Biochemistry: Fundamentals and Bioenergetics

Edited by

Meera Yadav

&

Hardeo Singh Yadav

Department of Chemistry
North Eastern Regional Institute of Science And Technology (NERIST)
Nirjuli
Itanagar-791109
Arunachal Pradesh
India

Biochemistry: Fundamentals and Bioenergetics

Editors: Meera Yadav and H.S. Yadav

ISBN (Online): 978-1-68108-847-1

ISBN (Print): 978-1-68108-848-8

ISBN (Paperback): 978-1-68108-849-5

BENTHAM SCIENCE PUBLISHERS LTD.

End User License Agreement (for non-institutional, personal use)

This is an agreement between you and Bentham Science Publishers Ltd. Please read this License Agreement carefully before using the ebook/echapter/ejournal (**"Work"**). Your use of the Work constitutes your agreement to the terms and conditions set forth in this License Agreement. If you do not agree to these terms and conditions then you should not use the Work.

Bentham Science Publishers agrees to grant you a non-exclusive, non-transferable limited license to use the Work subject to and in accordance with the following terms and conditions. This License Agreement is for non-library, personal use only. For a library / institutional / multi user license in respect of the Work, please contact: permission@benthamscience.net.

Usage Rules:

1. All rights reserved: The Work is 1. the subject of copyright and Bentham Science Publishers either owns the Work (and the copyright in it) or is licensed to distribute the Work. You shall not copy, reproduce, modify, remove, delete, augment, add to, publish, transmit, sell, resell, create derivative works from, or in any way exploit the Work or make the Work available for others to do any of the same, in any form or by any means, in whole or in part, in each case without the prior written permission of Bentham Science Publishers, unless stated otherwise in this License Agreement.
2. You may download a copy of the Work on one occasion to one personal computer (including tablet, laptop, desktop, or other such devices). You may make one back-up copy of the Work to avoid losing it.
3. The unauthorised use or distribution of copyrighted or other proprietary content is illegal and could subject you to liability for substantial money damages. You will be liable for any damage resulting from your misuse of the Work or any violation of this License Agreement, including any infringement by you of copyrights or proprietary rights.

Disclaimer:

Bentham Science Publishers does not guarantee that the information in the Work is error-free, or warrant that it will meet your requirements or that access to the Work will be uninterrupted or error-free. The Work is provided "as is" without warranty of any kind, either express or implied or statutory, including, without limitation, implied warranties of merchantability and fitness for a particular purpose. The entire risk as to the results and performance of the Work is assumed by you. No responsibility is assumed by Bentham Science Publishers, its staff, editors and/or authors for any injury and/or damage to persons or property as a matter of products liability, negligence or otherwise, or from any use or operation of any methods, products instruction, advertisements or ideas contained in the Work.

Limitation of Liability:

In no event will Bentham Science Publishers, its staff, editors and/or authors, be liable for any damages, including, without limitation, special, incidental and/or consequential damages and/or damages for lost data and/or profits arising out of (whether directly or indirectly) the use or inability to use the Work. The entire liability of Bentham Science Publishers shall be limited to the amount actually paid by you for the Work.

General:

1. Any dispute or claim arising out of or in connection with this License Agreement or the Work (including non-contractual disputes or claims) will be governed by and construed in accordance with the laws of the U.A.E. as applied in the Emirate of Dubai. Each party agrees that the courts of the Emirate of Dubai shall have exclusive jurisdiction to settle any dispute or claim arising out of or in connection with this License Agreement or the Work (including non-contractual disputes or claims).
2. Your rights under this License Agreement will automatically terminate without notice and without the

Bentham Science Publishers Ltd.
Executive Suite Y - 2
PO Box 7917, Saif Zone
Sharjah, U.A.E.
Email: subscriptions@benthamscience.net

CONTENTS

FOREWORD

It is immense pleasure to write the forward of the book titled "Biochemistry: Fundamentals and Bioenergetics" edited by Dr. Meera Yadav and Prof. H.S. Yadav, which is comprehensive and highly informative for the students. I also believe that teachers and scholars at every level and stage of their careers can enrich and strengthen their knowledge of biological chemistry by updates and practices presented in this book. The initial interest in the area of biological chemistry became intense as new tools and techniques of biotechnology came into existence. Biological chemistry is continuously and steadily progressing at the laboratory level and at the computational level. The authors and editors have chosen the specific topics and details that are important and relevant to the chemistry behind life sciences.

This book discusses a wide range of topics related to the fundamental and applied aspects of the chemistry behind life sciences. The book contains a range of topics, including the scopes and importance of biochemistry, the latest physical techniques to determine structures of biomolecules, detailed classification, structure and function of biomolecules like carbohydrates, lipids , amino acids, proteins, nucleic acids, enzymes, hormones as well as the thermodynamics of life sciences and bioenergetics and metabolism of biomolecules. It also deals with photosynthesis and respiration, oxidative phosphorylation, DNA replication, transcription and translation, and recombinant DNA technology.

The book has contributions from scientists, teachers and research scholars. The editors have done a commendable job in bringing and collecting and compiling a wide range of excellent papers shared by expert researchers . The collection of expertise and knowledge has been shaped to provide a unique piece of work in the form of a book. I am very much confident that the book should prove to be a very useful source of knowledge to the students, teachers, research scholars, scientists, engineers and doctors in the disciplines of life sciences, microbiology, biochemistry, biotechnology and engineering.

I express my sincere appreciation to the editors for their contributions and I am sure that this book will be very handy and widely used especially by the aspiring young generation students who wish to create a niche in the field of biological chemistry.

Saket Kushwaha
Rajiv Gandhi University
India

PREFACE

Based on progressive experimental achievements of biochemists and biologists, biochemical information is updated day by day and documented in the form of a book. Biochemistry is continuously and steadily progressing at the laboratory level to life sciences. We therefore have chosen specific topics and details that are important and relevant to understand the fundamentals of biological chemistry.

This textbook mainly aims to fulfill the requirement of undergraduates, postgraduates and research students having strong chemistry background with an ambition to enter into the biochemistry field. It is also helpful to instructors to get updates related to the field of biochemical sciences with little effort. The topics are explained, ranging from basics to a detailed knowledge in the area of biochemistry. To enable students to grasp the key points of chapters, keynotes have been included and a brief summary is given at the end of chapters.

We have attempted to integrate chemical concepts and details throughout the text. It includes the scope and nature of chemical forces, structural and mechanistic basis for the action of biomolecules, the thermodynamic basis for the folding and assembly of proteins and other macromolecules. Bearing specific functional groups, biomolecules are important intermediates for the synthesis of many chiral medicines and are widely used in the preparation of hormones, flavors, fragrances, liquid crystals and chiral auxillaries. These fundamental topics will help in understanding of all biological processes taking place. Our goal is to provide pricise and detailed view on specific topics concerning biological chemistry that will enable students to understand how the chemical features help to meet the biological needs.

Chemical insight often depends on a clear understanding of the structures of biochemical molecules. We have taken considerable care in preparing structures of stereochemically biomolecules. These structures should make it easier for the student to develop an intuitive feel for the shapes of molecules and comprehension of how these shapes affect reactivity *i.e* structure function relationship.

This edition of Biochemistry: Fundamentals and Bioenergetics offers a wide selection of high-quality supplements to assist students and instructors.We are optimistic to see the satisfaction of students and teachers aspiring to have clear concepts and knowledge in the area of biological chemistry. The editors and authors confirms that this book content has no conflict of interest.

Meera Yadav

&

Hardeo Singh Yadav
Department of Chemistry
North Eastern Regional Institute of Science And Technology (NERIST)
Nirjuli
Itanagar-791109
Arunachal Pradesh
India

List of Contributors

Archana Pareek	Department of Chemistry, NERIST, Nirjuli, Itanagar-791109 (AP), India
Dencil Basumatary	Department of Chemistry, North Eastern Regional Institute of Science And Technology (NERIST), Nirjuli, Itanagar-791109, Arunachal Pradesh, India
Hardeo Singh Yadav	Department of Chemistry, North Eastern Regional Institute of Science And Technology (NERIST), Nirjuli, Itanagar-791109, Arunachal Pradesh, India
Kamlesh Singh Yadav	DDU Gorakhpur University, Gorakhpur-273009, Uttar Pradesh, India
Meera Yadav	Department of Chemistry, North Eastern Regional Institute of Science And Technology (NERIST), Nirjuli, Itanagar-791109, Arunachal Pradesh, India
Nagendra Nath Yadav	Department of Chemistry, North Eastern Regional Institute of Science And Technology (NERIST), Nirjuli, Itanagar-791109, Arunachal Pradesh, India
Nene Takio	Department of Chemistry, North Eastern Regional Institute of Science And Technology (NERIST), Nirjuli, Itanagar-791109, Arunachal Pradesh, India
Nivedita Rai	Department of Chemistry, North Eastern Regional Institute of Science And Technology (NERIST), Nirjuli, Itanagar-791109, Arunachal Pradesh, India
Pratibha Yadav	Department of Chemistry, IIT Delhi, Hauz Khas New Delhi-110016, India
Saroj Yadav	Delhi University, Benito Juarez Road, South Campus, South Moti Bagh, New Delhi-110021, India
Shilpa Saikia	Department of Chemistry, North Eastern Regional Institute of Science And Technology (NERIST), Nirjuli, Itanagar-791109, Arunachal Pradesh, India
Sonam Tashi Khom	Department of Chemistry, North Eastern Regional Institute of Science And Technology (NERIST), Nirjuli, Itanagar-791109, Arunachal Pradesh, India

CHAPTER 1

Scope and Importance of Biological Chemistry

Nene Takio[1], **Meera Yadav**[1,*] and **Hardeo Singh Yadav**[1]

[1] *Department of Chemistry, North Eastern Regional Institute of Science And Technology (NERIST), Nirjuli, Itanagar-791109, Arunachal Pradesh, India*

Abstract: Biochemistry allows us to understand how various chemical processes in all living organisms interact and function to support life. It covers a wider range of scientific disciplines, which are sub-categorised into different branches. In this unit, the importance of biochemistry in medicine, health sectors, nutrition, and the living system has been discussed in detail. It also helps to understand biological phenomena of the environment and its conversation, genetic manipulations of genes *via* recombinant DNA technology and gene sequencing *via* human genome project. The knowledge of biochemistry has advanced tremendously and in forthcoming years, it has a potential role in unravelling the mystery of life processes.

Keywords: Biomarkers, Biopsy, Genome, Genotype, Transgenes, Xenobiotics.

INTRODUCTION

Biochemistry, as an interdisciplinary subject, includes a wide range of scientific disciplines like life sciences, forensics, chemical sciences, plant sciences, and medicine. It focuses on the chemical processes occurring within the living system at the molecular level. Therefore, there is a need for a biochemical approach because biochemistry attempts to understand the chemical composition, structure, biological functions and metabolism of biomolecules and in this process, it goes much deeper into the problem of life than any other branch of science. The term 'Biochemistry' was first coined in 1903 by a German chemist named Carl Neuberg. During the last two decades, knowledge of biochemistry has advanced tremendously and in forthcoming years, it is predicted to have a potential role in unraveling the mystery of the processes of life [1].

The concept of biochemistry is very old; its knowledge and understanding have been applied for exploring and investigating components of a living system for more than a thousand years. Modern biochemistry will help in a better understanding of enzymes, molecular biology and their functions in the body.

* **Corresponding author Meera Yadav:** Department of Chemistry, North Eastern Regional Institute of Science And Technology (NERIST), Nirjuli, Itanagar-791109, Arunachal Pradesh, India. E-mail: drmeerayadav@rediffmail.com

Meera Yadav & Prof. Hardeo Singh Yadav (Eds.)

Cells and tissues in the human body are made up of chemical elements like H_2, C, N, O, Ca and P, which play a pivotal role in overall body functions and form a vital part of biochemistry [2].

CENTRE OF BIOCHEMICAL REACTIONS

DNA is the core part where genetic material stores data, directs and controls all biochemical reactions. It directs the cell to release chemicals like enzymes to perform various mechanical functions like replication, synthesis, digestion, catalysis, *etc*., which occur in a regulated manner. The information is contained long sequences of nucleic acid subunits and each subunit is made up of four nucleotides. The sum of weak interactions between molecules affects the overall stability of the biological structures and functions. All the biochemical reactions follow the 2nd law of thermodynamics, stating that all systems with spontaneous reactions run "downhill," motion with an increase in entropy or randomness [3, 4].

BRANCHES IN BIOCHEMISTRY

Biochemistry is a diverse subject quite useful in all other branches of science. Nowadays, it has been sub-categorised into different branches to study different biological functions involving RNA and DNA, protein synthesis, cell membrane and much more. Some of these have been discussed below:

Enzymology

It is a study of properties and biological functions of enzymes like enzymatic activity, kinetics, enzyme-substrate complex, the kinetics of the reaction, enzymatic regulation, and transition state, *etc*. They fulfill a multitude of functions in living organisms. They are essential for signal transmission and cellular control, usually by kinases and phosphatases. They also produce movement with ATP, which hydrolyzes myosin to induce muscle contraction as well as moves cargo in and out of the cell. Other cell membrane ATPases are active transport ion pumps [5]. Enzymes like amylases and proteases present in the intestine participate in the digestive system and breakdown of large molecules like starch and proteins into smaller ones.

Factors Affecting the Enzyme Activity

Clinical enzymology is another sub-branch of biochemistry that deals with the studies of enzymes responsible for prolonged diseases and their diagnosis. Enzymes are highly specific and selective, therefore required in small quantities with high purity.The reaction rate is the maximum when an enzyme gets fully saturated with substrate, designated as Vmax. The affinity of an enzyme with substrate influence the relationship between the reaction rate and concentration of substrate normally represented as the Km (Michaelis-Menten constant) of an enzyme. For practical purposes, Km is defined as the concentration of substrate at which the enzyme achieves its half Vmax. A high Km value represents the low affinity of the enzyme with a substrate and to achieve Vmax, a higher concentration of substrate is required. The favoured kinetic properties of these enzymes are low Km and high Vmax for maximum efficiency at low enzyme and substrate concentrations, as shown in Fig. (**1**). Thus, to avoid contamination from incompatible materials, the enzyme source is selected with utmost care to get a purified enzyme. Enzymes have huge potential in the therapeutic application for treating cancer [6], such as Asparaginase, has proved to be efficient in treating acute lymphocytic leukaemia. Its action relies upon the fact that tumour cells have poor aspartate-ammonia ligase activity,which limits their potential to synthesize the typically non-essential amino acid L-asparagine [7]. Table **1** shows various applications of enzymes in clinical diagnosis.

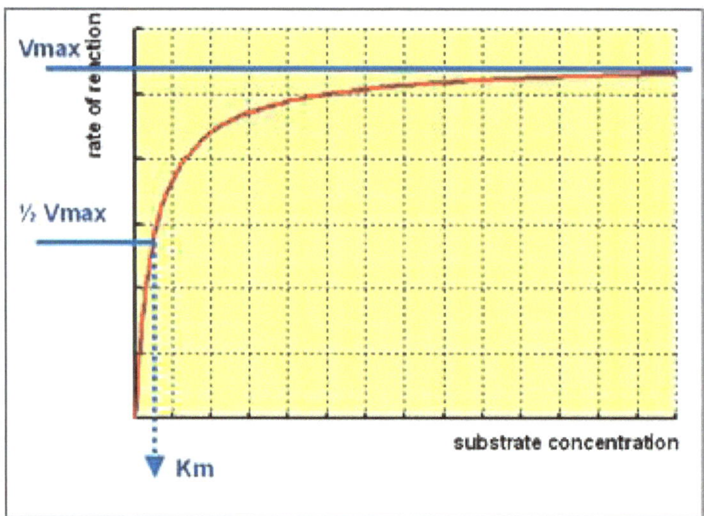

Fig. (1). Rate of reaction *vs* substrate concentration.

Table 1. Uses of enzyme for diagnostic purposes.

S.No.	Enzyme for diagnostic purpose	Estimation of	Method of estimation
1	Urease	Urea	Rapid Ureate Test(RUT)
2	Uricase	Uric acid	A colorimetric method
3	Glucose oxidase	Glucose	Glucose oxidase method

(Table 1) cont.....

4	Peroxidase	Glucose/Cholesterol	(GPO-PAP) method/ CHOD-PAP method
5	Hexokinase	Glucose	Hexokinase/G6PD method
6	Cholesterol oxidase	Cholesterol	CHOD/POD **method.**
7	Lipase	Triglyceride	GPO/PAP **method**
8	Horse raddish peroxidase	Specific proteins and antigens	ELISA
9	Alkaline phosphate	Bone and liver disorders	ELISA
10	Restriction endonuclease	Natural mutations like hemoglobinopathies and thalassemias.	Recombinant DNA technology
11	Reverse transcriptase	Gene expression	Polymerase chain reaction

Endocrinology

It is a study dealing with biosynthesis, signal process, storage and functions of hormones in living organisms. Hormones control metabolism, respiration, growth, reproduction, sensory perception, and movement and its imbalance in the body causes a wide range of medical conditions. Endocrinology deals with both hormones and the glands, also the tissues from where it is produced.

There are more than 50 different hormones produced in the human body, though they are present in small amounts and yet have a significant effect on physical function and development. Endocrine tissues include the adrenal glands, hypothalamus, ovaries, and testes. The endocrine system includes tissues such as the adrenal gland, hypothalamus, ovaries, and testes. The most common hormonal disorder found in women is polycystic ovary syndrome (PCOS) [8]. Hormonal imbalances can be caused by genetic or environmental factors. Endocrinologists generally deal- with the subsequent conditions like:

- diabetes
- osteoporosis
- menopause
- metabolic disorders
- thyroid diseases
- excessive or insufficient production of hormones
- some cancers
- short stature
- infertility

Molecular Biology

This study aims to understand the molecular and chemical processes that occur in living organisms from a molecular perspective. Here you will find detailed information on classical, biochemical and metabolic cycles and also learn about the integration and degradation of molecules *in vivo*. Molecular biology helps us understand the chemical properties of molecules *e.g.* cell metabolism. Chemical reactions occurring in the body are beneficial in sustaining life. Reproduction, structural restoration, and autonomic response to stimuli involve a number of intracellular processes. Molecular biochemists study two major types of metabolism: catabolism and anabolism. Catabolism is the process by which matter is broken down and energy is released through the respiration of cells whereas the anabolic process uses energy to make various components inside the cell.

In addition to biomolecules, molecular biochemistry also deals with the study of viruses. The virus can only develop inside the host cell, making it a form of pseudo-life. They can influence different parts of molecules, from protein synthesis to cell membrane transport and also infect all other organisms, including plants or animals. More than 5,000 varieties of viruses have been described by molecular biochemists worldwide and they have given the term 'virology' to the study of viruses.

Molecular Genetics and Genetic Engineering

It deals with genetic modification and the processes involving gene insertion, gene silencing, gene expressions, mutation and various properties. The goal of this study is to overcome the limits of genetic manipulation by transferring genes from one species to another species or by splicing the unwanted genes. The purpose of this study is also to model the effects of genes. Genetic engineering is used to alter the genetic make-up of cells, thus exchanging properties inside and over species to create better or new living things. The new DNA can be inserted into the host's genome by first isolating it, copying the ancestral stimulus material, creating a DNA sequence using nuclear sequencing techniques, or synthesizing the DNA and then inserting it into the host's body. Genes may be removed, or "knocked out", using a nuclease [9]. The different methods for knockout are (1) Gene silencing (2) Conditional knockout (3) Homologous recombination (4) Gene editing and (5) Knockout by mutation. Overcoming obstacles, boundaries between species, for example, the genome of one species can be integrated into another to create new species. One of its main goals is to obtain current administrative entities and gene expression, that is, to obtain epigenetic code. It is the foundation of some other disciplines of life sciences, especially biotechnology.

Applications of Genetic Engineering

Agriculture

An important use of recombinant DNA technology is to modify the genotype of crops to increase crop productivity, nutritional value, protein abundance, immunity, and reducethe use of fertilizer. Recombinant DNA technology and tissue culturegenerate high-yielding grains, legumes and vegetables. Some plants can grow their own fertilizers, while others are genetically engineered to make their own pesticides. Examples are Bt cotton, Bt brinjal. *etc*. Fig. (**2**) shows the increasing rate of Bt cotton production in India. Several varieties have been genetically modified which can bind directly to the atmospheric nitrogen, to avoid dependence on fertilizers. There are certain genetically evolved weed killers which are not specific to weeds alone but kill useful crops also. Glyphosate is a commonly used weed killer which simply inhibits a particular essential enzyme in weeds and other crop plants.

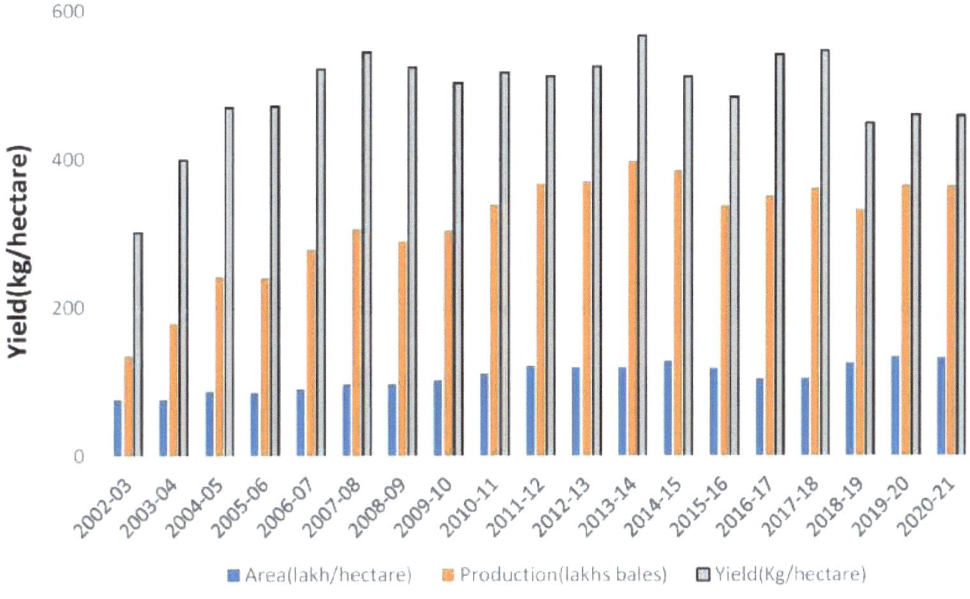

Fig. (2). Bt cotton production in India.

Medicine

Genetic Engineering was found to be quite popular to treat genetic diseases.It plays an important role in the manufacture of drugs. Microorganisms and herbal materials are currently being manipulated to produce many useful drugs vaccines, enzymes and hormones at low cost. Gene therapy is perhaps the most innovative and promising aspect of genetic engineering, allowing individuals with defective genes to insert healthy genes directly.

It is quite useful for the production of vaccines and artificial hormonesfor the treatment of diseases as shown in Fig. (**3**) for the production of insulin by recombinant DNA technology.

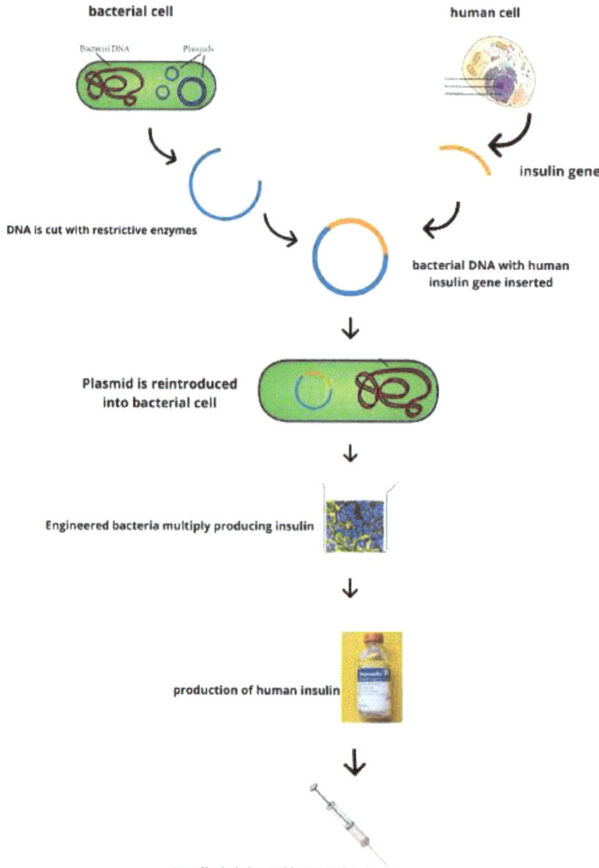

Fig. (3). Human insulin production by recombinant DNA technology.

Energy Production

It has immense potential for energy production. With the help of this technology, it is now possible to produce bioengineered crops or biofuels which in turn produce biomass that can be used as fuel or oil, alcohol, diesel or other energy products. As a result, the waste can be converted to methane. Genetic engineers try to transfer the cellulose gene to the right organism, which can convert wastes like sawdust and cornstalk into sugar and then alcohol.

Industries

Nowadays through this technology, wide varieties of chemicals are being produced in the industries. Synthesis of Glucose from sucrose can be achieved by enzymes extracted from genetically modified organisms. Genetically modified strains of bacteria and cyanobacteria have been developed that are capable of synthesizing large amounts of ammonia for fertilizer production at acheaper rate. Microorganisms have been developed which convert cellulose to sugar and sugar into ethanol.It can also be used to track the deterioration of wastes, petroleum products, naphthalene, and other industrial wastes.

Structural and Metabolic Biochemistry

The purpose of this study is to provide clear information about the biological architecture like proteins and nucleic acids (DNA and RNA) and understand different metabolic pathways at the cellular level.

Chemical reactions are mostly synchronised and occur in sequences called metabolic pathways, each of which is catalysed by a particular enzyme. These pathways are classified according to the reactions that lead to material breakdown or synthesis [9].

WHY UNDERSTANDING BIOCHEMISTRY IS IMPORTANT?

Biochemistry is a scientific discipline that explains that chemical elements are vital for structural components like carbohydrates, lipids, proteins, and nucleic acids, which are involved in metabolic activity. Biochemistry gives valuable insights into the complex molecular relationships that make life sustainable [10]. It also helps to understand the processes associated with aging and cell death. It transfers knowledge for a better understanding of signaling processes of energy changes and to carry out scientific and technological research. Therefore, it is important to understand the importance of biochemistry and its extensive application in our daily activities [11].

Importance of Biochemistry in Medicine

Drug Designing

Structural Biochemistry has been essentialfor the production of new medicines. Medicines are currently being studied using biochemistry methods such as Xray

Crystallography. Modern biochemistry techniques are generally used to explain the function of the enzyme by understanding the folding and bending of the molecule. The European Federation for Medicinal Chemistry says ***"Biochemistry is a guide to drug discovery and forits Application"*.** For example, Morphine is a drug that reduces pain in terminal cancer.The most basic goal in drug development is to predict how a particular molecule is attached to a target, and how strongly it will bind. Molecular mechanics are mostly used to measure the strength of the intermolecular interaction between a small molecule and its biological target. These methods are also used to predict the structure of small molecules and to model the structural change of a target that can occur when small molecules bind to it [12 - 16]. There are two major types of drug design:

1. **Ligand based or indirect drug design** that depends on the knowledge of how a molecule binds with another target molecule of interest. It can be built on the knowledge that it works with a model of a biological target, which can be used to design new molecular entities that interact with the target.
2. **Structure based or direct drug design** uses data information of the 3D structure of biological sample using x-ray crystallography or NMR spectroscopy method, which helps in predicting the binding affinity and selectivity of the target molecule.

Diagnosis

Clinical biochemistry is a branch of medicine which deals with the detection and treatment of associated disorders in a patient by using various biochemical methods. For example, according to the symptoms described by the patient, the physician may prescribe medicine and test to detect diseases.

Nutrition

Many diseases occur due to a deficiency of vital minerals in our body. Hence, a good knowledge of biochemistry is required to overcome deficient nutrients and better functioning of the body. Nutritional biochemistry is the study of nutrition and is composed of various studies of food nutrients and their function and chemical components in humans and other mammals. Specifically, human nutrition refers to the use, absorption, and elimination of essential chemicals found in foods and beverages that help the body produce energy and support its growth and development. Nutrients boost the immune system ofthe body to fight diseases effectively.

Importance of Biochemistry in Agriculture

Gout – It is a form of inflammatory arthritis due to the deposition of uric acid in joints, tendons and tissues.

Agricultural biochemistry deals with agricultural production, food processing, monitoring and remediation of the environment. The study emphasizes the relationships between plants, animals and bacteria and their environment [17]. Some important areas where the knowledge in agricultural biochemistry is highly useful include:

1. Assessment of thenutritional value of grains, poultry, cattle and pulses.
2. Production and processing of improved genotypes.
3. Elimination and inactivation of harmful non-nutritional factors in food grains by reproduction and chemical treatments. *e.g.* BOAA in Lakhdal, Trypsininhibitors of soybean, Aflatoxins of groundnut.
4. Preservation of food, processing and post-harvest physiology and nutritional quality of fruit, crops and vegetables.
5. Biochemistry of resistance to disease and insects.

Importance of Biochemistry in Nutrition

Nutrition entails a healthy diet that can prevent diseases, reduce disease conditions and promote health. Biochemical studies help us determine the optimal amount of nutrients for good health, and the nutritional value of food and drink can also be determined by different biochemical tests. Food consists of nutrients that are categorised according to their role in the body: energy-producing macronutrients (carbohydrates, proteins and fats), essential micronutrients (vitamins, minerals and water) and many other ingredients. Although micronutrients do not provide energy for the body to make fuel, they are essential for the proper functioning of the body's metabolic and regulatory activities, as shown in Table **2**.

Table 2. Recommended dietary allowance (RDA) of important nutrients for an adult man, weighing 70kg.

Nutrients RDA
Carbohydrates 400 g
Fats 70 g
Proteins 56 g
Essential fatty acids 4 g
Vitamin A 1000 μg

(Table 2) cont.....

Vitamin D 5 μg
Vitamin E 10 μg
Vitamin K 70 μg
Calcium 800 mg
Iron 10 mg

Nonessential nutrients, such as flavonoids, phytoestrogens, carotenoids, probiotics, also have important properties for good health. The regular consumption of a variety of foods provides energy and nutrients which are important for an individual's health and well-being. The recommended daily food and their nutritional values are shown in Table **3**.

Table 3. Importance of minerals.

Minerals	Sources	Functions	Deficiency/Excess	Daily Requirement
Sodium	Processed food, salt	Osmotic skeleton of extracellular fluid	>10g intake/d increase B.P	5g/d
Magnesium	Green leafy vegetables, refined grains, milk, fruits	Metabolism of sodium and potassium	Irritability, tetany, hyperreflexia, hypo-reflexia	340mg/day
Zinc	Meat, milk, fish	Component>300 enzymes, Metabolism of glucides and peptides, synthesis of insulin, maintain the integrity of the immune system	Growth failure, sexual infantilism in adolescents, loss of taste, delayed in wound healing, spontaneous abortions, congenital abortions	12mg/day- Men 10mg/day-Women

The biochemistry of nutrition is the backbone for understanding the composition and function of food and nutrients in the body. Nutrients function as a cofactor for enzymes, hormonal components and metabolic processes and participate in oxidation/reduction reactions.

Nutrients are important for body growth, sexual development, reproduction, psychological wellbeing, energy level and the normal functioning oforgan systems in the body, albeit needed in small quantities.

Importance of Biochemistry in Pathology

The ultimate application of biochemistry is for the health and welfare of mankind.Clinical biochemistry or chemical pathology is a necessary laboratory service for clinical practice. The results of biochemical tests conducted can help diagnose the disease and early treatment. Biochemical screening is important in diagnosing diseases like diabetes mellitus, jaundice, myocardial infarction, arthritis, pancreatitis, rickets, cancer, acid-base imbalance, and so on. It is widely used for testing in clinical laboratories. Diagnosis creates a list of different diagnoses based on history and clinical examination. Based on this list, tests can be selected to include or exclude as many variations as possible.

Importance of Biochemistry in Pharmacy

Biochemistry has an eminent importance in pharmacy.The pharmaceutical industry relies heavily on biochemistry, as the systems in the bodywork with a wide variety of chemicals. Biochemistry works with hormones, enzymes, proteins, and cell interactions to understand what types of chemicals are needed to correct any imbalance without adversely affecting other chemicals produced in the body [18]. The important areas where biochemistry plays a major role are:

Drug Constitution

Biochemistry examines the composition of drugs, the possibility of theirdegradation at different temperatures and it also helps to change the chemistry of medicine to improve performance and reduce side effects, *etc.*

The Half-life

This test is performed on biochemical drugs to determine how long the drug lasts when kept at a high temperature.

Drug Storage

The required storage conditions can be estimated by using biochemical tests. For example, many enzymes and hormones that participate in drug delivery deteriorate over time due to temperature or oxidation and improper storage, *etc.*

Drug Metabolism

This helps in understanding how drug molecules get metabolized by enzymes through various biochemical reactions, which may help in avoiding medications with side effects [19].

Importance of Biochemistry in Plants

Biochemistry helps in explaining the chemical reactions that have taken place in the plants and how we can optimise them to improve our productivity. Some of thesehas been discussed below:-

Photosynthesis

This is one of the chemical reactions in plants that will help us to understand how carbohydrates are produced with the help of sunlight, CO_2 and water.

Different Sugars

There are a number of carbohydrates produced in plants. Each one of them hasdifferent structures and functions, thus understanding its biochemistry means understanding its physical and chemical properties.

Plants Secondary Metabolites

Biochemistry provides us with knowledge about the mechanism of the plant forming various products such as tannins, resins, alkaloids, gums, enzymes, and phytohormones [19].

RECENT TRENDS IN BIOCHEMISTRY

The advancement in biochemistry is ever-growing due to its exponential growth and so are its applications in various disciplines of sciences. New techniques are being introduced, leading to the development of extraordinary and medicinally useful molecules, to modify hereditary characteristics of plants and animals, to diagnose new diseases and ultimately finding new ways of curing them. In recent years many such discoveries and inventions have been made to understand the mystery of the living systems as well as to study biochemical phenomena of the environment for its conservation [20]. Some of the blooming topics have been discussed below which would have a tremendous role to remold the future in the years to come:

- **Human genome project**
- **Environmental biochemistry**
- **Gene therapy**

Human Genome Project

The Human Genome Project (HGP) was initiated in the year 1984 but was formally launched in October 1990 and completed in the year 2003. HGP was an international research project to classify the sequence of the human genome and the genes contained in it. The HGP's purpose is to classify all of the approx.

30,000 genes in human DNA but the project failed to get sequences of all human DNA. Only the 'euchromatic' regions which make up 92% of the human genome was sequenced [21].

The sequence of the three billion chemical base pairs that make up human DNA is calculated and this information is stored in the database. It develops data processi ng methods, transfers relevant technology to the private sector and tackles the ethical, legal and social problems that might occur in the project. Molecular medicine with HGP improves the diagnosis of disease, early detection of genetic predisposition to the disease, thus it can help in design drug, gene therapy and drug monitoring systems, and to form custompharmacokinetic drugs [22]. The ability to use the gene to treat a disease called gene therapy has captured the imagination of the biomedical community and has huge potential to treat or cure inherited and acquired diseases.

Application and Benefits of Human Genome Project (HGP)

The sequencing of the human genome and other species' geno mes is expected to significantly improve our understanding and interpretation of biology and medicine. Some of the advantages of HGP are:

- Identification of human genes and their functions.
- Understanding of polygenic disorders *e.g.* cancer, hypertension, and diabetes.
- Improved diagnosis of diseases
- Development of pharmacogenomics.
- Genetic basis of psychiatric disorders
- Improved knowledge on mutations.
- Comparative genomics
- A better understanding of developmental biology
- Improvement in gene therapy

Environmental Biochemistry

Due to rapid modernization and exploitation of natural resources, there are rapid climatic changes taking place as well as an increase in environmental pollution. Therefore, Environmental biochemistry as a new discipline primarily deals with the metabolic (biochemical) responses and adaptations in man (or other organisms) due to environmental factors. It studies the microbial metabolism of contaminants with a focus on metabolite elucidation and its reactions. Every day we encounter or interact with different environmental pollutions that become highly poisonous when ingested or get absorbed in the body as shown in Fig. (**4**) the bio-magnification of pollutants. So studying the biochemistry of pollutants it is possible to study the behaviour, transformation of pollutants and how they can affect the biological functions of the body.

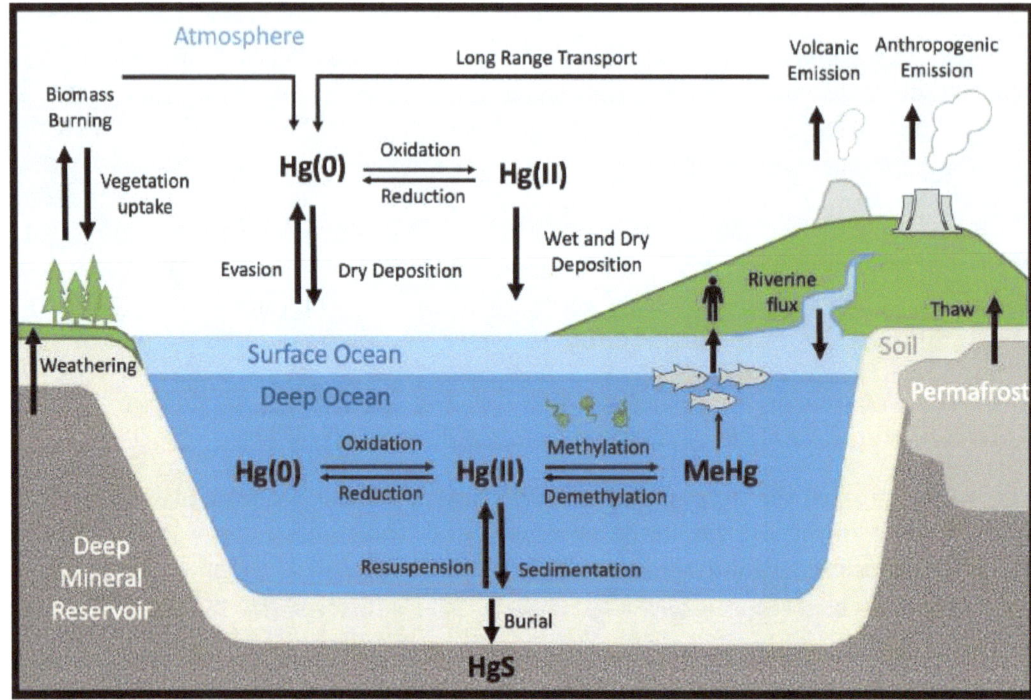

Fig. (4). Biomagnification of mercury metal.

At the molecular level, the main steps that make up catalytic enzymes in the metabolism of pollutants are analyzed in terms of genetic, kinetic, and structural criteria. The knowledge gained will help us to develop tools and methods for improving and advancing the techniques for purification of wastewater, bioremediation process, formation and selection of microbes with certain physical properties and development of environmental friendly process to promote green chemistry.

Via biochemistry, we have been able to establish that some xenobiotics such as PCB, dioxinand DDT disrupt the normal function of the body by mimicking body hormones [21].

Biomarkers

Biomarkers or biological markers are observable measures of any biological condition or state. These are used to indicate exposure to xenobiotics present in the environment and species or their effect.

A biomarker can itself be an external substance (*e.g.*, asbestos particles or NNK from tobacco), or a type of body-processed external substance (metabolite) that can usually be quantified. Biomarkers are major molecular or cellular events that associate specific environmental relationships with health outcomes. They play a significant role in understanding the relationships between exposure to environmental pollutants, chronic human disease development and detection of subgroups that are elevated risk of disease. There has been considerable progress in identifying and validating new biomarkers which can be used in population-based environmental disease studies as shown in Fig. (**5**), presenting various biomarkers and their applications.

Molecular Biomarker	Sources	Applications
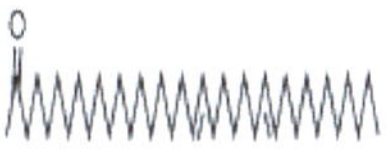 Alkenones	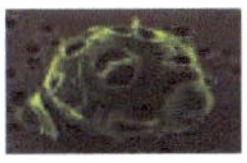 Haptophyte algae	U^{K}_{37} - sea surfaces and temperature $\delta^{13}C$ – Paleo CO_2 δD -Hydrography, salinity
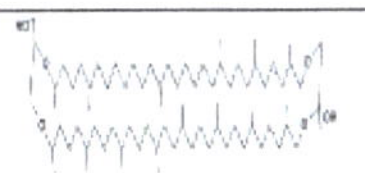 Isoprenoidal GDGTs	 Thaumarchaeota	TEX_{86} – Sea surfaces and lake temperatures MI – Anaerobic oxidation of methane
 Long chain Diols	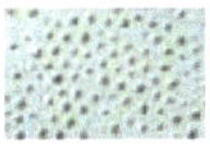 Eustigmatophyceae	DIX – Sea surface temperatures
 Plant Waxes	 Higher plant species	Land plants organic matter. $\delta^{13}C$ – Changes in carbon cycle/reservoirs δD -P/E hydrography, paleotopography
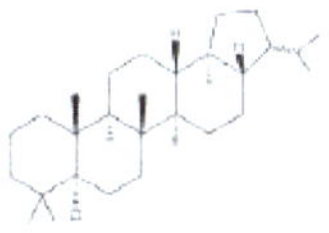 Hopanes	 Soil Bacteria	$\delta^{13}C$ – Changes in methanogen populations

Fig. (5). Biomarkers and their applications.

Biosensors

These sensors have become very common in recent years, and they are applicable in various fields listed below:

- General medical examination
- Testing metabolites
- Disease diagnosis
- Insulin therapy
- Professional psychotherapy and disease detection
- Military
- Livestock and Veterinary use
- Drug reformation
- Industrial Production and monitoring
- Monitor environmental friendly emission

Latest Developments in Environmentally Friendly Biosensors

For the detection and monitoring of various environmental contaminants, biosensors including immunosensors, aptasensors, genosensors, and enzymatic biosensors have been documented using antibodies, aptamers, nucleic acids, and enzymes as recognition elements [23]. Table **4** summarises recent biosensors used for monitoring the environment [23].

Table 4. Summary of recent biosensors for environmental monitoring.

Analyte/Pollutant Detected	Biosensor Type	Recognition Element	Electrode/Sensing Material
Paraoxon	Electrochemical (amperometric)	Enzyme (AChE 1)	Gold SPE 2 and cysteamine SAM 3
	Electrochemical (voltammetric)	Enzyme (butyrylcholinesterase)	SPE 2 with carbon black nanoparticles
	Optical (colorimetric)	Enzyme (AChE 1 and ChO 4)	Iodine-starch
	Electrochemical (amperometric)	Enzyme (AChE 1)	GCE 5 and gold nanorods

(Table 4) cont.....

Methyl parathion	Electrochemical (impedimetric)	Enzyme (hydrolase)	SPE 2 with Fe_3O_4 and gold nanoparticles
	Electrochemical (amperometric)	Enzyme (AChE 1)	Graphite and macroalgae
	Electrochemical (impedimetric)	Enzyme (AChE 1)	Carbon paste electrode and reticulated spheres structures of $NiCO_2S_4$
	Electrochemical	Enzyme (AChE 1)	Carbon paste electrode with chitosan, gold nanoparticles, and Nafion
	Optical	Sphingomonas sp. cells	Microplate with silica nanoparticles and PEi 6 hybrid
Chlorpyrifos	Electrochemical (impedimetric)	Enzyme (tyrosinase)	SPCE 7 and IrOx nanoparticles
	Electrochemical (voltammetric)	Enzyme (AChE 1)	Boron-doped diamond electrode with gold nanoparticles and carbon spheres
	Electrochemical (voltammetric)	Aptamers (#1)	Carbon black and GO 8/Fe_3O_4
	Electrochemical (amperometric)	Enzyme (AChE 1)	GCE 5 with NiO nanoparticles-carboxylic graphene-Nafion
Dichlorvos	Optical (fluorescence)	Enzyme (AChE 1 and ChO4)	QD 9 and acetylcholine
	Electrochemical (voltammetric)	Enzyme (AChE 1)	Platinum electrode with ZnO
	Electrochemical (impedimetric)	Enzyme (AChE 1)	Ionic liquids-gold nanoparticles porous carbon composite
Acetamiprid	Optical (colorimetric)	Aptamers (#2)	Gold nanoparticles
	Electrochemical (impedimetric)	Aptamers (#3)	Gold nanoparticles, MWCNT 10, and rGO 11 nanoribbons
	Electrochemical (impedimetric)	Aptamers (#3)	Silver nanoparticles and nitrogen-doped GO 8
	Electrochemical (impedimetric)	Aptamers (#3)	Platinum nanoparticles

(Table 4) cont.....

Atrazine	Electrochemical (voltammetric)	Antibodies (monoclonal)	Gold nanoparticles
	Electrochemical (FET 17)	Antibodies (monoclonal)	SWCNT
	Electrochemical (impedimetric)	Aptamers (#4)	Platinum nanoparticles
	Electrochemical (amperometric)	Phage/antibody (monoclonal) complex	Magnetic beads functionalized with protein G
Pirimicarb	Electrochemical (voltammetric)	Enzyme (laccase)	Carbon paste electrode with MWCNT 10
	Electrochemical (amperometric)	Enzyme (AChE 1)	Prussian blue-MWCNT 10 SPE 2
Carbofuran	Electrochemical (voltammetric)	Enzyme (AChE 1)	IrOx-chitosan nanocomposite
	Electrochemical (amperometric)	Enzyme (AChE 1)	GCE 5 with GO 8 and MWCNT10
	Electrochemical (amperometric)	Enzyme (AChE 1)	GCE 5 with NiO nanoparticles-carboxylic graphene-Nafion composite
Carbaryl	Electrochemical (impedimetric)	Enzyme (AChE 1)	Gold electrode with cysteamine SAM 3
	Electrochemical (impedimetric)	Enzyme (AChE 1)	Interdigitated array microelectrodes with chitosan
	Electrochemical (amperometric)	Enzyme (AChE 1)	MWCNT 10 and GO 8 nanoribbons structure
	Electrochemical (amperometric)	Enzyme (AChE 1)	Porous GCE 5 with GO 8 network
Legionella pneumophila	Optical (SPR 12)	Nucleic acids (#5)	Gold substrate with streptavidin-conjugated QD 9
	Optical (SPR 12)	Antibody (polyclonal)	Gold substrate with protein A SAM 3
	Electrochemical (amperometric)	Antibody (polyclonal)	SPCE 7 with Fe_3O_4@polydopamine complex
	Optical (SPR 12)	Antibody (polyclonal)	Gold gratings substrate

(Table 4) cont.....

Escherichia coli	Optical (SPR 12)	Polymerizable form of histidine	Gold substrate
	Piezoelectric (QCM 13)		
	Electrochemical (capacitive)	Polymerizable form of histidine	Gold electrode
	Optical (electrochemiluminescence)	Antibodies (polyclonal)	GCE 5 with polydopamine imprinted polymer and nitrogen-doped QD 9
Bacillus subtilis	Electrochemical (amperometric)	Antibodies (polyclonal)	Gold electrode with SWCNT 14
Hg^{2+}	Optical (evanescent-wave optical fibre)	Nucleic acids (#6)	Optical fibre platform
	Optical (fluorescence)	DNA	MOF 15 (UiO-66-NH_2)
	Electrochemical (voltammetric)	Nucleic acids (#7)	Gold substrate with vertically aligned SWCNT
	Optical (SERS 16)	Nucleic acids (#8)	SWCNT 11 and $CoFe_3O_4$@Ag substrate
Pb^{2+}	Optical (fluorescence)	DNAzymes (#9)	Carboxylated magnetic beads
	Optical (fluorescence)	DNAzyme (#10)	Graphene QD 9 and gold nanoparticles
	Optical (fluorescence)	Aptamers (#11)	Micro-spin column
Brevetoxin-2	Electrochemical (impedimetric)	Aptamers (#12)	Gold electrodes with cysteamine SAM 3
	Electrochemical (voltammetric)	Cardiomyocyte cells	Microelectrode array with platinum nanoparticles
Saxitoxin	Electrochemical (voltammetric)	Cardiomyocyte cells	Microelectrode array with platinum nanoparticles
	Optical (interferometry)	Aptamers	-
Microcystin	Electrochemical (impedimetric)	Antibodies (monoclonal)	Graphene
	Electrochemical (voltammetric)	Antibodies (monoclonal)	Gold electrodes with MoS_2 and gold nanorods
	Electrochemical (voltammetric)	Enzyme (protein phosphate 1)	SPE 2
Okadaic acid	Optical (SPR 12)	Antibodies	Gold electrode with a carboxymethylated surface
	Electrochemical (FET 17)	Antibodies (monoclonal)	Graphene
	Optical (fluorescence)	Antibodies (monoclonal)	Carboxylic acid-modified magnetic beads and CdTe QD 9

(Table 4) cont.....

Domoic acid	Optical (SPR 12)	Antibodies	Gold electrode with a carboxymethylated surface
	Electrochemical (FET 17)	Antibodies (monoclonal)	SWCNT 14
	Optical (SPR 12)	Antibodies	Glass side chip with gold surface
Bisphenol A	Optical (fluorescence)	Aptamers	Gold nanoparticles
	Optical (evanescent-wave optical fibre)	Aptamers (#13)	Optical fibre surface
	Optical (fluorescence)	Aptamers (#14)	Molybdenum carbide nanotubes
Nonylphenol	Electrochemical (FET 17)	Antibodies (monoclonal)	SWCNT 14
17β-estradiol	Photo-electrochemical	Aptamers (#15)	CdSe nanoparticles and TiO_2 nanotubes
	Electrochemical (voltammetric)	Antibodies	Gold electrode with MPA 18 SAM 3
	Electrochemical (capacitive)	Antibodies	Gold electrode with MUA 19 SAM 3

Note: 1 AChE: acetylcholinesterase, 2 SPE: screen-printed electrode, 3 SAM: self-assembled monolayer, 4 ChO: choline oxidase, 5 GCE: glassy carbon electrode, 6 PEi: polyethyleneimine, 7 SPCE: screen-printed carbon electrode, 8 GO: graphene oxide, 9 QD: quantum dots, 10 MWCNT: multi-walled carbon nanotubes, 11 rGO: reduced graphene oxide, 12 SPR: surface plasmon resonance, 13 QCM: quartz crystal microbalance, 14 SWCNT: single-walled carbon nanotubes, 15 MOF: metal-organic framework, 16 SERS: surface enhancement Raman spectrum, 17 FET: field-effect transistor, 18 MPA: 3-mercaptopropionic acid, 19 MUA: 11-mercaptoundecanoic acid.

Gene Therapy

Gene therapy is the process of gene manipulation and insertion into cells to treat diseases. Newly introduced genes will encode proteins and repair the defect found in a genetic disorder. It involves gene manipulations in animals or humans in order to correct a disease, and maintain individual healthy. Gene therapy in theory is the ultimate solution for the treatment of genetic disease.

It involves isolating a particular gene, making its copies, injecting it into the target tissue cells to create the desired proteinas shown in Fig. (**6**). It is absolutely essential to verify that the gene is not harmful to the patients and is appropriately expressed. As of now due to several limitations, gene therapy has not progressed the way it should despite intensive research. But a breakthrough may come anytime, and this is only possible with persistent research. And a day may come when almost every disease will have gene therapy, as one of the treatment modalities. Hence, gene therapy can revolutionize the practice of medicine.

Gene therapy is categorized into two types: germline and somatic, which apply to humans.The purpose of germline gene therapy is to insert transgenic cells into the germ cell and somatic cell as shown in Fig. (**7**). This therapy cure the treated person and the gametes. Toinsert a new gene directly into a cell, scientists use a bioengineered medium called a "vector" for gene delivery.

For example, viruses have a natural tendency to transmit genetic material to cells, and so can be used as vectors. However, before a virus can be used to transfer therapeutic genes into human cells, it is modified to suppress its ability to cause infectious disease. Gene therapy can be used to alter cells within or outside the body. When performed within the body, a doctor injects the vector that contains the gene directly into the portion of the body that has faulty cells. In gene therapy that is used to change cells outside the body, a patient may take blood, bone marrow, or other tissue and isolate particular cell types in the laboratory. These cells are fed into the vector containing the desired gene. The cells are left in the laboratory to multiply and are then injected back into the patient, where they proceed to multiply and ultimately produce the desired result.

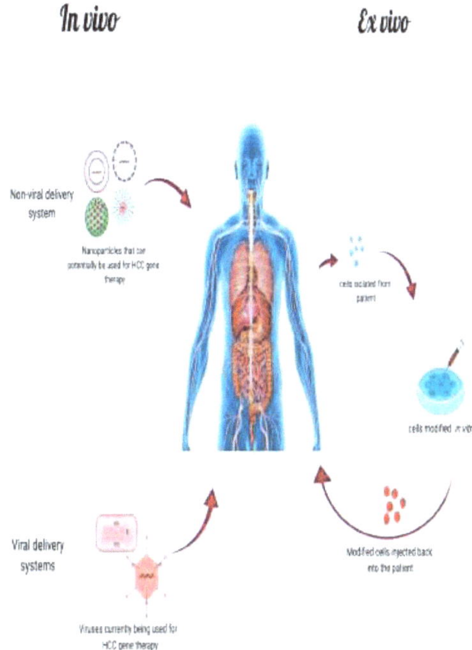

Fig. (6). Types of gene therapy: **a)** *Ex vivo* **b)** *In vivo*.

CANCER

The proliferation of the body cells is under strict regulation under normal circumstances. Sequentially, the cell differentiates, divides, and dies. In cancer, the regulation of cell growth and development is lost, leading to excessive proliferation and spreading of viral cells as shown in Fig. (**7**). The retroviral replication in host cells. Cancer is the second-most killer disease in developed countries. It is estimated that more than 20 percent of deaths in the United States account for cancer. Some signs and symptoms, or screening tests may detect cancer. Usually, it is then further tested by medical imaging and biopsy confirmation.

Survival rates differ by the type of cancer and the stage at which it is diagnosed. These days, the tumor markers are employed to detect the presence of cancers such as carcinoembryonic antigen (CEA) and Alpha-fetoprotein (AFP).

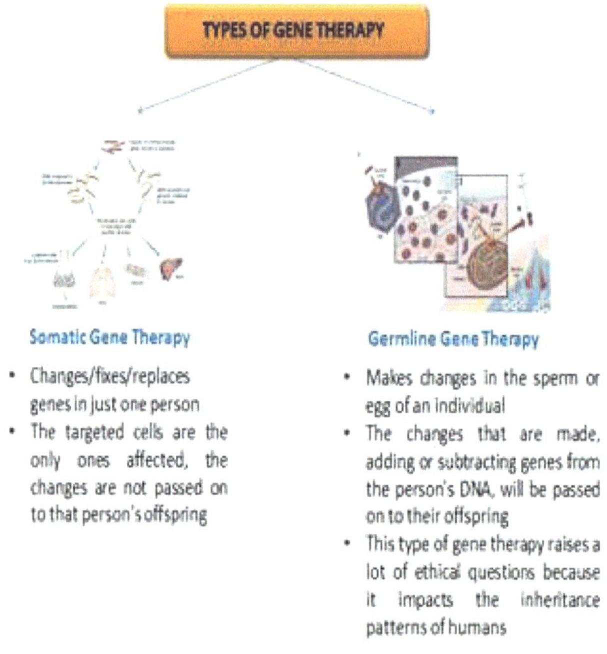

Fig. (7). Types of Gene Therapy.

Chemotherapy is another treatment process that employs certain anticancer drugs to patients. The effectiveness of anticancer drugs is inversely proportional to the size and number of tumor cells. As of now, the best way to combat the killer disease is to prevent the usage of carcinogenic products and have a healthy lifestyle because the survival rate of a person decreases if detected late. Perhaps the developments in these fields provide the most promising prospects of significantly reducing the cancer mortality rate.

Some of the emerging technologies and therapies that will revolutionize cancer care are:

a. **Immunotherapy**
Immunotherapy is a cure for cancer intended to improve the body's natural ability to combat cancer by restoring the role of the immune system. A healthy immune system kills about 10,000 cancerous and precancerous cells a day on average. When the immune system becomesweakened and the process of destroying these harmful cells can no longer be done, hence cancer grows. Immunotherapy is a cancer treatment designed to boost the body's natural ability to fight cancer by restoring functions of the immune system. A healthy immune system destroys an average of 10,000 cancerous and precancerous cells a day. When the immune system becomes compromised and can no longer handle the task of removing these dangerous cells, cancer develops. Many of these cancer survival mechanisms can be overcome with the right Immunotherapy treatment [24].

b. **Immunomodulators and Vaccine**
Checkpoint inhibitors medications improve the body's defenses, and antitumor sensitive vaccines are anticipated to help doctors fight advanced cancer with greater effectiveness.

c. **Oncologic Liquid Biopsy**
Liquid biopsies are capable of detecting cancer cells or tumour DNA that shed tumours in the blood.

d. **CRISPR: Cancers and Gene Editing**
In 2016, CRISPR, or Clustered Frequently Interspaced Short Palindromic Repeats, made a number of headlines. The technology enables scientists to modify genes with unparalleled ease as indicated in Table **5**.

e. **Patient-specific Studies to Make Drug Discovery More Effective**
Individualized monitoring and awareness of patient subpopulations help researchers predict the response of individual patients to therapies.

f. **Nanotechnology and the Treatment of Cancer**
Researchers use nanotechnology to selectively deliver therapeutics to cancer cells and to increase the therapeutic effectiveness of such treatment [25].

Table 5. Major highlights of Human Genome Project.

The draft represents about 90% of the entire human genome. It is believed that most of the important parts have been identified.
The remaining 10% of the genome sequences are at the very ends of chromosomes (*i.e* telomeres) and around the centromeres.
The human genome is composed of 3200 Mb *i.e* 3.2 billion base pairs.
Approximately 1.1 to 1.5% of the genome codes for proteins.
Approximately 24% of the total genome is composed of introns that split the coding regions (exons) and appear as repeating sequences with no specific functions.
The number of protein-coding genes is in the range of 30,000-40,000.
An average gene consists of 3000 bases, the sizes however vary greatly. Dystrophin gene is the largest known human gene with 2.4 million bases.
Genes and DNA sequences associated with many diseases such as breast cancer, muscle diseases, deafness and blindness have been identified.
About 100 coding regions appear to have been copied and moved by RNA-based transposition (retro-transposons).
Repeated sequences constitute about 50% of the human genome.
A vast majority of the genome (97%) has no known functions.
Between humans, the DNA differs only by 0.2% or one in 500 bases.

KEYWORDS

Genome

Chromosomes with haploid set found in gamete or microorganism.

Biopsy

It is a diagnostic procedure usually conducted by a surgeon, radiologist or cardiologist involving the extraction of sample cells or tissues for analysis to assess the existence or nature of a disease.

Genotype

It is the part of the genetic makeup of an organism or individual's cell, which defines one of its features.

Xenobiotics

It is a chemical substance contained within an organism that is not created naturally or supposed to be present inside the organism. It may also include substances present in concentrations far higher than normal.

Transgenes

It is a gene or genetic material that has been naturally passed from one organism to another, or through any of a variety of genetic engineering techniques. The introduction of a transgene probably changes an organism's phenotype.

SHORT-ANSWER TYPE QUESTION

1. What are biomolecules? Give examples.
2. What is the term half-life means in pharmacy?
3. What happens when the human body is infected with cancer cells? Name the common markers used to detect tumor cells..
4. Explain the role of biochemistry in pathology.
5. How biochemistry helps to overcome deficient nutrients and minerals in our body?
6. Mention the application of biochemistry in agriculture.
7. Differentiate carbohydrates, fats and proteins with examples.
8. Give full forms of DNA, RNA, ATP and NADPH and their role in the living system.
9. What are the types of drugs used for the treatment of cancer?
10. What are the scopes of biochemistry?
11. What do you understand by myocardial infarction?

LONG-ANSWER TYPE QUESTIONS

1. Why there is a need for a biochemical approach?
2. Explain Human Genome Project in detail. Mention its various applications that will revolutionize the medicine world.
3. Why survival rates vary when infected with Cancer? How the deadly disease can be overcomed with emerging technologies?
4. Explain the importance of biochemistry.
5. What do you understand by Environmental Biochemistry? Explain in detail.
6. What are various branches in biochemistry?
7. Explain the life cycle of cancer-causing retrovirus in the tumor cell.
8. How gene therapy is useful in treating diseases?
9. Describe the inter-relation of biochemistry with other branches of science.
10. What do you mean by nanotechnology? How the technology will be useful for the treatment of deadly diseases in the years to come?

MULTIPLE CHOICE QUESTIONS

1. Which of the following is an example of polygenic disorder?
   a. Diabetes
   b. Sickle-cell Anemia
   c. Hemophilia
   d. Polycystic kidney
2. The tumor markers employed to detect the presence of cancer is a
   a. BOAA
   b. Carcinoembryonic Antigen
   c. Trypsin
   d. Nucleoproteins
3. The gene manipulation in humans and animals to correct or cure disease is called
   a. Gene therapy
   b. Immunotherapy
   c. Psychotherapy
   d. Chemotherapy
4. Vitamin C deficiency leads to
   a. Anemia
   b. Scurvy
   c. Ricket
   d. Marasmus
5. Given below are carbohydrates EXCEPT
   a. Starch
   b. Chitin
   c. Glycogen
   d. Cholesterol
6. How many bonds carbon form with other elements?
   a. 5
   b. 6
   c. 4
   d. 3
7. Which of the following elements is the LEAST abundant in the living system?
   a. Oxygen
   b. Nitrogen
   c. Phosphorus
   d. Sodium
8. The cohesion of water is caused by:
   a. Covalent bonds
   b. Hydrogen bonds
   c. Ionic bonds

d. Hydrophobic bonds

9. Name the process involving gene insertion and manipulation into cells to correct diseases?
   a. Stem cell therapy
   b. Gene therapy
   c. Live vector vaccines
   d. Molecular cloning
10. Which one of the following is the most current gene therapy trial target:
   a. Cancer
   b. HIV
   c. SCID deficiency
   d. Cystic Fibrosis
11. Which one of the following is used as a vehicle for gene therapy to carry a healthy gene?
   a. Bacteria
   b. Plastic capsule
   c. Powder balls
   d. Viruses
12. How many numbers of chromosomes humans have?
   a. 44
   b. 46
   c. 42
   d. 50
13. The human genome is composed of:
   a. all of our genes
   b. Only DNA
   c. DNA and RNA
   d. responsible for our characteristics
14. The primary goal of HGP was:
   a. to determine the sequence of base pair of DNA
   b. to identify human defective genes
   c. to differentiate eukaryotic and prokaryotic genes
   d. to sequence chromosomes genes
15. What are genes?
   a. working units of DNA
   b. a thread-like structure in the cell
   c. are linear and carry information
   d. made up of proteins and amino acids
16. The most widely present element in the living system?
   a. Zinc, Hydrogen, Potassium, Calcium
   b. Carbohydrates, Lipids, Proteins, Nucleic Acids

c. Fat, Muscles, Organs, Bones
d. Calcium, Lithium, Potassium, Nitrogen

17. Purified wastewater that is released into rivers or lakes is known as:
    a. Sewage
    b. Effluent
    c. Septic sludge
    d. Storm runoff
18. What is produced when we burn fossil fuel?
    a. Oxygen
    b. H_2O
    c. Acid rain
    d. Carbon Dioxide
19. Macronutrients mean the nutrients that are needed in large amounts.
    a. True
    b. False
    c. None of these
20. In diffusion, molecules move form an area of __________concentration to and area of __________ concentration.
    a. Low to High
    b. Low to low
    c. High to high
    d. high to low
21. Which one is the function of aldosterone?
    a. increase in blood volume
    b. decrease in blood volume
    c. increase in blood sugar
    d. decrease in blood sugar
22. Common symptoms of hyperthyroidism are?
    a. Diarrhoea
    b. Urinary frequency
    c. Weight gain
    d. Heat tolerance
23. What is DNA made up of?
    a. pentose sugar
    b. hexose sugar
    c. triose sugar
    d. heptose sugar
24. The structural study of the nuclear receptor is achieved by
    a. Ultrasound
    b. X-ray crystallography
    c. Nanoindentation

d. Colour Doppler

25. Which would NOT be a correct match?
    a. Keratin- amino acid
    b. DNA- nucleic acid
    c. Phospholipid- fatty acids
    d. Cellulase –monosaccharide
26. An unsaturated fatty acid is made up of what kind of bonds?
    a. Single
    b. Double
    c. Hydrogen
    d. Ionic
27. A feature of nucleic acids not found in lipids is
    a. Oxygen
    b. Phosphorus
    c. Potassium
    d. Hydrogen
28. Which stage of vehicle emission norms are applied currently in India in Internal Combustion Engine?
    a. BS V
    b. BS IV
    c. BS III
    d. BS II
29. Which of the following is not a greenhouse gas,
    a. CH_4
    b. N_2O
    c. O_3
    d. H_2
30. Which of the following toxic gas is dissolved in human blood faster than O_2?
    a. SO_2
    b. CO
    c. O_3
    d. N_2O

CONSENT FOR PUBLICATION

Not Applicable.

CONFLICT OF INTEREST

The authors confirm that this chapter contents have no conflict of interest.

ACKNOWLEDGEMENTS

Authors are thankful to Department of Chemistry, North Eastern Regional Institute of Science and Technology, Nirjuli, Itanagar for providing facilities to accomplish the work.

REFERENCES

[1] Biochemistry and its definition. Available from: www.biochemistry.org

[2] Biochemistry KT. 2ndedition.. New Delhi: Wiley Eastern Ltd 1990.

[3] Voet D, Voet JG, Pratt CW. Principles of Biochemistry. 4thedition. NJ: John Wiley & Sons Inc 2013.

[4] Cliffsnotes. The Scope of Biochemistry. Types of Biochemical Reactions. Houghton Mifflin Harcourt 2016. Available from: https://www.cliffsnotes.com/study-guides/biology/ biochemistry/the-scope--f-biochemistry /types-of-biochemical-reactions

[5] Nelson DL, Cox MM. Lehninger Principles of Biochemistry. 7th ed., W.H. Freeman and Company 2017.

[6] Anand C, Debashree D, Prashant S, Vikash KM, Sushil KK, Nanoemulsion: The emerging contrivance in the field of nanotechnology. Nanosci Nanotechnol Asia 2018; 8(2): 146-71. [http://dx.doi.org/10.2174/2210681207666170418123826]

[7] Chaplin M. Enzyme Technology. Medical applications of the enzyme. [updated 6th August 2014].Available from: https://www1.lsbu.ac.uk/water/enztech/medical.html

[8] Brazier Y. What is endocrinology? Medical News Today [updated on 12th October 2017].Available from: https://www.medicalnewstoday.com/articles/248679.php

[9] Wikibooks.org. Structural Biochemistry/Metabolism. [updated 18th July 2018]. Available from: https://en.wikibooks.org/wiki/Structural_Biochemistry/Metabolism

[10] Saha D. Biochemistry: scope and its importance. Available from: www.biologydiscussion.com

[11] Maldonado FS. Biochemistry-The importance of basic sciences in dentistry. Rev. Mex de Ortod 2013; 17(2): 74-5.

[12] Wikipedia.org. Drug design [updated 2nd October 2019]. Available from wikipedia.org/wiki/Drug_design

[13] Neda A, Fatemeh S, Tahereh MI. Quantitative Structure-Property Relationship (QSPR) investigation of camptothecin drugs derivatives. Comb Chem High Throughput Screen 2018; 21(7): 533-42. [http://dx.doi.org/10.2174/1386207321666180927102836] [PMID: 30264675]

[14] Daryoush J., Fatemeh S. QSPR models to predict thermodynamic properties of cycloalkanes using molecular descriptors and GA-MLR method. Curr Computeraided Drug Des 2020; 16(1): 6-16. [http://dx.doi.org/10.2174/1573409915666190227230744]

[15] Roop K, Poornima D, Anil Kumar V, Abha B. Design, synthesis, and quantum chemical calculations of 2,6-Diphenylspiro[cyclohexane-1,3′-pyrido[1,2-a]pyrimidine]-2′,4,4′-trione through DFT approach. Lett Org Chem 2019; 16(1): 983-95.

[16] Abhishek P, Dilpreet S, Neena B. Aprepitant loaded self-microemulsion preconcentrates for enhanced solubility and bioavailability: a design. Curr Nanomed 2018; 8(2): 156-73. [http://dx.doi.org/10.2174/2468187308666180426120823]

[17] Agriculture Information Bank. Scope & Importance of Biochemistry in Agriculture [updated 21st January 2018] Available from: . https://agriinfo.in/scope-and-importance-of-biochemist-y-in-agriculture-1547/

[18] Importance of biochemistry in the pharmaceutical industry. Available from: http://www.answers.com/Q/Importance_of_biochemistry_in_pharmaceutical_industry

[19] Applications of biochemistry in daily life –medical – agriculcultural – pharmacy [updated 10th April 2017]. Available from: https://azchemistry.com/applications-of-biochemistry

[20] Owais M. Recent trends in biochemistry and biotechnology. J BiochemBiotechnol 2017; 1(1): 6-7.

[21] Satyanarayana U, Chakrapani U. Biochemistry. 4th edition. New Delhi: Elsevier Health Sciences APAC 2013.

[22] Ejelonu A. What is human genome project- why is it important to society? [updated 2nd May 2018] Available from: http://serendipstudio.org /biology /b103/ f01/ web1/ ejelonu. html 2001.

[23] Justino CIL, Duarte AC, Rocha-Santos TAP. Recent progress in biosensors for environmental monitoring: a review. Sensors (Basel) 2017; 17(12): 2918. [http://dx.doi.org/10.3390/s17122918] [PMID: 29244756]

[24] Envita Medical Centre. Advanced immunotherapy options are the future of cancer treatment. 2018.www.envita.com

[25] The Medical Futurist. 14 technologies that will shape the future of cancer care [Updated on 21st March 2016]. Available from: www.medicalfuturist.com

CHAPTER 2

Structure of Water: Acid-Base and Buffers; Hydrogen Bonding

Nivedita Rai[1], **Shilpa Saikia**[1], **Meera Yadav**[1,*] and **Hardeo Singh Yadav**[1]

[1] *Department of Chemistry, North Eastern Regional Institute of Science And Technology (NERIST), Nirjuli, Itanagar-791109, Arunachal Pradesh, India*

Abstract: Water, being a universal solvent, makes up to 70% of the weight of living organisms. The amphoteric nature of water is explained by Lewis and Bronsted acid-base theory. Hydrogen bonding is crucial in many chemical processes. It is accountable for having water's unique solvent capabilityand also for hydrophobic effect. This chapter discusses in detail the hydrogen bonding between neighbouring water molecules and theirrole in physical and chemical changes in the water molecule. It also discusses the relationship between pressure and temperature in water *via* a phase diagram.

Keywords: Acids, Bases, Buffer, Hydrogen bonding.

INTRODUCTION

Water (H_2O) is the basic need of every single living being, making up to 70% or a greater amount of the weight of most life forms. Water comprises little polar V-shaped particles with molecular formula H_2O. It is viewed as the widespread universal solvent for reasons including its basic, synthetic, and physical properties. These properties bring about numerous extraordinary attributes of water [1].

* **Corresponding author Meera Yadav:** Department of Chemistry, North Eastern Regional Institute of Science And Technology (NERIST), Nirjuli, Itanagar-791109, Arunachal Pradesh, India; E-mail: drmeerayadav@rediffmail.com

Meera Yadav & Prof. Hardeo Singh Yadav (Eds.)

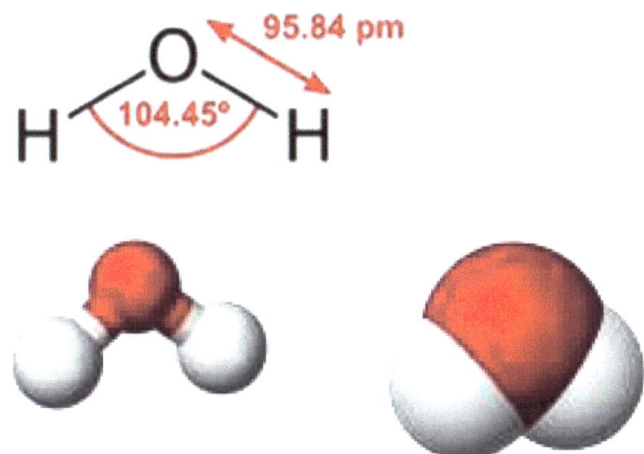

Water is a very small bent molecule, consisting of two hydrogen atoms connected to a heavier oxygen atom. Water molecule possesses a neutral charge, but polarity lies with the center of positive and negative charges located oppositely. Each hydrogen atom has a core comprising a unit positively charged and proton encircled by a 'cloud' of a unit negatively charged electron. Oxygen atom similarly has a core comprising of eight positively charged protons and eight uncharged neutrons encircled by a 'cloud' of eight negatively charged electrons.

The water molecule contains ten paired electrons in five 'orbitals'; one pair of the electron is connected with the oxygen atom, the other two pairs of electrons are associated as 'outer' electrons on the oxygen atom and the remaining two electron pairs form each of the two similar hydroxyl (O-H) covalent bonds.

Water molecules are polar in nature and form a hydrogen bond with other water molecules. The polarity of water permits it to isolate particles of salts and get strongly bonded to other polar substances, for example, acids, alcohols, *etc*., consequently dissolving them.

Hydrogen bonding of water is responsible for its various unique properties, like the density of its solid form (ice) is less compared to its liquid form; it generally possesses a high boiling point of 100 °C and has a high molar heat capacity.

Water is amphoteric in nature, *i.e.* it has both an acidic and a basic character, and self-ionizes to H^+ and OH^- ions. At a fixed temperature, the product of the concentrations of H^+ and OH^- ions is constant. Since water is an excellent solvent, it is rarely pure. There are numerous other compounds that are not soluble in water, like oils, fats and other nonpolar substances [2].

Names	
IUPAC name	Water, oxidane
Other names	Hydrogen hydroxide(HH or HOH), Hydrogen oxide, Di-hydrogen monoxide(DHMO), Hydrogen monoxide, Di-hydrogen oxide, Hydric acid, Hydroxic acid, Hydrol, μ-oxido dihydrogen

Water nomenclature table

PHYSICAL AND CHEMICAL CHARACTERISTIC OF WATER

- Water has no taste and has no odour at the standard condition of temperature and pressure. It is colorless when present in small quantities, however, the color of water and ice appears to a very light blue. Additionally, ice also appears colorless and the vaporized form of water essentially appears invisible as a gas.
- It is translucent, and as daylight can enter the aquatic plants, they can survive within the water. UV light isrelatively less absorbed.
- The shape of the water molecule is nonlinear V-shaped. The electronegativity of the oxygen atom of the water molecule is higher than the electronegativity of the hydrogen atoms and hence there is a small -ve charge on the oxygen atom and +ve charge on the hydrogen atom. Therefore water is a polar molecule having a net dipole moment. Likewise, water can form a remarkably large number of intermolecular hydrogen bonds (four) for a particle of its size. Because of these unique factors of water, there is a strong attractive force between the molecules of water as a result of which water develops a very high capillary force and high surface tension. With capillary movement, water goes up against the force of gravity through small capillary tubes. All vascular plants, for example, trees, depend on this property.
- It is a very good solvent and hence it is considered a universal solvent. Substances thatcan dissolve in water, *e.g.* sugars, acids, alkalis salts and few gases such as oxygen and carbon dioxide (carbonation) are called hydrophilic which means water-loving in nature and those substances which don't dissolve well in water (*e.g.* oils and fats), are called as hydrophobic which means water-hating in nature.
- All the biomolecules which are found in the cells are also soluble in water.

For example, DNA, proteins, polysaccharides, *etc.*

- Water in its pure form conducts very low electricity but conductivity can be greatly improved by dissolving a small quantity of ionic solid such as NaCl.
- The boiling point of water depends upon the barometric pressure. At the top of Mt. Everest, for example,water boils at 68°C, while at sea level, it boils at

100°C. Contrarily, water somewhere down in the sea close to geothermal vents can arrive at temperatures of several degrees and stay fluid *i.e.* in the liquid state.

- The molar specific heat capacity of water is the second-highest after ammonia, and also hasahigh heat of vaporisation (40.65 kJ·mol^{-1}), all due to the strong hydrogen bond between the molecules. These two unique properties of water allow it to balance the climate of the earth through the buffering of broad temperature variation.
- The highest water density is at 39°F (or 3.98°C to be exact) thus changing the properties of water *i.e.* Ice has the peculiar property of being less dense than its liquid state. It extends to occupy an additional volume of 9% in its solid-state, representing the liability of ice floating on fluid water [3].

ACID AND BASE CONCEPT

Properties of Acids and Bases According to Robert Boyle

In the year 1661, Robert Boyle stated the properties of acids as follows:

Acids are

- Sour in taste.
- Corrosive in nature.
- Change the color of certain vegetable dyes, such as litmus, from blue to red.
- Lose their acidity *i.e.* they become neutral when they react with alkalies.

The word "acid" comes from the Latin word "acidus" which means "sour," and also refers to the sharp odour and sour taste of various acids. Boyle also stated the properties of alkalies as follows.

Alkalies are

- slippery in nature
- change the colour of litmus from red to blue
- become less alkaline when they react with acids [4].

The Arrhenius Definition of Acids and Bases

According to Arrhenius "acids form hydrogen ions in aqueous solution while Arrhenius bases form hydroxide ions."

Later, in the year 1884, Svante Arrhenius recommended that salts such as sodium chloride (NaCl) get dissociated when dissolved in water and produce the charged particles called ions.

$$NaCl(s) \rightarrow Na^+(aq) + Cl^-(aq)$$

After three years, Arrhenius elaborated the above theory and added that acids are neutral in nature and when they are dissolved in water, they get ionised to give hydrogen (H^+) ions and a corresponding -ve ion. According to the definition of Arrhenius, HCl (hydrogen chloride) is an acid because when it is dissolved in water, it ionizes to give hydrogen ions (H^+) and chloride (Cl^-) ions.

$$HCl(g) \rightarrow H^+(aq) + Cl^-(aq)$$

Arrhenius asserted that bases are also neutral compounds and when they are dissolved in water, they get ionised or dissociated to give either hydroxide (OH^-) ions or a corresponding positive ion. For example, sodium hydroxide (NaOH) acts as an Arrhenius base as it gets dissociated in water and gives the hydroxide (OH^-) ions and sodium (Na^+) ions.

$$NaOH(s) \rightarrow Na^+(aq) + OH^-(aq)$$

Examples of Arrhenius acid: HCl, HCN, and H_2SO_4

Examples of Arrhenius base: NaOH, KOH, and $Ca(OH)_2$

Limitations of Arrhenius Theory

This theory suggests it must contain the particular ions all together for a substance to release either H^+ or OH^- ions. However, this theory cannot explain the nature of ammonia (NH_3) (a weak base) which, when dissolved in water, produces hydroxide (OH^-) ions, but there is no OH^- ion present in ammonia.

HCl gets neutralized by using the solution of NaOH and ammonia. In both cases, a colourless solution is obtained that can be crystallized to produce a white salt *i.e.*, either NaCl or NH_4Cl. The two reactions are very similar. The equations are given as follow:

$$NaOH(aq) + HCl(aq) \rightarrow NaCl(aq) + H_2O(l)$$

$$NH_3(aq) + HCl(aq) \rightarrow NH_4Cl(aq)$$

In the case of sodium hydroxide, there is a reaction between the H^+ ions which are released from the acid and the OH^- ions of sodium hydroxide (NaOH) –following the Arrhenius theory. However, in the case of ammonia, there is no OH^-ion. When ammonia reacts with water, it produces ammonium ions and hydroxide ions as shown below:

$$NH_3(aq) + H_2O(l) \rightleftharpoons NH^{+4}(aq) + OH^-(aq)$$

The reaction given above is reversible. Approximately 99 percent of ammonia remains as ammonia molecules in a typical dilute ammonia solution. However OH^-ions are also present as explained by Arrhenius theory. The same reaction also occurs between NH_3 and HCl, both present in the gaseous form.

$$NH_3(g) + HCl(g) \rightarrow NH_4Cl(s)$$

In the example above, the solution does not contain H^+ ions or OH^- ions – as there is no solvent or solution. The theory of Arrhenius does not take the above reaction into account as an acid-base reaction, despite the fact that it gave the same product when the two molecules were put together in the solution. Taking into account these controversial cases, attempts were made later on to remove defects in these theories about acids and bases in a new and different manner.

Bronsted Definition of Acids and Bases

Brønsted-Lowry's theory of acid and bases had taken the Arrhenius theory one step forward and removed its flaws. According to this theory, there is no need for a substance to carry H^+ ions or OH^- ions to be classified as an acid or base. For instance, let us consider the following reaction:

$$HCl(aq) + NH_3(aq) \rightleftharpoons NH_4^+(aq) + Cl^- (aq)$$

In the above chemical equation, ammonia (NH_3) accepts a proton from hydrochloric acid (HCl) to forms an NH_4^+ ion called ammonium ion, carrying a positive charge and Cl^- ion known as chloride ion carrying a negative charge. So, according to the Bronsted-Lowry concept HCl is acid as it gives a proton and ammonia is a base that accepts a proton. In the above equation, Cl^- (chloride ion) acts as the conjugate base of the acid HCl ((hydrogen chloride) and ammonium ion NH_4^+ act as the conjugate acid for the base NH_3.

The above explanation is given in terms of an equilibrium expression as shown in the Fig. (**1**) below.

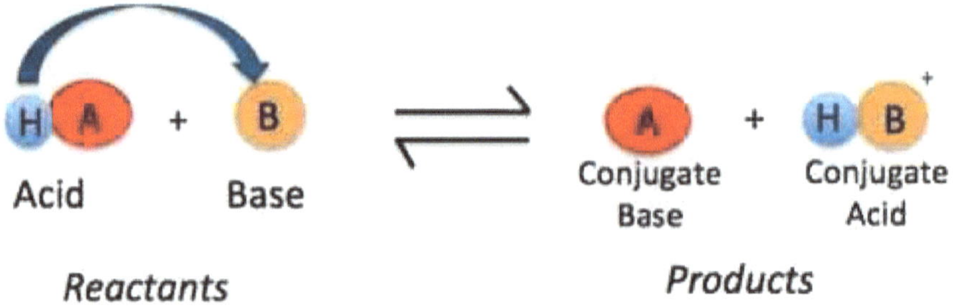

Fig. (1). Bronsted-Lowry Acid-Base Reaction.

The equilibrium sign, indicates that the reaction takes place both in forward and backward directions. From the above figure, it is clear that the HA which is taken as acid, gives or loses a proton and converts into its conjugate base A^- and B acts as a base that receives that proton and converts into its conjugate acid as HB^+.

Conjugate Acids and Bases

A significant outcome of the concept of equilibrium is that each acid (here commonly taken as HA) has a base form (A−), and the other way around. In the base, equilibrium for conjugate acid of base B is HB^+. These equilibria are related to the equilibrium for a given acid and base as given below:

$$\mathbf{H_2O\ (l) \rightleftharpoons H^+(aq) + OH^-(aq)}$$

With equilibrium constant,

$$K_w = [H^+][OH^-]$$

At 25°C, the value of K_w *i.e.* equilibrium constant is found to be 1.00×10^{-14}. The value of K_w is determined by taking the product of the acid constant (K_a) and base dissociation constants (K_b).

The conjugate acid strength and conjugate base strength depend on the strength of its parent acid or base and is inversely proportional to it. Any base or acid is in fact additionally a type of conjugate base or conjugate acid; these terms are generally utilized to differentiate the species present in solution (for example, acetic acid act as a conjugate acid of the acetate anion, which behaves as a base and the acetate ion acts as the conjugate base of acetic acid which behaves as an acid) [5].

The Acid-Base Chemistry of Water

The chemistry of water is illustrated by the phenomenon of equilibrium which takes place between neutral molecules of water and its dissociated ions.

$$\mathbf{2\ H_2O(l) \rightleftharpoons H_3O^+(aq) + OH^-(aq)}$$

The equilibrium constant of the above-given reaction can be expressed as follows-

$$K_c = \frac{[H_3O^+]\,[OH^-]}{[H_2O]^2}$$

The above expression is known as equilibrium constant, but at equilibrium, it does not consider the large difference in the concentrations of water molecules and its dissociated ions (H_3O^+ and OH^-).

The ability of water to conduct the electricity indicates that pure water at room temperature possesses 1.0 x 10^{-7} moles of each hydronium (H_3O^+) ion and hydroxide (OH^-) ions in one litre.

$$[H_3O^+] = [OH^-] = 1.0 \text{ X } 10^{-7} \text{ M}$$

At 25°C temperature, the molar concentration of neutral water molecules is 55.35 mol.

$$\frac{0.9971 \text{ g } H_2O}{1 \text{ mL}} \text{ X } \frac{1000 \text{ mL}}{1 \text{ L}} \text{ X } \frac{1 \text{ mol } H_2O}{18.015 \text{ g } H_2O} = 55.35 \text{ mol} \frac{H_2O}{L}$$

The concentration ratio of hydrogen (H^+) ion or hydroxide (OH^-) ion and the neutral water molecule is found to be 1.8 x 10^{-9}.

$$\frac{1.0 \text{ X } 10^{-7}}{55.35} = 1.8 \text{ X } 10^{-9}$$

It follows that at room temperature, just around 2 ppb (parts per billion) of molecules of water separates into their respective ions. The water molecule concentration is much higher than the concentrations of its dissociated ions (H_3O^+

and OH^-) that the value of equilibrium is almost constant.

Rearrangement for the expression for the equilibrium constant of the dissociated water molecules can be given by the following equation.

$$[H_3O^+][OH^-] = K_c X [H_2O]^2$$

In the above equation, the product of H_3O^+ and OH^- ions is replaced by an equilibrium constant K_w known as water dissociation equilibrium constant:

$$[H_3O^+] [OH^-] = K_w$$

At room temperature, the concentration of both the ions *i.e.* $H_3O^+$and OH^- ions is 1.0×10^{-7} M, respectively. The calculated value of K_w (dissociation constant) at 25°C is thus found to be 1.0×10^{-14}.

$$[1.0 \times 10^{-7}] [1.0 \times 10^{-7}] = 1.0 \times 10^{-14}$$

Despite the fact that Kw is defined as the dissociation of the water molecule, above mentioned expression of the equilibrium constant is also recommended for acid-base solutions. Without considering the sources of the dissociated ions of water *i.e.* OH^- ions and H_3O^+ ions, at equilibrium, the calculated value of the product of the concentration of these two ions at room temperature is always found to be 1.0×10^{-14}.

Lewis Acid-base Concept

In the year 1923, G. N. Lewis proposed yet another point of view on response among H^+ and OH^- ions of water. According to the theory of Bronsted, the hydroxide (OH^-) ion is the dynamic species and it forms a covalent bond by gaining an H^+ ion from the solution. According to the concept of Lewis acid-base theory, the hydrogen (H^+) ion is an active species and it forms the covalent bond by gaining two electrons from the hydroxide (OH^-) ion.

$$H^+ + :\ddot{O}—H^- \longrightarrow H—\ddot{O}—H$$

According to the model of Lewis acid-base, bases are known to donate a pair of electrons and acids are known to accept that pair of electrons. **Lewis acid** can hence be defined as a species that can gain a pair of non-bonding electrons *e.g.* the hydrogen (H^+) ions. In another way, we can say that a **Lewis acid is an acceptor**

of electron pair or electron pair acceptor. Similarly, a **Lewis base** can be defined as a species that can donate or give a pair of non-bonding electrons *e.g.* the OH^- ion. Hence we can say that **a Lewis base is a donor of electron pair or an electron-pair donor**.

The benefit of the hypothesis of Lewis is the manner with which it supports the concept of the oxidation-reduction process in the reaction. The reactions of oxidation and reduction essentially include the exchange of electrons from one species to another. During this process, there is a net change in the oxidation number of the reactive species which are involved in the reaction.

$$Na\cdot + :\ddot{\underset{..}{Cl}}\cdot \longrightarrow [Na^+][:\ddot{\underset{..}{Cl}}:^-]$$

The Lewis hypothesis recommends that sharing of electrons happens when an acid reacts with a base, as a result of which there is no change in the oxidation states or oxidation number of any of the atoms. There are several chemical reactions that can be classified into these classes. There are two ways to exchange a pair of electrons between the species (1) electrons are moved from one atom to another atom (2) atoms of the molecule come closer to each other so that they can perform the sharing of electrons.

The important benefit of the theory of Lewis acid-base is how it increases the number of acids resulting in the increase in the number of acid-base reactions. According to the hypothesis of Lewis, an acid is any molecule or ion that is capable of gaining or receiving a pair of valenceelectrons (non-bonding electrons). Al^{3+} ions are assumed to form bonds with six molecules of water in order to give an intricate complex ion.

$$\mathbf{Al^{3+}(\mathit{aq}) + 6\ H_2O(\mathit{l}) \rightleftharpoons Al(H_2O)_6{}^{3+}(\mathit{aq})}$$

The above reaction is an example of a Lewis acid-base. According to the Lewis structure of water, it contains pairs of nonbonding electrons and hence it is considered as a Lewis base. Al^{3+} ion's electronic structure has vacant 3*s*, 3*p*, and 3*d* orbitals which are used to accommodate the pairs of nonbonding electrons which are given by neighbouring molecules of water.

$$\mathbf{Al^{3+} = [Ne]\ 3\mathit{s}^0\ 3\mathit{p}^0\ 3\mathit{d}^0}$$

In the above reaction $Al(H_2O)_6{}^{3+}$ ion is formed, when the surrounding water molecules which behave as a Lewis base, donate six pair of electrons to the aluminium ion (Al^{3+}) which acts as a Lewis acid and form an **acid-base complex** or **complex ion**.

The theory of Lewis acid-base can clarify easily the reason behind the reaction of BF3 molecules with ammonia. The shape of a BF_3 molecule is trigonal-planar. The boron atom is present in the centre of the molecule having sp^2 hybridization, and has an empty $2p_z$ orbital. The vacant orbitals of boron can accept electrons and hence the BF_3 molecule behaves as an acceptor of electron-pair or also called Lewis acid. The vacant $2p_z$ orbital can be used to accommodate the nonbonding electrons donated by a Lewis base resulting in the formation of a covalent bond. Therefore BF_3 molecule reacts with NH_3 which acts as Lewis base forming an acid-base complex having a filled orbital of all the atoms which are occupied by valence electrons. The theory of Lewis acid-base may also explain why nonmetal oxides for example CO_2 get dissolved in water to form acids, like H_2CO_3(carbonic acid).

$$\mathbf{CO_2(\mathit{g}) + H_2O(\mathit{l}) \rightleftharpoons H_2CO_3(\mathit{aq})}$$

In the above reaction, the H_2O molecule behaves as a donor of an electron-pair or as a Lewis base. The carbon atom of the CO_2 molecule acts as an electron-pair acceptor. The carbon atom does not need the double bond with its two oxygen atom as it receives a pair of electrons from the surrounding molecule of water atoms as shown in the figure below,

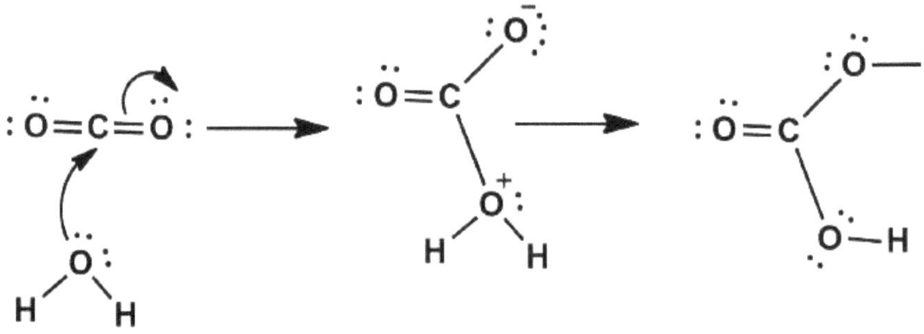

If water is added to CO_2 an intermediate is formed. One of the oxygen atoms holds a positive charge in its intermediate form and another bears a negative charge. One H^+ ion is transferred as a result of which the molecule comprises all the oxygen atoms as neutral charge. Hence the formation of a product resulting from the reaction of CO_2 and water is a carbonic acid, H_2CO_3.

pH Scale

Water in its pure form behaves as a weak acid as well as a weak base. Water produces a few numbers of H_3O^+ and OH^- ions which indicates that it is an aqueous solution of both strong acids and strong bases.

$$\mathbf{H_2O(l) + H_2O(l) \rightleftharpoons H_3O^+(aq) + OH^-(aq)}$$

Base acid acid base

The concentrations of hydronium (H_3O^+) ion and hydroxide (OH^-) ions in water arecarefully determined by measuring the ability of water to conduct electricity. At room temperature, these ions have a concentration of 1.0 x 10^{-7} moles per liter.

$$[H_3O^+] = [OH^-] = 1.0 \times 10^{-7} \text{ M (at 25°C)}$$

On adding a strong acid to water, there occurs a rapid increase in the concentration of the H_3O^+ions.

$$\mathbf{HCl\ (aq) + H_2O(l) \rightleftharpoons H_3O^+(aq) + Cl^-(aq)}$$

Simultaneously, there is a decline in the concentration of hydroxide (OH^-) ions as the hydronium (H_3O^+) ions generated in the reaction are used in the neutralization of the OH^- ions in water.

$$\mathbf{H_3O^+(aq) + OH^-(aq) \rightleftharpoons 2\ H_2O(l)}$$

The value of the product of the concentration of two ions *i.e.* H_3O^+ and OH^- ions is constant, regardless of the fact that how much acid or base is being added to water. The resulting product of the concentration of these ions is found to be 1.0 x 10^{-14} in pure water at 25°C

$$\mathbf{[H_3O^+][OH^-] = 1.0 \times 10^{-14}}$$

In an aqueous solution, the concentrations of the OH^-ions and H_3O^+ ions are very large because of which is very difficult to work with. The Danish biochemist S. P. L. Sorenson in the year 1909 reported the H_3O^+ ion concentration on a logarithmic scale and called its pH scale. Since the concentration of the H_3O^+ ion is often lower than 1, a negative number is the log of such concentration. To avoid constant use of -ve numbers, he described pH as the negative log of the concentration of H_3O^+ ion.

$$\mathbf{pH = -\log [H_3O^+]}$$

The pH concept confined the range of concentration of H_3O^+ ion into a scale thatis much more convenient to handle. As the concentration of H_3O^+ ion decline from around 100 to 10^{-14}, the pH of the solution isincreased from 0 to 14.

If the H_3O^+ ion concentration in pure water is 1.0 x 10M, at the temperature of 25°C, the pH value of pure water is 7.

$$\mathbf{pH = -\log [H_3O^+] = -\log (1.0 \times 10^{-7}) = 7}$$

If the pH value of a solution is less than 7, then it is called acidic solution and if the pH value of the solution is greater than 7, then it is called a basic solution [6].

HYDROGEN BONDING IN WATER

Polar molecules such as molecules of water bear a slightly negative charge at one end (the oxygen atom in water) and a slight positive charge on the other end (the hydrogen atoms in water) as shown in Fig. (**2**). When molecules of water are close to each other, their ends possessing the positive and negative charge get attracted to the oppositely charged ends of neighbouring molecules. The force of attraction among the molecules is shown as a dotted line called a hydrogen bond. Each water molecule is bonded to four other water molecules as shown in Fig. (**2**).

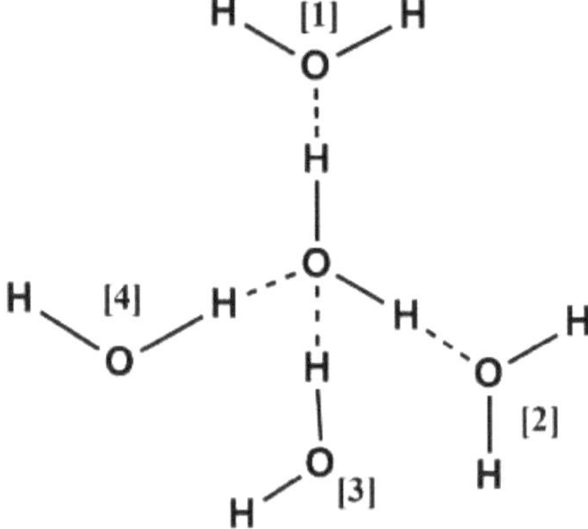

Fig. (2). Bonding in the water molecule.

Some of the essential and special properties of water are due to the presence of hydrogen bonds between water molecules. The obsession with hydrogen bonds holds water fluid over a broader temperature spectrum that is seen with any other molecule of its size.

The energy required to beat the attractive forces among the water molecules *i.e.* the hydrogen bonds, allow water to possess a strong heat of vaporization, there by requiring a large amount of energy to turn the liquid water into a vaporising state.

The hydrogen bond is a weak bond and the strength of a covalent bond is just 5 percent. At points where various hydrogen bonds are formed between two atoms (or parts of a similar particle), the relationship that follows is sufficiently strong and very stable [7].

TYPES OF HYDROGEN BOND

Hydrogen bonding is an interesting occurrence. It is responsible for changes in the physical properties of any substances. Hydrogen bonding is a bond between hydrogen, which is a highly electropositive element and a highly electronegative element such as nitrogen, oxygen and fluorine. A hydrogen bond is not an actual chemical bond in the true sense. It's an attraction between two, unlike poles. Hence, it is a dipole-dipole interaction. There are two types of hydrogen bonds, namely intermolecular hydrogen bond and intramolecular hydrogen bond.

If there is hydrogen bonding between two atoms of two different molecules, then it is called an intermolecular hydrogen bond.

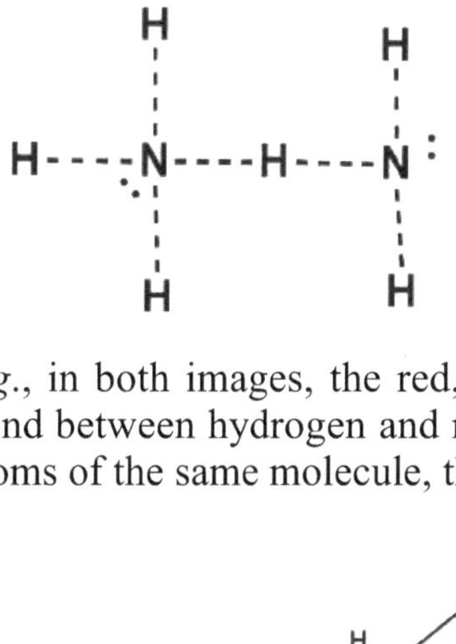

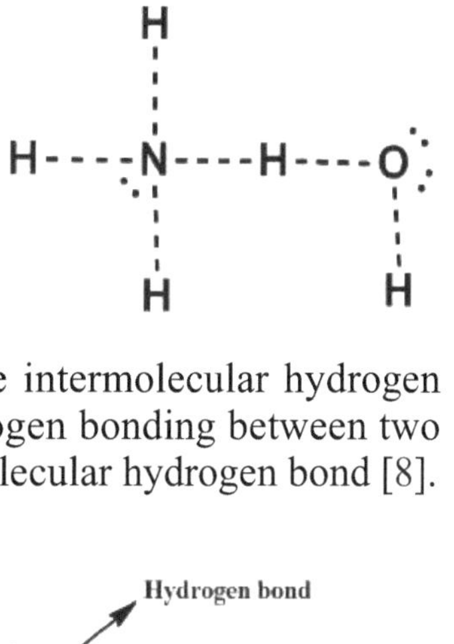

e.g., in both images, the red, dotted line indicates the intermolecular hydrogen bond between hydrogen and nitrogen. If there is hydrogen bonding between two atoms of the same molecule, then it is called an intramolecular hydrogen bond [8].

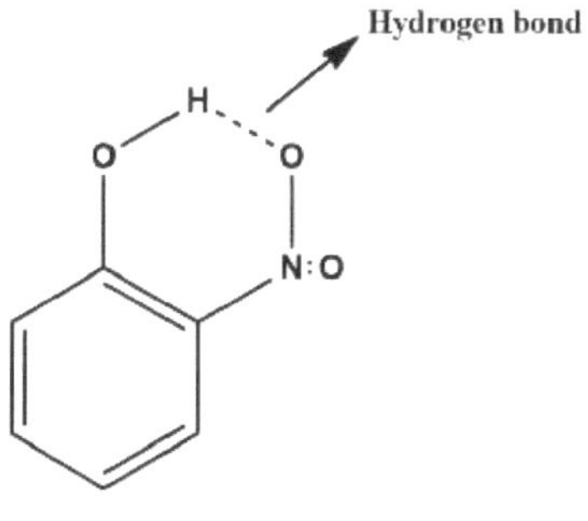

Hydrogen bond

Hydrogen bond

O-nitro phenol

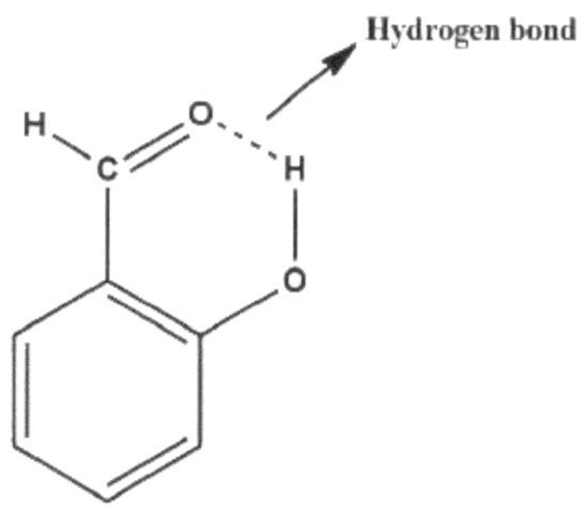

Salicyladehyde

E.g. in this image, the dotted line indicates an intramolecular hydrogen bond between oxygen and hydrogen.

Another example is shown below,

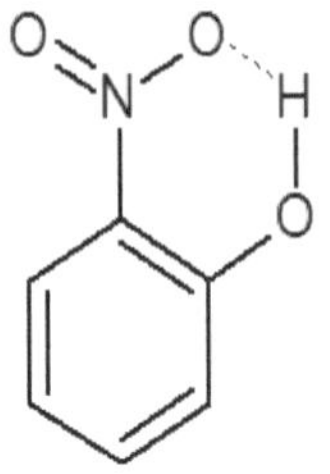

o-nitrophenol (intramolecular hydrogen bonding)

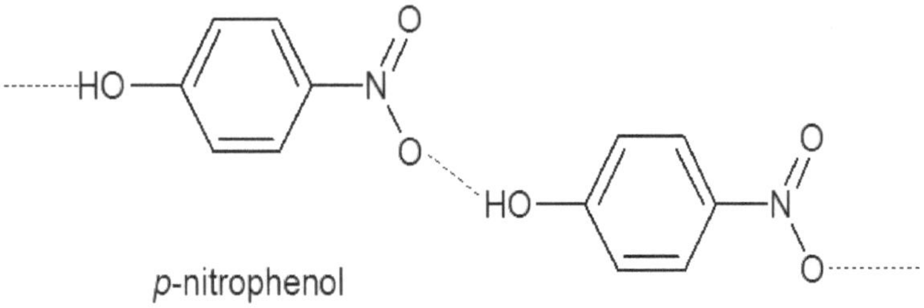

p-nitrophenol (intermolecular hydrogen bonding)

Intermolecular H-bonding increases intermolecular attraction hence increases the boiling point, viscosity and surface tension. Whereas, intramolecular H-bonding tends to decrease the surface area of the molecule hence decreases the boiling point and surface tension [9].

HYDROPHOBIC

The word 'hydrophobic' came from the Greek roots hydro- (meaning 'water') and -phobia (meaning 'fearing'). The word 'hydrophobic' relates to the certainty that nonpolar substances do not dissolve in water molecules. Universal solvent water is a polar molecule, which means that it sustains a partial net charge among its atoms. Oxygen, being an electronegative atom, attracts the electrons associated with each bond closer to its nucleus, thus generating a more negative charge on it. Therefore, any substance with a charge, either negative or positive, can interact with water molecules. So basically, hydrophobic particles are atoms that don't have a charge, which means they are non-polar. These particles don't have any charge-to-charge connections that will permit them to associate with water. Hydrophobic materials don't disintegrate in water or in any arrangement that contains a generally watery condition. Waxes are made up of hydrophobic atoms that are utilized industrially and organically as a result of their capacities to oppose connecting with water. Commercial waxes repel water to cause beading [10].

HISTORY

In the year 1781, Henry Cavendish proved that water was made from oxygen and hydrogen. Initially, the water molecule wasdecayed into hydrogen and oxygen, by the process of electrolysis, in the year 1800 by two English scientists, William Nicholson and Anthony Carlisle. Later on in the year 1805, the two scientists named as Joseph Louis Gay-lussac and Alexander von Humboldt told that water molecule consists of two parts of hydrogen and one part of oxygen. Gilbert Newton Lewis segregated the primary example of pure heavy water in 1933.

The properties of water have been commonly used to describe various temperature scales. Eminently, the scales of Kelvin, Rankine, Celsius and Fahrenheit were distinguished by the freezing and breaking points of water. The more rare scales of Delisle, Reaumur, Newton and Romer were also described. The triple point of water is the most utilized standard point today [11].

NOMENCLATURE

The recognised IUPAC (the International Union of Pure and Applied Chemistry) name of water is oxidane or simply it is called water. The name Oxidane is used for the mononuclear parent hydride used by substituent classification for naming water subordinates. Generally, these subordinates have other suggested names like the name hydroxyl is suggested for the – OH group over oxidanyl. For this reason, the IUPAC expressly refers to the word oxane as inadmissible, as it now tetrahydropyran is known as the word of a cyclic ether.

Hydrogen oxide is the basic term used for water. It closely resembles related compounds, for example, H_2O_2 (hydrogen peroxide), H_2S (hydrogen sulfide), and D_2O (deuterium oxide), commonly called heavy water. The name of water is taken hydrogen monoxide by using the terms for ionic binary compounds, but this isn't one of the names published by IUPAC. Another rarely used name of water is dihydrogen monoxide. Some other systematic names of water are hydroxylic acid, and hydrogen hydroxide, which utilizes the names of acid and base. Generally, none of these interesting names are used. In terms of IUPAC, the hydrogen (H^+) and hydride (OH^-) ions of the water molecule are called hydron hydroxide. Generally, the term water substance is being used instead of hydrogen oxide (H_2O) when it is not mentioned whether one is talking about the fluid form of water, vapour form of water, some type of ice, or a segment in a blend or mineral [12].

SUMMARY

The chemical formula of water is H_2O. Water is colorless, odorless and tasteless liquid in its natural state. Water is amphoteric in nature. Arrhenius acid is any species that gets ionized to give the hydrogen (H^+) ions, in water and Arrhenius base is any substance that gives hydroxide (OH^-) ions, when dissolved in an aqueous solution. As per the concept of the Bronsted-Lowry acid-base model, acid is a species that contributes hydrogen ion (H^+) or also called as proton and a base is a species that accepts the hydrogen (H^+) ion or proton. pH (capability of hydrogen) is a logarithmic scale used to determine the extent of acidity or basicity of a watery solution. The hydrogen bond in water is a powerful attraction between neighboring water molecules involving one hydrogen atom situated between the two oxygen atoms.

LONG-ANSWER TYPE QUESTIONS

1. Calculate the pH value of Pepsi if the given concentration of the H_3O^+ ion in the solution is 0.0045 M.
2. For the given acid or base, write the chemical equation of the reactions with water when they donate or accept one hydrogen ion.
   a) Hydrocyanic acid b) Carbonate ion c) Ammonium ion d) Sulfuric acid e) Acetic acid
3. Write a short note on the chemical and physical properties of water.
4. What are the 'pH' of pure water and that of rainwater? Explain the difference. How do you increase or decrease the pH of pure water?
5. Explain the phenomena ‘Hydrophobicity’.
6. Explain the Lewis model of acid-base chemistry.
7. What is the difference between a Lewis acid or base and a Brønsted Lowry acid or base?
8. What is required to happen in Lewis acid-base reactions?
9. Draw the Lewis structures of the reactants and product of each of the following equations, and identify the Lewis acid and the Lewis base in each equation:

   a) $CO_2 + OH^- \longrightarrow HCO_3^-$
   b) $B(OH)_3 + OH^- \longrightarrow B(OH)_4^-$
   c) $I^- + I_2 \longrightarrow I_3^-$
   d) $AlCl_3 + Cl^- \longrightarrow AlCl_4^-$ (use Al-Cl single bonds)
   e) $O^{2-} + SO_3 \longrightarrow SO_4^{2-}$

10. Boric acid H_3BO_3, is a Lewis acid.
    (a) Write an equation for its reaction with water.
    (b) Predict the shape of the anion thus formed in the reaction.

SHORT-ANSWER TYPE QUESTION

1. Name and describe three acid-base base theories.

2. What's the difference between an acidic and alkaline buffer?
3. What is the difference between the strength and concentration of an acid?
4. State appropriate reason for the following;
   a) Crude water provides electricity while the distilled water does not.
   b) Dry HCl gas does not turn blue litmus red while the dilute hydrochloric acid does.
   c) Typically a milkman applies a very small amount of baking soda during thesummer season to fresh milk.
   d) Acid is added to water for the dilution of acid and not water to acid.
   e) Ammonia is a base but contains no hydroxyl group.
5. State the chemical properties on which the following uses of baking soda are based:
   (i) as an antacid
   (ii) as a soda acid fire extinguisher
   (iii) to make bread and cake soft and spongy.
6. Name the natural source of the following given acid
   (i) Citric acid. (ii) Lactic acid
   (iii) Oxalic acid. (iv) Tartaric acid.
7. The pH of four unknown solutions A, B, C and D isas follows 11, 7, 5 and 2.What is the nature of the solution.
8. Write the name of the bases which are highly soluble in water. Also, give an example.
9. How is the decay of tooth-related to pH? How can it be prevented?
10. Why does bee sting cause pain and irritation? How the rubbing of baking soda on the sting area is gives relief?
11. "Sodium hydrogen carbonate is a basic salt". Justify the statement. How it can be converted into washing soda?
12. What is rock salt? Mention its colour and the reason because of which it has this colour.
13. What is the pH range of our body? Explain how the use of antacids gives relief from acidity. Give the name of one such antacid.
14. Fresh milk has a pH value of 6. How does the pH value will change as it gets converted into curd? Explain your answer.
15. A milkman adds a very small amount of baking soda to fresh milk. Why does this milk take a longer time to set as curd?
16. What is the nature of toothpaste? How do they help to prevent tooth decay?

MULTIPLE CHOICE QUESTIONS

1. Water is a

(a) Polar solvent

(b) Nonpolar solvent

(c) Amphipathic solvent

(d) Nonpolar uncharged solvent

2. The polar molecule can readily dissolve in water. This is because

(a) The polar molecule can form a hydrogen bond with water

(b) The polar molecule can replace water-water interaction with more energetically water solute interaction

(c) Polar charged water can interact with the charge of a polar molecule

(d) All polar molecules are amphipathic in nature

3. Which of the following compound can be Bronsted acid and Bronsted base

(a) H_2O

(b) NH_3

(c) HSO^{4-}

(d) OH^-

4. Which of these compounds are Bronsted-Lowry acids? There may be more than one answer.

(a) HCl

(b) NaCl

(c) H_2O

(d) NH_3

5. Which of these compounds are Bronsted-Lowry bases? There may be more than one answer.

(a) HBr

(b) CH_3O^-

(c) NH^{2-}

(d) HCl

6. Which list of acids and bases is not correct?

(a) Bronsted-Lowry Acid: HCl, H_2O, CH_3COOH

(b) Bronsted-Lowry Base: -OH, TsOH, -H

(c) Lewis Acid: $AlCl_3$, HCl, H_2O

(d) Lewis Base: -OH, $-NH_2$, $CH_2=CH_2$

7. A substance that donates a pair of electrons to form a coordinate covalent bond is called

(a) Lewis acid

(b) Lewis base

(c) Bronsted-Lowry acid

(d) Bronsted-Lowry base

8. A species that is able to accept a proton is called

(a) Acid

(b) Base

(c) Neutral compound

(d) Cation

9. Substances that react with both acids and bases are called

(a) Neutral

(b) Conjugate bases

(c) Amphoteric substances

(d) Conjugate acids

10. Bronsted-Lowery acid in reaction $H_2O + NH_3$ $NH^{4+} + OH^{-}$ is

(a) H_2O

(b) NH_3

(c) OH^{-}

(d) NH^{4+}

11. Mg^{2+}, Ca^{2+} and Sr^{2+} ions form stable complexes with crown ethers. This involves interactions between:

(a) soft acid and hard base

(b) soft acid and soft base

(c) hard acid and hard base

(d) hard acid and soft base

12. $BeCl_2$ is an example of a:

(a) Lewis base

(b) Bronsted base

(c) Bronsted acid

(d) Lewis acid

13. Which of the following compounds reacts with water, the other compounds being soluble or sparingly soluble in water?

(a) $Ca(OH)_2$

(b) CaH_2

(c) $CaBr_2$

(d) $Ca(NO_3)_2$

14. Why water is considered a polar molecule?

(a) The hydrogen end is slightly negative and the oxygen end is slightly positive

(b) The hydrogen and oxygen are covalently bonded

(c) Because it is hydrophobic

(d) Because it is hydrophilic

15. Attractions between water molecules are called

(a) Covalent bonds

(b) Ionic bonds

(c) Polar bonds

(d) Hydrogen bonds

16. $Ca(OH)_2$ is an example of an Arrhenius base. What can be inferred from the given statement?

(a) $Ca(OH)_2$ accepts OH^- ions when dissolved in water.

(b) $Ca(OH)_2$ releases H^+ ions when dissolved in water.

(c) $Ca(OH)_2$ releases OH^- ions when dissolved in water.

(d) $Ca(OH)_2$ accepts H^+ ions when dissolved in water.

17. Which of the following is an example of an Arrhenius base?

(a) KCl

(b) HNO_3

(c) NH_3

(d) KOH

18. Which of the following represents the correct Arrhenius definition of bases?

(a) These are the substances that accept H^+ ions when dissolved in water.

(b) These are the substances that release OH^- ions when dissolved in water.

(c) These are the substances that release H^+ ions when dissolved in water.

(d) These are the substances that accept OH^- ions when dissolved in water

19. Which of the following contain(s) intramolecular hydrogen bonding?

I. NH_3 II. HF III. HCOH IV. $CH_3CH_2CH_2OH$

(a) I only

(b) I, II, III, IV

(c) II and IV

(d) II and I

CONSENT FOR PUBLICATION

Not Applicable.

CONFLICT OF INTEREST

The authors confirm that this chapter contents have no conflict of interest.

ACKNOWLEDGEMENTS

Authors are thankful to Department of Chemistry, North Eastern Regional Institute of Science and Technology, Nirjuli, Itanagar for providing facilities to accomplish the work.

REFERENCES

[1] Chaplin M. Water molecule structure. [updated 16th June 2019] Available from: http://www1.lsbu.ac.uk/water/water_molecule.html

[2] Lowry TM. The uniqueness of hydrogen. J Soc Chem Ind 1923; 42(3): 43-7. [http://dx.doi.org/10.1002/jctb.5000420302]

[3] Myers R. The Basic of Chemistry. Connecticut: Greenwood Press 2003; pp. 157-61.

[4] Brent L. Acids and Bases. New York: Crabtree Publishing Company 2009.

[5] Hulanicki A, Masson MR. Reactions of Acids and Bases in Analytical Chemistry. England: Ellis Horwood. New York: Halsted Press 1987.

[6] Bates RG. Determination of pH; Theory and Practice. New York: Wiley 1973. [http://dx.doi.org/10.1149/1.2403829]

[7] National Center for Biotechnology Information. PubChem Database. Hydron; hydroxide. [Updated 1st February 2020] Available from: https:// pubchem.ncbi.nlm.nih.gov/compound/22247451

[8] Chaplin M. Hydrogen bonding in water. [updated 18th December 2019]. Available from: http://www1.lsbu.ac.uk/water/water_hydrogen_bonding.html

[9] Lewis GN, MacDonald RT. The concentration of H_2 Isotope. J Chem Phys 1933; 1(6): 341-4. [http://dx.doi.org/10.1063/1.1749300]

[10] Chandler D. Interfaces and the driving force of hydrophobic assembly. Nature 2005; 437(7059): 640-7. [http://dx.doi.org/10.1038/nature04162] [PMID: 16193038]

[11] Greenwood NN, Earnshaw A. Chemistry of the Elements. 2nd ed., Oxford: Butterworth-Heinemann 1997.

[12] Leigh GJ, Favre HA, Metanomski WV. Principles of Chemical Nomenclature: A Guide to IUPAC Recommendations. Blackwell Science 1998.

CHAPTER 3

Electrostatic and Van Der Waals Forces

Nene Takio[1], **Dencil Basumatary**[1], **Meera Yadav**[1,*] and **Hardeo Singh Yadav**[1]

[1] *Department of Chemistry, North Eastern Regional Institute of Science And Technology (NERIST), Nirjuli, Itanagar-791109, Arunachal Pradesh, India*

Abstract: Non-covalent intermolecular interactions are the fundamental forces which defines all the shape and structural stability of the biomolecules. Hydrogen bonds hold strands of DNA together and also responsible for determining the three-dimensional structure of folded proteins, whereas electrostatic and vander waals interactions are found to be a dominant factor in determining conformation of biomolecules. Noncovalent forces are individually weak relative to covalent bonds, but the cumulative effect of such interactions in a protein or nucleic acid can be very significant. They all play a fundamental role in fields as diverse as supramolecular chemistry, structural biology, polymer science and nanotechnology.

Keywords: Covalent and non covalent forces, Dipole-Dipole interactions, Ionic bonds, Thermophiles: Supramolecules, Vander waals contact distance.

INTRODUCTION

Molecular interactions, also known as non-covalent interactions, are attractive or repulsive forces involving atoms and molecules. These are significant for the determination of structural biomolecules, drug designing, material science, sensors, nanotechnology, and chemical sciences. In the biological processes, non-covalent interactions allow protein and ribosomal RNAs folding to form globular structures, and also help the formation of the DNA double-helical structure. These stabilize the assembly of proteins with DNA, RNA and other proteins, and help in the formation of phospholipids in the membrane. Hence, understanding the forces behind the interactions between biomolecules is essential, providing a better perspective on how a living system functions.

FUNDAMENTAL FORCES IN BIOMOLECULES

Covalent bonds with high energy formation are considered to be much stronger than the bonds in biomolecules, but still, they manage to hold the complex mole-

* **Corresponding author Meera Yadav:** Department of Chemistry, North Eastern Regional Institute of Science and Technology (NERIST), Nirjuli, Itanagar-791109, Arunachal Pradesh, India. E-mail: drmeerayadav@rediffmail.com

Meera Yadav & Prof. Hardeo Singh Yadav (Eds.)

cular structure with great ease since all these small non-covalent forces have a very strong cumulative effect on bimolecular stability. A double-stranded helical DNA with thousand base pairs held by different stacking forces is considered, as shown in Fig. (**1**).

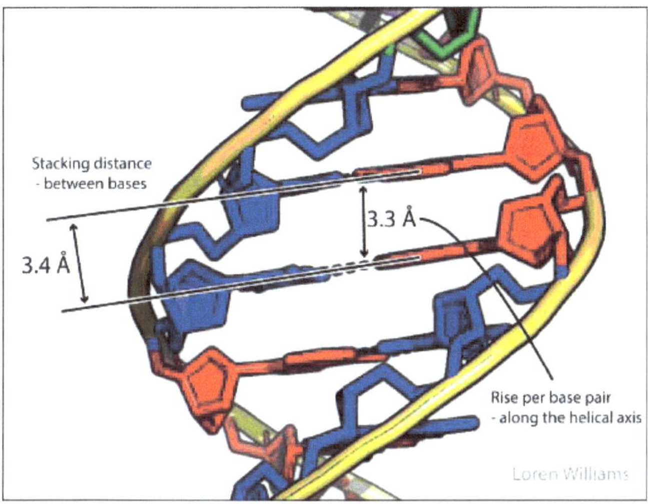

Fig. (1). Base pair stacking in B-DNA. Short range repulsion sets a distance of 3.4 Å between base pairs. Major stacking forces- Hydrophobic Interactions Van der Waals forces. https://ww2.chemistry.gatech.edu/~lw26/structure/molecular_interactions/mol_int.html.

The base pair energy of the average base pair, which is about 0.5 kJ/mol, is not very large, but the energy range of 1000 base pairs is 500 kJ/mol, which is equal to the range of many covalent bonds. It also has important implications for the mobility of individual base pairs. These can be easily opened when whole molecules are held together. These weak and uncoordinated forces play an important role in DNA replication, 3-dimensional protein synthesis, enzyme-substrate complex formation, and also in molecular signals identification. Infact, all biological compositions and processes are interdependent on covalent and non-covalent interactions. Some of the forces responsible for the structural determination of proteins have been discussed below (Fig. **2**).

These include:

Covalent Bonds

1. The breakdown of protein-peptide bonding with the respective amino acids can be achieved by acidic or alkali treatment or by enzymatic treatment. *e.g.*, Proteases.
2. A disulfide bridge is formed between cysteines and cystine. (Cysteine has

another HS- and -SH that forms the disulfide-SS- bridge). The bridge is destroyed by reduction with β-mercaptoethanol and cysteine is reformed [2].

Non-Covalent Bonds

1. Hydrogen bonds – Biological components like proteins are made up of hydrogen bonds thatplay an important role in stability because these are away from water (which disrupts them). They can be broken down by overheating.
2. Van der Waals forces – These are short dipole-dipole (δ+ & δ-) interactions between closely distances atoms. The bonds can be broken down by denaturing agents like heat, chemicals, *etc.*
3. π-π overlap – These are delocalized in rings with π electron clouds and easily break down by heat.
4. Electrostatic or ionic interactions and salt bridges in biomolecules.

These also get easily disrupted by changes in ionic strength or pH.(*e.g.*, positive residues include Arg, His: while negative residues include Asp, Glu, Cys). They all differ in structure, arrangement, strength, and specificity. Also, all the bonds behave differently in the presence of water [1].

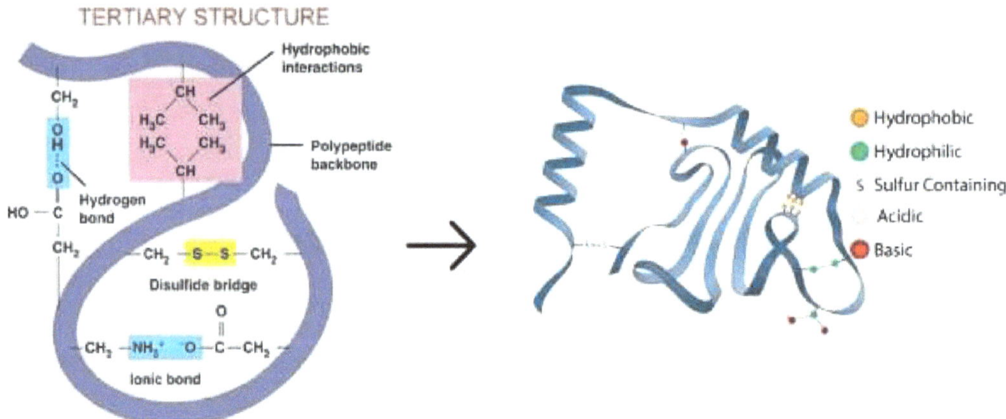

Fig. (2). Different forces in protein stability.

HYDROGEN BONDING

Hydrogen bonding is considered a strong bond that participates in many important chemical processes. This is also responsible for the hydrophobic effect and unique solvent capabilities. This bond formed is between uncharged molecules as well as charged ones. The atom that hydrogen is most strongly bonded to is called H donor, while the other atom is said to be an H acceptor with a partial negative charge that attracts hydrogen. These are the strongest dipole-dipole interactions but much weaker than covalent bonds. These have 1–3 kcal mol^{-1} (4–13 $kJmol^{-1}$) energy compared to approximately 100 kcal mol^{-1}(418 kJ mol^{-1}) of carbon-hydrogen covalent bonds. Hydrogen bonds in biomolecules are more specialized than other weak bonds because these require specific complement groups that donate or accept hydrogen. The significant characteristic of hydrogen bonds is that they are extremely directional. These are an essential part of the biomolecules, such as proteins, nucleic acids and carbohydrates [2]. These bring complementary strands of DNA together and help to determine the three-dimensional structure of folded proteins, antibodies and enzymes.

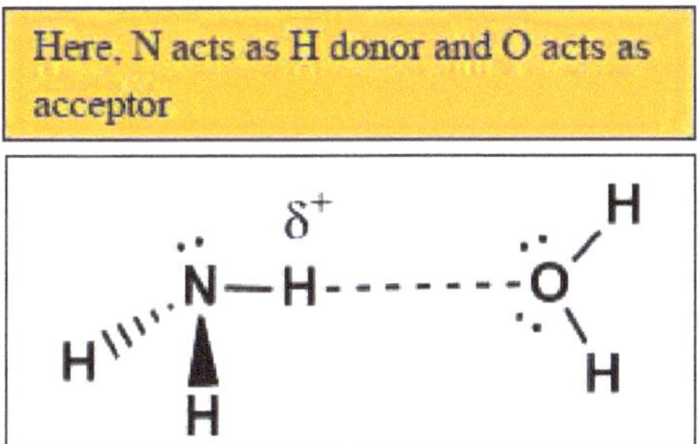

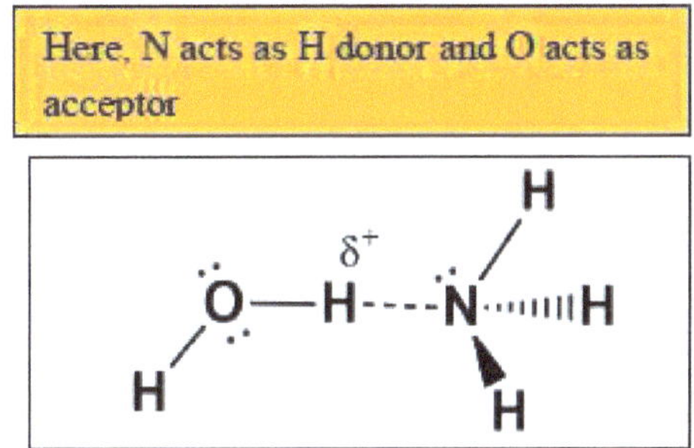

Here, N acts as H donor and O acts as acceptor. Hydrogen bonds are essential for the solubility of compounds and important non-covalent interaction between biomolecules in the cellular environment. The macromolecules fold and organize into three-dimensional structures through intermolecular and intramolecular H bonds, as well as other non-covalent interactions. The double-stranded helical structure of DNA is formed due to the presence of Hydrogen bonding between bases of the complementary strand.

Carbohydrates like starch and glycogen form hydrogen bonding with water due to high solubility. Also, the presence of H bonding in amino acids and globular proteins makes them soluble in water. According to their ability to form an H bond, solvents are categorized as polar and non-polar solvents, where polar solvents form hydrogen bonding with solute molecules, whereas non-polar solvents do not form any bond with water. Examples of polar solvents like water, liquid NH_3, ethanol, acetic acid, *etc.* These are further classified into less polar and high polar solvents based on theirpolarity degree *e.g.* Liquid NH_3 is a less polar solvent compared to water. Hydrogen bonding also participates in many important biochemical reactions. In the form of nutrients, macromolecules are digested in the stomach by various agents; including the H-bond between the substrate exist in the enzyme active site and an important amino acid residue. Residues of important amino acids exist in enzyme active sites interacting with the substrate in a variety of reactions, affecting the catalysis that results in the formation of the product. The cytosol is also a water resource and the various proteins and the dissolution of the biomolecules in the cytosol is only possible by forming hydrogen bonds with the water molecules of the cytosol.

HYDROGEN BONDING AND ITS ROLE IN DNA

DNA is composed of four nitrogen bases: guanine, cytosine, adenine and thiamine. Guanine pairs with cytosine whereas, adenine complements thiamine, bind to each other *via* hydrogen bonds shown in Figs. (**3** and **4**). These hydrogen bonds between complementary nucleotides hold two DNA strands in the double helix. The nitrogenous bases also form hydrogen bonds with the external environment like water. Such internal, as well as external hydrogen bonding, is not strong but the presence of millions of hydrogen bonds in DNA makes it a stable molecule. The helical structure of the DNA is formed *via* the interaction of hydrogen bonds of phosphate groups present on its complementary base pairs. They also sequester the bases present on the interior of the double-helical DNA structure. Therefore, hydrogen bonds between bases enhance the hydrophobic effect of stabilizing DNA. In DNA structure, the hydrophobic bases are the present interior of the helix, hydrophilic part is arranged exterior interacting with the water.

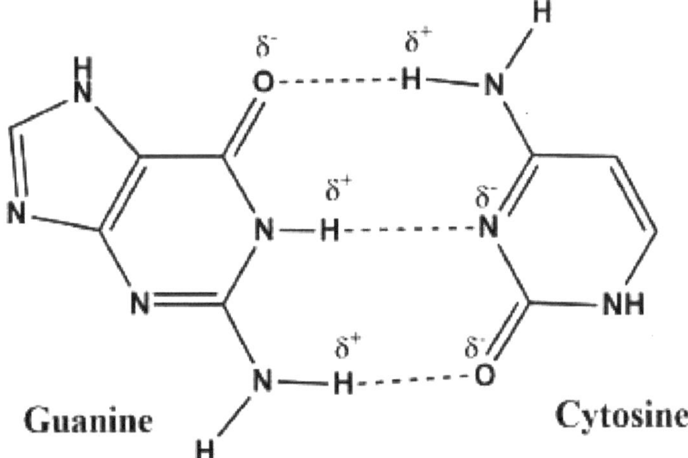

Fig. (3). Hydrogen bonding in base pairs.

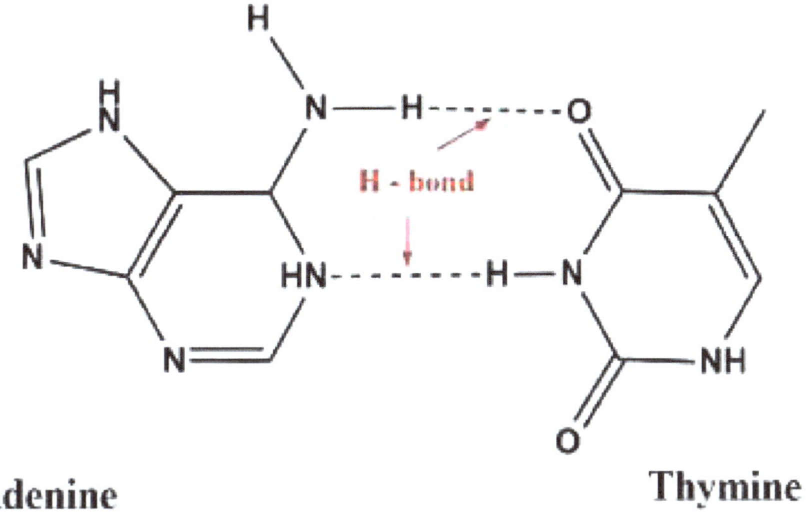

Adenine Thymine

Fig. (4). Hydrogen bonding in complementary base pairs.

Though hydrogen bonding is a weak interaction, its additive effect stabilize DNA molecule andcomplementary base pairs. The bases are maintained by bonding energy of hydrogen ranging from 1 to 5 kcal/mol (4 to 21 kJ/mol).

HYDROGEN BONDING IN PROTEINS

Hydrogen bonds are abundant in the secondary structure of the protein but rare in tertiary conformations. The secondary structure was found to have interactions between the nitrogen - hydrogen bond pairs and the neighbouring polypeptide backbones, which contain oxygen atoms (see Fig. **5**). Since both N and O are electronegative in nature, the hydrogen atoms attached to the nitrogen in the polypeptide backbone form hydrogen bonds with the oxygen atoms in the second chain, and *vice versa*. Although weak in nature, these bonds provide greater stability to the formation of secondary proteins because they repeat themselves. In the tertiary structure of the protein, interactions occur in the polypeptide chain between a functional group called hydrophobic interaction. These interactions are formed between water hydrogen molecules around the hydrophobe and further improve the conformation.

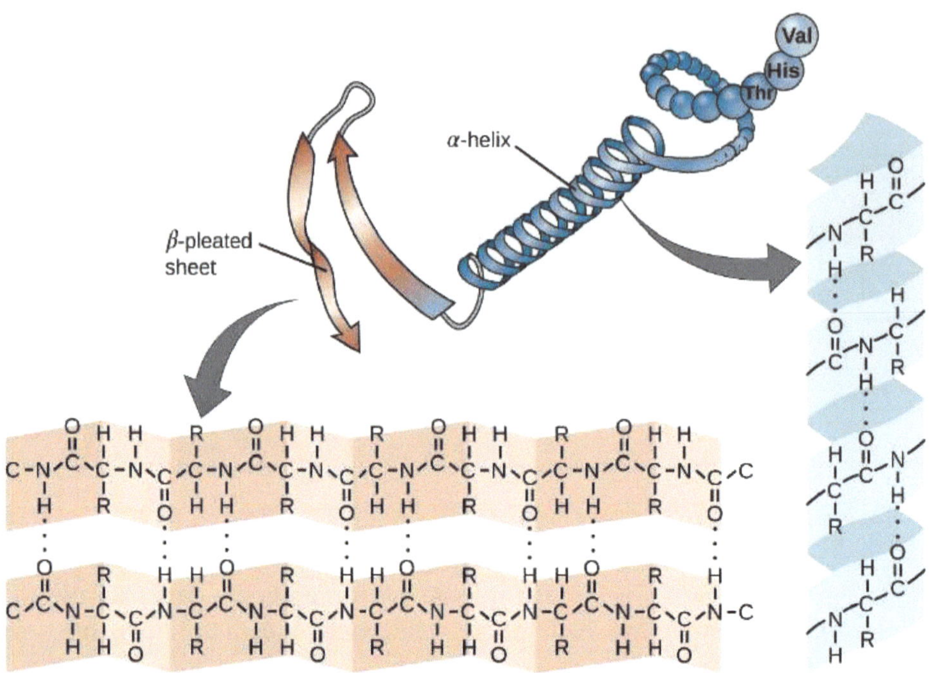

Fig. (5). The Secondary Structure of DNA is due to Hydrogen Bonding.

ELECTROSTATIC INTERACTIONS

Electrostatic interactions are formed between charged species which can be either attractive or repulsive. This is shown by the following equation

$$F = \frac{kq_1q_2}{r^2}$$

In this, k is Coulomb's constant ($k = 8.9875 \times 109 N\ m^2C^{-2}$), q_1 and q_2 are the magnitudes of charges, r is the distance between the charges. The interactive force between charges is attractive if q_1 and $q_2$are opposite in sign, which means they have different charges.

It is considered one of the strongest non-covalent interactions covering greater distances than any other molecular forces. In Fig. (**6**), it was found that electrostatic interactions are a necessary factor in determining the conformation of the protein, which are important for ligand-receptor binding and play a vital role in the stability and functioning of biomolecules. Electrostatic or ionic bonds of crystals and salts ionize in the presence of water [3].

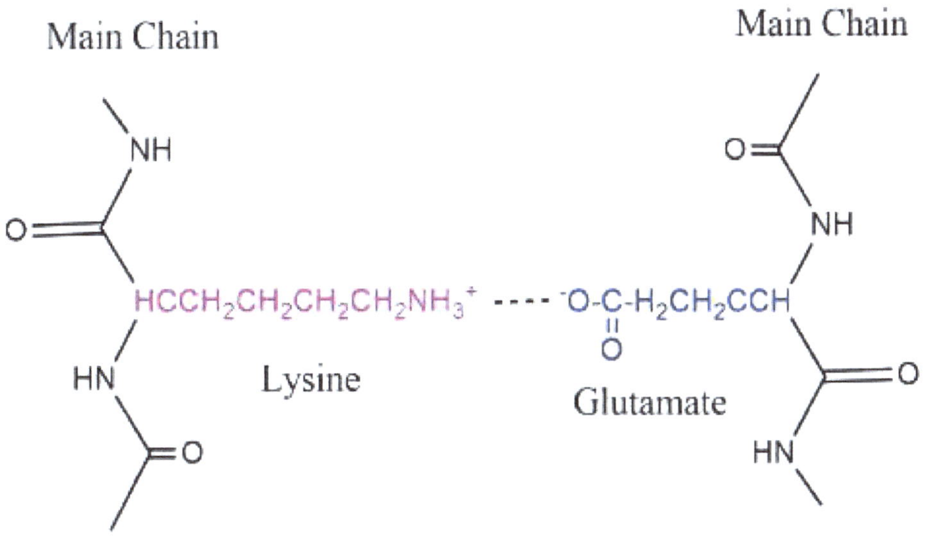

Fig. (6). Electrostatic interactions between positive charge lysine and negative charge glutamate carboxyl group of amine group in proteins.

Salt bridges- These are non-covalent forces between two ionized sites that are made up of two components: one is hydrogen bond and the other is electrostatic interaction. For instance, enzymes either have anionic and cationic sites that bind reactants of opposite charge. The forces of attraction between protein's functional groups with opposite charge are =known as salt bridges.

The salt embedded inside the hydrophobicity of the protein is not affected by water, making it stronger than its surface. Such interactions can be termed Ion-pairing. Electrostatic interaction is also associated with the mutual repulsion of uniformly charged ion groups. Such charged repulsion can affect the formation of individual biomolecules when interacting with other similarly charged molecules. Electrostatic interactions are the main stable forces between phosphate group with magnesium and oxygen ions of RNA molecule as shown in Fig. (7). Many magnesium ions are involved in RNA and DNA structure. During RNA and protein folding, as well as the DNA annealing process, the electrostatic interactions depend on the concentration of salt and pH [4]. Proteins residues have negatively charged ionisable sidechains like aspartic acid, glutamic acid and also have positively charged ones like lysine, histidine, and arginine. The polypeptide chain with amide and carboxyl-terminal can ionize either with positive or negative charges.

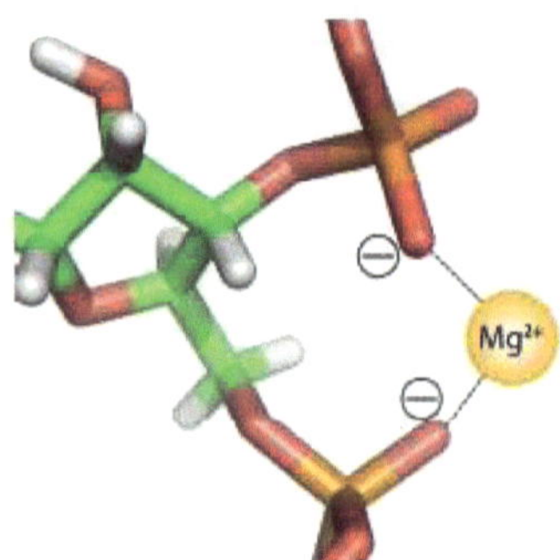

Fig. (7). Electrostatic interactions.

The actual pKa value of the folded group of proteins can be diverting by its own value which depends on the electrostatic field produced. The net charge of any protein is the sum of positive and negative ionized amino acid groups and cofactor ions, and it depends on the pH of its solvent. The pH value of the protein with no net charge is called the isoelectric point, which is represented by the PI.

All amino acids have isoelectric point ranges from 2.8 to 10.8. For example, an acidic amino group like aspartic acid has an IP value of 2.8, it is considered a neutral salt at pH 2.8, which forms negative ions with charges -1and -2 at pH greater than 2.8.

In neutral pH, most amphoteric proteins have cationic and anionic groups at specific locations in the amphoteric with fold structures. An important polar group in a polymer is a peptide bond with a broad dipole moment. Polar groups and ions present in the same or other molecules interact with each other in a polar solvent. In general, pI values for proteins and polypeptides are calculated by classical isoelectric focusing (CIEF) in slab gels or free solution. Mix a uniform protein solution with a particular polymer to purify proteins using the isoelectric focusing process. This polymer has some properties that ensure that the protein forms a pH gradient when the electric field is spread across the solution. As the pH gradient grows, the protein molecules move simultaneously in the solution until these molecules reach the isoelectric point of the protein. Therefore, this accepted method purifies complex protein mixtures by helping proteins reach their respective isoelectric points. In the electrophoresis method, an electric current is passed into a separate mixture of amino acids, where positively charged amino acids go to the negative electrode, and amino acids with a negative charge to a positive electrode, and conversely, since the amino acids in PI do not move, they form separate bands on the filter paper or thin layer Plate.

ROLE IN PROTEIN FOLDING

Examination ofthe distribution of charge on protein surfaces shows that charged groups are, normally surrounded by opposing signals and form ion pairs, which contribute to transport stability. The intramolecular electrostatic energy of a single pair of ions is either exposed to solvents or buried within the protein, which varies from 0.5 to 5 kcal/mol. The entropy of a rigid salt bridge is the factor that opposes and decreases the free energy of folded protein. The cooperative network of salt bridges contributes significantly to protein stability, which is often observed on the surface of thermophilic proteins. The acid residues occur at the polarity of the helical dipole at the N-terminal end of the helix, and the main residues occur at the negative pole. These are propeller covers. These interactions do not depend on the context, but only on the first or second turn of the helix, which, regardless of the length of the α-helix, contributes significantly to the stability of the protein, not exceeding 2 kcal/mol. Unfavourable electrostatic reactions can destabilize the conformation of protein folding. Hydrophobic core unpaired charge and isolated hydrogen bonds are energy expensive due to their high self-energy. Secondary structures such as α- helix and sheets of protein core can generally be presented as simple framework structures because the central chain does not have a pair of backbone structures of hydrogen donors or acceptors.

MOLECULAR RECOGNITION

Molecular recognition is an important tool in the field of modern biology and biochemistry. Many biological phenomena involve the identification of specific molecules of individual biopolymers. Interprotein interactions are important for the transmission of biological signals and form a combination of multi-factorial proteins commonly found in cell organs, cytoplasmic cells and cell membranes. The molecular level of a protein helps in the identification of different molecules. Hydrogen bonds and salt bridges usually have specific electrostatic properties between molecular surfaces. In the protein interface,approx. 0.88 hydrogen bonds per 100 Å feet ofaccessible surfaces are found, with 0-5 intersubunit bridges between the protein dimers.

VANDER WAALS INTERACTIONS

London scattering forces or VanderWaal forces are weak attractions between atoms that are close to each other. These are non-specific interatomic attractions and come into play when two free atoms are separated at 3-4 Å apart. The interaction depends on the electronic charge distribution around the atom and how it fluctuates over time. The attraction between two atoms increases when these come closer until they are separated by a Vander Waal contact distance, then these experience repulsion due to potential energy. There is a cloud of electrons around every molecule, so even if the atom is neutral, it can repel. When these molecules approach each other, repulsive energy is generated between molecules, and as the molecules separate from each other, the potential energy generated by the repulsive force decreases. For example, the contact gap between oxygen and carbon atoms shown in Figs. (**8** and **9**). For example, the contact distance between O and C is 3.4 Å is obtained by adding the contact radii of the O and C atoms by adding 1.4 and 2.0, respectively [5].

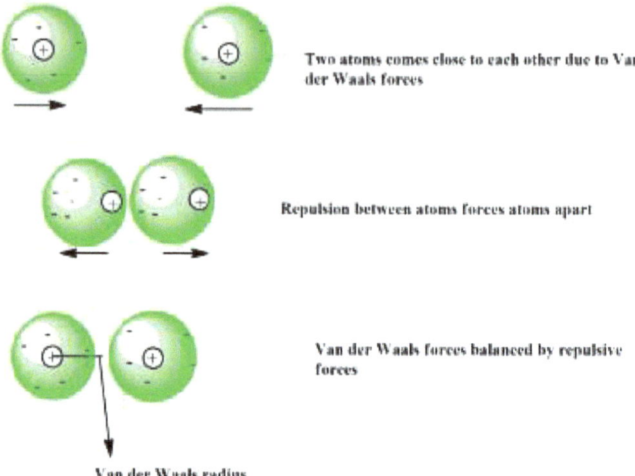

Fig. (8). Schematic diagram of Van der Waals forces of attraction.

Atom	VDW Radius (rin Å)
H	1.2
C	1.7
N	1.6
O	1.5
P	1.8
S	1.8

Fig. (9). Van der Waalscontact radii of atoms and repulsion between two atoms.

Atom	VDW Radius (rin Å)
H	1.2

(Table 9) cont.....

Atom	VDW Radius (rin Å)
C	1.7
N	1.6
O	1.5
P	1.8
S	1.8

The energies involved with Vander Waals interactions are very small; the usualinteractions contribute from 0.5 to 1.0 kcal-mol^{-1} (from 2 to 4 kJmol^{-1}) per atom pair. However, when a large number of atoms come closer within Vander Waals's contact distance, the cumulative effect of all the atoms can be very significant (Fig. **9**). Van der Waals contact radii of atoms and repulsion between two atoms.

As seen in Fig. (**10**) above, the most desirable energy is the Van der Waals contact distance. Due to electron-electron repulsion, the energy keeps on increasing as atoms pass closer together than this distance. They are very weak forces; hence due to the presence of such forces, crystalline substances have low melting points and are generally soft in nature. The more symmetrical the atom, the greater the Vander Waal forces. These forces are more important for compounds than the atoms and molecules in elements. The Forces of Vander Waals also play a crucial role in various fields such as supramolecular and physical chemistry, structural biology, polymer science, nanotechnology, surface science and condensed matter physics [6]. Vander Waalsforces also help in defining physical properties like solubility in the polar and non-polar solvent of many organic compounds as well as molecular solids.

Vander Waals forces are divided into three types:

- Dipole-dipole forces
- Dipole-induced dipole forces
- London dispersion forces

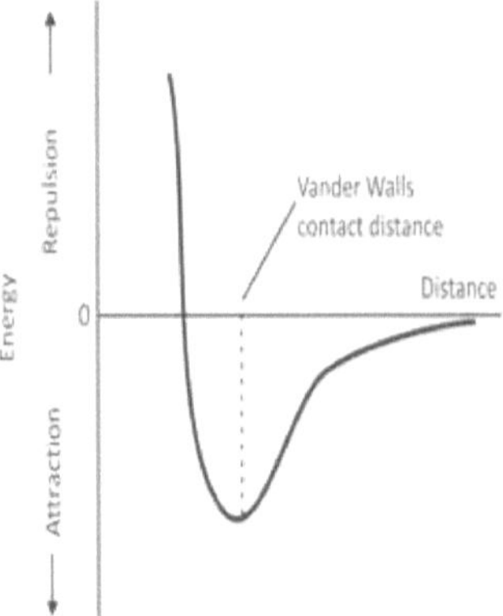

Fig. (10). Energy diagram ofvan der Waals interaction between two atoms and its contact distance.

Dipole-dipole/Keesom Interactions

Dipole-dipole force is generated between molecules with the permanent dipole. Like HCl, chlorine is more electronegative than hydrogen, so its molecule has a permanent dipole. These permanent built-in dipoles attract molecules to each other. Molecules with permanent dipoles have a higher boiling point than molecules with only time-varying dipoles. For example, the boiling points of ethane, CH_3CH_3, and fluoromethaneas 184.5 K and 194.7.

Both have equal electron numbers and sizes. The high boiling point of fluorine is due to the high electronegativity of fluorine and the large permanent dipoles of the molecule.

Dipole Induced Dipole/Debye Forces

It is a weak interaction where a dipole is induced by polar molecules to non-polar molecules. These induced dipoles arise when a permanent dipole molecule repels the electrons of another molecule. A permanent dipole molecule causes mutual interactions between the adjacent excited dipoles. The energies between excited and permanent dipoles do not depend on temperature unlike Keesom dipole interactions, since the excited dipole can move and rotate freely around a non-polar molecule. It is much weaker than dipole-dipole interactions.

For instance, the oxygen molecule is a non-polar molecule [7] but induced a dipole when a polar molecule like water interacts with the O_2 molecules.

London Dispersion Forces

London or dispersion forces arise from non-immediate dipole moments of both atoms and molecules. Such polarizations may be caused either by a polar molecule or by the repulsion of the negatively charged electron cloud in non-polar molecules. As a consequence, London interactions are caused by the random electron density changes in the electron cloud.

The dispersion force (London) is the most important factor, since all materials are polarizable, whereas the Keesom and Debye forces need permanent dipoles. In a symmetric molecule like hydrogen, the electrons are in continuous motion. At any given moment, they will be at one end of the molecule, making that end of the molecule δ^- while the other end of the molecule becomes δ^+. At any given moment, the electrons reverse the polarity of the molecule by moving to the opposite end. This leads to the fluctuation of dipole moments of symmetrical molecules. For example, if a non-polar molecule approaches the polar molecule of temporary polarity. The positive end of the left-hand molecule absorbs the electrons as the right-hand molecule approaches. This triggers the induced dipole in the approaching molecule. The molecules are arranged in such a way that the δ^+ end of one is drawn to the δ^- end of the other. As long as the molecules are close together, this coordinated movement of electrons will take place over a large number of molecules. This diagram illustrates how the entire lattice of molecules could be kept together in a solid with the help of Van der Waals's dispersion powers.

HYDROPHOBIC INTERACTION

Hydrophobic interactions represent the relationship between water and hydrophobic substances (molecules that are poorly soluble in water). Hydrophobic substances are nonpolar molecules and usually have long carbon chains that do not interfere with water molecules. A mixture of fat and water is a good example of this particular interaction. The general belief is that water and fat do not bind because the van der Waals forces acting on the water and fat molecules are too weak. Here, that's not the case. The behaviour of the fat droplet in the water is more related to the entropy and enthalpy of the reaction than to its intermolecular power.

Causes of Hydrophobic Interactions

Non-polar molecules such as fat molecules stick together rather than diffusing into an aqueous medium because fat and water molecules have limited contact interaction. The image above shows that hydrophobes have less contact with water since a total of 16 water molecules interact before conjugation and only 10 actually after the interaction. Hydrophobic interactions are the primary drivers of macromolecule folding, substrate binding of enzymes, and membrane formation that define the boundaries of cells and their internal compartments.

Thermodynamics of Hydrophobic Interactions

As a hydrophobe enters an aqueous environment, the hydrogen bond between water molecules is broken and the hydrophobic is allowed to enter the system, but the water molecules do not combine with hydrophobic molecules. This is called an endothermic reaction, as heat is released into the system when the bond is broken. Water molecules are bent by the presence of hydrophobes, forming new hydrogen bonds, and creating ice-like cages known as clathrate cages around hydrophobes. This approach makes the system (hydrophobic) more organised and thus decreases the total entropy of the system by $\Delta S < 0$. Changes in system enthalpy (ΔH) may be negative, zero, or positive, as new hydrogen bonds may partially or fully compensate for broken hydrogen bonds as hydrogen enters the substance. However, due to the large changes in entropy (s), enthalpy changes are not important to determine the reaction's spontaneity.

Formation of Hydrophobic Interactions

The fusion of hydrophobes and water molecules is not spontaneous; however, there are spontaneous hydrophobic interactions between hydrophobes. When hydrophobes come together and interact with each other, enthalpy increases (ΔH is positive) as some of the hydrogen bonds that make up the clathrate cage will be broken. The entropy will increase by breaking down a portion of the clathrate frame (ΔS is positive)

Accordingly, itwill be $\Delta H = +ve(small)$

$\Delta S = +ve(large)$

Strength of Hydrophobic Interactions

Hydrophobic interactions are more powerful than other weak interactions, such as Vander Waal Interactions, hydrogen bonding, *etc.* Its strength depends on different factors, such as:

Temperature

As temperature rises, the intensity also rises, but at high temperature, denaturation occurs.

Number of Carbon Atoms on the Molecules

Higher the number of carbon atoms present in the biomolecules, the more will be the strength of the interactions.

Shape of Hydrophobes

Organic aliphatic molecules have stronger interactions than aromatic compounds. The carbon chain and its branches minimize the hydrophobic effect of this molecule, whereas the linear carbon chain generates the strongest hydrophobic interaction. As carbon branches form a rigid barrier, it is difficult for the two hydrophobes to interact closely with each other to reduce contact with the water [8].

Hydrophobic Interactions in Protein Folding

The proteins are bundled together and kept together by different mechanisms of molecular interaction. Molecular interactions provide thermodynamic stability of complex, hydrophobic, and disulfide protein bonds. Hydrophobic interactions not only influence the primary structure of the protein but also affect secondary and tertiary structural changes [9]. Globular proteins achieve a clear, compact native conformation in water due to their hydrophobic effects. When a protein is properly folded, it usually has a hydrophobic core as it becomes hydrated with water around the system. This is necessary since it forms the charged core of the protein and can cause channeling within the protein. It has been found that hydrophobic interactions affect the correlations near the native state, although they do not affect the same temporal properties of structural fluctuations around the native state. As shown in Fig. (**11**), hydrophobic interactions have been shown to affect proteins, although it has been found that proteins can have the most stable conformation and folding [10].

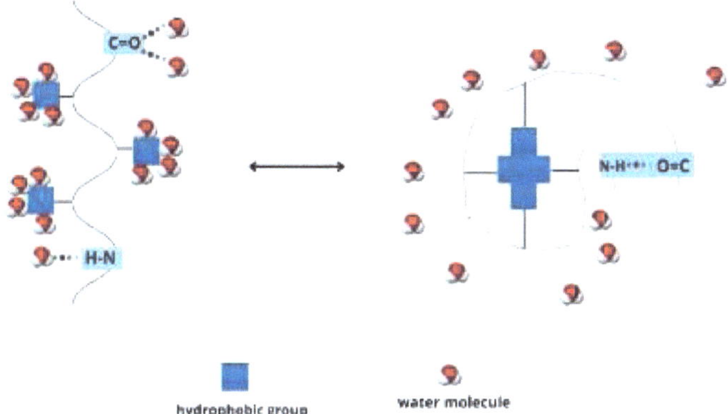

Fig. (11). The image showing the folding of proteins due to hydrophobic interactions (.http://www.cryst.bbk.ac.uk).

It is the main driving force behind protein folding. Basically, the protein chain is twisted in such a way that the risk of hydrophobic residues in the chain is reduced.

This causes the hydrophilic (polar) side chain residues to lie outside the molecule [11, 12].

Zwitterions

Zwitter ions are amino acids that are a double charged free solution. Their net charge depends on the pH of the solution and each amino acid has an isoelectric point, with no charge. Zwitter ions are molecules with functional groups, at least one of which is positively charged and the other negatively charged. The net charge for a whole atom is zero. Amino acids are the most accepted example of Zwitter ion. These include amino groups (basic) and carboxyl groups (acids). Since the NH_2 group is the strongest base, it takes H^+ from the -COOH group and leaves behind Zwitter ions (*i.e.* the amine group reduces carboxylic acid).

Zwitterion (neutral) is the normal form of amino acids in solution. There are two types, anion, and cation, depending on the pH. This describes the normal behavior of diprotic acids. The two dissociation stages are controlled by two acidity constants K_1 and K_2. When amino acids are dissolved in water, zwitterion interacts with H_2O molecules and acts as both acids and bases. However, unlike simple zwitterion compounds, which can only form cationic or anionic components, zwitterion has both ionic states at the same time. Below the isoelectric point (PI) is a positive net charge (+ve), above which is a negative net charge (-ve). When amino acids are part of a polypeptide/protein, they lose their NH_2 and OH groups, resulting in only the charge of the side chains.

Proteins can have their own electron points, which depend on the number and type of residues of various amino acids.

Glycine vs. Carbonic Acid

Glycine (NH_2-CH_2-COOH) is the simplest amino acid and abbreviate by HGly, or shorter by HA with A = Gly-.it has the shortest side chain R = H. The three species are:

	$[0] = [H_2A^+]$	$= [H_2Gly^+]$:	NH_3^+-CH_2-COOH	(glyciniumcation)
	$[1] = [HA]$	$= [HGly]$:	NH_3^+-CH_2-COO^-	(*neutral* zwitterion)
	$[2] = [A^-]$	$= [Gly^-]$:	NH_2-CH_2-COO^-	(glycinate*anion*)

When compared with carbonic acid the two acidity constants are:

	glycine:	$pK_1 = 2.35$	$pK_2 = 9.78$
	carbonic acid:	$pK_1 = 6.35$	$pK_2 = 11.33$

The three groups dependence on pH (abbreviated by [j] = [0] [1], [2],) is shown in the form of the corresponding ionization fractions aj = [j]/CT:

Titration Curves

The titration curves demonstrate what happens to glycine when one adjusts the pH by adding either a strong acid (HCl) or a strong base (NaOH):

In the left Figure of Fig. (**12**), four CT doses of glycine are taken into account. Calculations are carried out using analytical formulas. The points in the right Figure are numerical calculations with acquisitions.

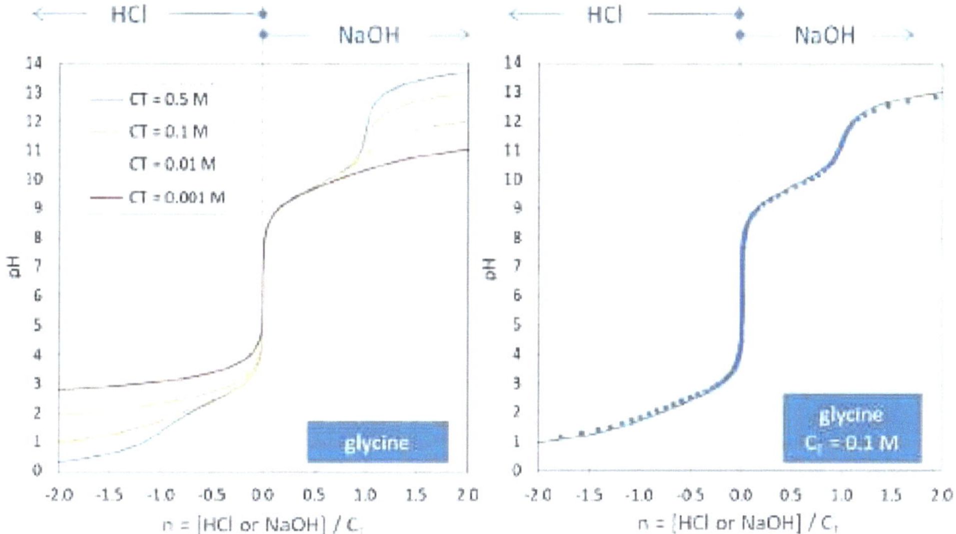

Fig. (12). Figure shows titration curve of glycine in the presence of strong acid or base.https://www.aqion.de.

Buffer Capacity and Buffer Intensity

The mathematical definition of buffering capacity and strength is the same for conventional acids (except for displacement z = 1). The diagrams below illustrate the buffer capacity (blue titration curve) as well as the corresponding buffer strength β (green) and its derivative dβ/dpH (red). This is achieved in two scenarios: exponentially high concentration of glycine and CT= 500 mm. The small dots represent zero dβ/dpH, which indicates a change in the intensity of the buffer and the direction of the titration curve. The blue titration curve in the back left plot is similar to the previous plot, except that the x and y axes are swapped.

ROLE OF FORCES IN BIOMOLECULES

There are different stabilization energy in covalent interactions. Typical Vander Waal interactions can be disturbed by 4-8 kJ energy radiation where hydrogen bonds can be broken with an input of 20 kJ/mol. In an aqueous solvent at 25°C, the strength of thermal energy become equal to these weak interactions. Although non-covalent interactions are individually weaker than covalent bonds, the cumulative effect of many of these interactions on proteins or nucleic acids can be very large, as shown in Fig. (**13**). For example, many enzymes bound to their substrate contain many hydrogen bonds and one or more ionic interactions, as well as hydrophobic and Vander Waal interactions. Each of these weak links contributes to a net reduction in free energy emissions; this free binding energy is released while stabilizing the binding system. The most stable (native) structure of most macromolecules is the structure that maximizes the potential for low binding. Antibodies rely on the cumulative effects of many weak interactions. The energy released when an enzyme does not covalently bind to its substrate is a major source of the enzyme's catalytic power. Certain physical forces also play a role in the interaction of proteins with surfactant molecules [14]. These weak interactions promote the binding of hormones or neurotransmitters to their protein cell receptor, as shown in Table **1** below.

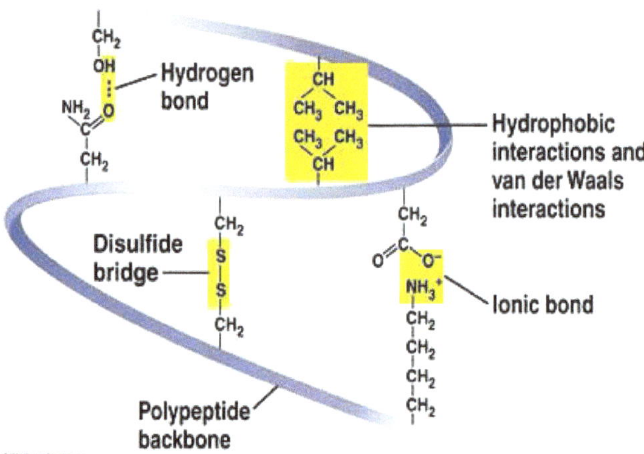

Fig. (13). The combined forces of different interactions and bonds in protein folding.

Table 1. Weak interactions between proteins.

Weak Interaction	Stabilization Energy (kJ/mol)
Hydrogen bonds	8-21
Electrostatic Interactions	21-42
Hydrophobic Interactions	4-8

(Table 1) cont.....

Weak Interaction	Stabilization Energy (kJ/mol)
Van der Waals forces	4

SUMMARY

• Molecular interactions of force are attractive or repulsive forces that hold biomolecules together in three dimensions and are important in various areas of protein structural folding, drug designing, materials and physical science, sensors, nanotechnology, separation, *etc.*

• Hydrogen bonding is important in many chemical processes. It is also responsible for the unique solvent capabilities of water and hydrophobic effects. They are the strongest dipoles - dipoles interactions but are much weaker than covalent bonds and strongly directional.

•Electrostatic reactions are probably the strongest non-covalent forces and can cover greater distances than other non-covalent reactions. These interactions are an important factor in determining the structure of biomolecules, play an important role in ligand-receptor binding, and also responsible for the stability and function of macromolecules.

• Vander Waal force is a weak interaction, with very low energy occurs between molecules very close to each other. They provide around 0.5 to 1.0 kcal mol^{-1} (2 to 4 $kJmol^{-1}$) in each pair of nuclear atoms

• Noncovalent interactions are individually weaker than covalent bonds, but the overall effect of many of these interactions on proteins or nucleic acids can be much greater. The energy released when the enzyme-substrate binds non-covalently is the key source of catalytic power for an enzyme.

KEYWORDS

Vander Waals Contact Distance

This is the minimum distance at which the force becomes repulsive instead of attractive when the two atoms come close to each other, Hence there is a balance between the force of attraction and repulsion.

Dipole-dipole Interactions

It is an attractive force between the positive and negative end of two different polar molecules. It has a strength result range from 5kJ to 20 kJ per mole.

Thermophiles

Thermophiles are living organisms that are found in extremely high temperatures, between 41-122 °C.

Supramolecules

Stable system consisting of two or more molecules organized by intermolecular (noncovalent interactions) is called supramolecular chemistry, and the molecules involved in such interactions are called supermolecules.

SHORT-ANSWER TYPE QUESTION

1. Why molecular interactions in biomolecules are important?
2. Which is the weakest force among non-covalent interactions and why?
3. What is hydrogen bonding? Why they are highly specific?
4. What causes hydrophobic interactions?
5. Explain Vander Waals's force of attractions.
6. Mention the factors responsible for increasing the strength of hydrophobic interactions.
7. What are polypeptide bonding and disulphide bridges?
8. How repulsion and attractions occur in Vander Waals interactions of biomolecules?
9. What are salt bridges?
10. Define the energy of electrostatic interaction.
11. What causes hydrophobic interactions in biomolecules?
12. What are polar and non-polar solvents?
13. What are hydrophobes?
14. Give different bonds found in DNA & RNA.
15. Explain electrostatic attraction in RNA.
16. What is Zwitter ion?
17. What do you mean by the isoelectric point of an amino acid?

LONG-ANSWER TYPE QUESTIONS

1. Explain the role of forces in biomolecules in detail.
2. Explain the energy associated with Vander Waals forces in detail.
3. How electrostatic interactions are the dominant factor in determining the conformations in biomolecules.
4. How hydrogen bonds form? Mention its importance.
5. What are the fundamental forces in biomolecules? Mention its key role in maintaining the stability of the molecules.
6. What are the types of Vander Waals interaction?

7. How hydrophobic interactions help in protein folding?
8. Explain the thermodynamics of hydrophobic interactions.
9. Explain hydrogen bonding in DNA?
10. Differentiate between dispersion forces, Keesom forces and Debye forces in detail.
11. How Zwitter ion formation takes place in biomolecules.
12. How pH affects Zwitter ion formation.

MULTIPLE CHOICE QUESTIONS

1. Which of these interactions are highly directional?

a) Vander Waal

b) Hydrogen bonding

c) Hydrophobic

d) Electrostatic

2. The functional groups of proteins are held together by an oppositely charged extension called as

a) Anionic bridges

b) Cationic bridges

c) Salt bridges

d) London forces

3. When an electronegative atom and an electron-deficient atom interact, the force of interaction between them is

a) Hydrogen bond

b) Vander Waal interaction

c) Covalent bond

d) Ionic bond

4. The force of attraction between partial positive and negative end between two molecules is

a) Ionic interaction

b) Vander Waals force

c) Electrostatic force

d) Dipole-dipole interaction

5. The bond formed between two identical atoms who share electron pairs and exert force on each other is

a) Coordinate bond

b) Ionic bond bond

c) Covalent bond

d) Non-covalent bond

6. The strength between the ionic bond of two atoms……….than covalent bond

a) stronger

b) equal

c) weaker

d) none of them

7. Which one of the following t is not an example of chemical bonding?

a) Coordinate bond

b) Covalent bonding

c) Metallic bonding

d) Hydrogen bonding

8. Which one of the following isalso known as instantaneous dipole-induced dipole interaction?

a) Vander Waals interaction

b) Ionic bond

c) Hydrogen bond

d) Covalent bonding

9. Choose the correct statement.

a) The number of electrons in a molecule affects the strength of Vander Waals interaction

b) Vander Waals forces help the formation of layers in graphite

c) As we go down the group the boiling point of noble gases increases

d) Due to Vander Waals interaction, water has a high boiling point

10. Which are the weakest forces in chemical bonding?

a) Dipole-dipole interaction

b) Vander Waals forces

c) London dispersion forces

d) Hydrogen bonding

11. Select the molecule with zero dipole moment.

a) NH_3

b) CO_2

c) CCl_4

d) CH_3Br

12. Choose the incorrect statement about water?

a) Water molecule is bonded to four others molecule during ice formation

b) The presence of hydrogen bonding is responsible for high boiling in water

c) Diamond and ice have a similar structure

d) At 1° C Celsius, water has maximum density

13. Select the most contributing factor in hydrogen bonding.

a) Covalent bond

b) Vander Waals forces

c) Ionic bond

d) Hydrophobic interactions

14. Which bond is formed between the attraction of temporary and their induced dipoles?

a) Ionic

b) Covalent

c) Metallic

d) Hydrogen

e) Dipole-induced dipole

15. Select forces formed between two Nitrogen atoms.

a) Dispersion forces

b) Covalent bonding

c) Hydrogen bond

d) Ionic bonding

16. Select molecules thatonly have hydrogen bonding.

a) AsH_3, CH_3NH_2

b) CH_4, AsH_3, H_2Te

c) CH_3NH_2, HF

d) HF, H_2Te

e) AsH_3, H_2Te

17. What criteria a solute must have to soluble in a water solvent?

a) Ionic or polar in nature

b) Nonpolar in nature

c) Have high atomic mass

d) Dipole-dipole interaction

18. HF has ahigher boiling point than CH_4 because

a) Ion-dipole interactions in CH_4.

b) CH_4 is more polarizable

c) HF is more polarizable.

d) CH_4 is polar.

e) Hydrogen bonding

19. Which of the following is the most abundantly found in biomolecules like protein, nucleic acids?

a) Covalent bond

b) Ionic bond

c) Coordinate bond

d) Hydrogen bond

20. Given below following information:

CF_4, Molecular Weight 87.99 with Boiling Point = 182°C

CCl_4, Molecular Weight 153.8 with Boiling Point = 123°C

Which one of them has a higher vapourpressure at the same temperature and in a liquid state?

a) CCl_4

b) CF_4

c) CF_4 and CCl_4 will have an equal vapour pressure

d) Cannot be determined with a given condition

21. Why ICl has ahigher boiling point compared to Br_2 though both have the same number of electrons?

a) ICl has ahigher molecular mass

b) Cl in ICl is more polarizable than Br of Br_2

c) ICl molecules have hydrogen bonding

d) Br_2 is less polar than ICl

e) Br is less electronegative than Cl

22. I_2 is most soluble in which of the following solvent?

a) H_2O

b) CCl_4

c) Ethanol

d) Vinegar

23. Why HF has a higher boiling point than HCl?

a) HF molecules have high hydrogen bonding

b) Low dipole moment in HF

c) HCl is not polar

d) Cl in HCl is more electronegative than F in HF.

24. Why F_2 and Cl_2 are gases and Br_2 is liquid in room condition?

a) Molecular size is dependent on the polarity

b) Dipole-induced dipole forces increasewith molecular size.

c) Dipole-dipole forces increase with molecular size.

d) Dispersion interactions increase with molecular size and polarity increases with molecular size.

25. Select the most polar molecules given below.

a) H_2

b) HCl

c) HBr

d) HI

26. In which of the following solvents, Iodine would be highly soluble?

a) Equally soluble in H_2O and CCl_4

b) H_2O

c) impossible to determine the relative solubilities

d) CCl_4

e) equally insoluble in both of them

27. Why boiling point of H_2O is about 200°C higher than H_2S and H_2Se?

a) Due to low molecular weight.

b) Least polarity when compared to H_2S and H_2Se

c) The H-S and H-Se is weaker than the O-H bond

d) Lightest compared to H_2S and H_2Se

e) Strongest intermolecular attraction in water than H_2S and H_2Se

28. Which phase transition of a given substance would release most energy?

a) Liquid to solid

b) Gas to solid

c) Liquid to gas

d) Gas to liquid

e) Solid to gas

29. Why solid copper sulphate dissolves faster in hot water?

a) Hot water has higher kinetic energy than cold water

b) High dipole-dipole interactions between molecules of water and $CuSO_4$.

c) Instantaneous dipole-induced dipole forces between the Cu^{2+} and the SO_4^{-4} Ions.

d) Formation of hydrogen bonding between OH^- of ions and Cu^{2+} ions

30. In which of the following solvent, Solid I_2 will be highly soluble?

a) Ethanol

b) Impossible to determine the relative solubilities

c) CCl_4

d) Equally soluble in H_2O and CCl_4

e) H_2O

31. In covalent bonds, electrons are shared

a) independently

b) by force

c) mutually

d) by temperature

32. Strength of intermolecular as compared with intramolecular forces is

a) More

b) Less

c) Same

d) Double

33. Which of the following has a permanent dipole moment?

a) CCl_4

b) CO_2

c) H_2O

d) CH_4

34. Molecule having covalent bond is

a) NaCl

b) HCl

c) $CaCl_2$

d) KCl

35. Which among is the strongest attractive force?

a) Electrostatic force

b) Hydrogen bonding

c) Covalent bonding

d) Dipole-dipole interactions

e) Ionic bonding

36. Which are the leading intermolecular forces in the tertiary structure of proteins?

a) Electrostatic forces

b) Hydrophobic interaction

c) Hydrophilic interaction

d) Disulphide bonds

e) hydrogen-bonding

37. The ease with which the charge distribution in a molecule can be distorted by an external electrical field is called the __________.

a) Electronegativity

b) Hydrogen bonding

c) Polarizability

d) Volatility

e) Viscosity

38. Which forces make the water denser than ice?

a) Electrostatic bonding

b) Hydrophobic interaction

c) Hydrophilic interaction

d) Hydrogen bonding

e) Ionic interaction

39. What do you mean by the volatility of liquid?

a) Not flammable

b) High viscosity

c) Increase hydrogen bonding

d) Decrease hydrogen bonding

e) High intermolecular attraction

40. What is the backbone of the formation of synthetic polymer in industries?.

a) Carbon-carbon intermolecular bonding

b) Formation of disulphide bond

c) Repeating subunits of monomers to form a macromolecules

d) Strong covalent bonds hold together large polymeric groups

41. Whichcondition predominates when a solid compound melts into liquid?

a) Increase in internal energy of the solid

b) Decrease in internal energy

c) Change in temperature

d) Temperature is constant

42. Which of the following molecules experience dipole-dipole interactions between two different molecules?

a) H_2O

b) CO_2

c) NH_3

d) Cl_2

43. Which type of molecule can easily be polarized?

a) A large and nonpolar molecule

b) None of these

c) A large and polar molecule

d) Small and polar molecules

e) Small and non-polar molecules

44. London forces predominate in which of the given molecules?

a) CH_3OH

b) NH_3

c) H_2S

d) CH_4

e) I_2

45. Why I_2 is solid and Cl_2 is gaseous at room temperature?

a) Due to increasing London dispersion forces as we go down the group

b) Less surface area of I_2

c) Increase in hydrogen bonding

d) Increase interactions between covalent and non-covalent bonding in I_2

46. Which intermolecular force is prevalent in CH_3 –NH-CH_3?

a) Covalent bonding

b) Hydrogen Bonding

c) Ionic Bonding

d) Dipole-dipole interaction

47. Which of the given molecule does not have hydrogen bonding in their

structure?

a) 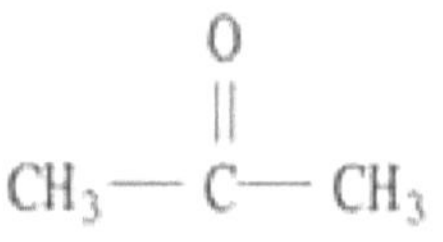 $CH_3-\overset{\overset{\large O}{\|}}{C}-CH_3$

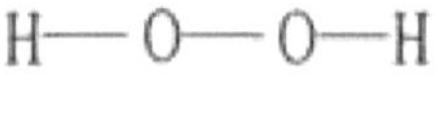

b) $H-O-O-H$

c) $CH_3-\underset{\underset{\large H}{|}}{\overset{\overset{\large H}{|}}{C}}-O-H$

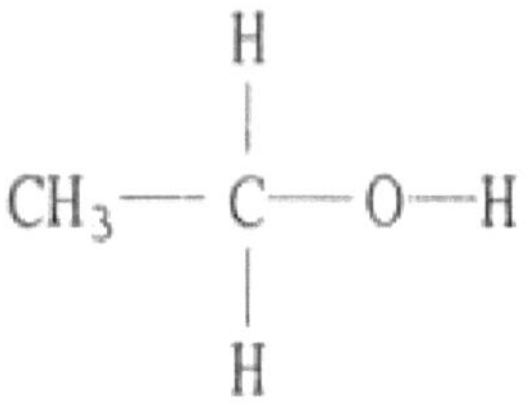

d) 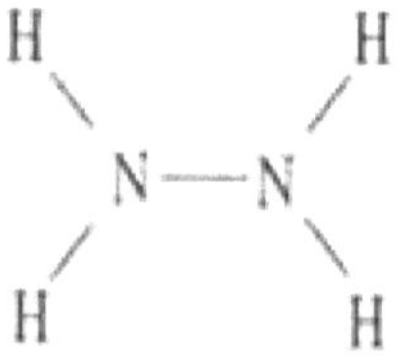 H_2N-NH_2 (H, H on each N)

48. Which of the given moleculeshashydrogen bonding as their predominantintermolecular forces?

a) I_2

b) H_2O

c) $C_6H_{13}NH_2$

d) CH_3OH

e) CO_2

49. Which intermolecular force predominates when there is an interaction between acetic acid and ethanol?

a) hydrogen bonding

b) Dipole-dipole interaction

c) Ion-dipole interaction

d) London dispersion forces

e) None of them

50. Choose the correct statement about Zwitter ions

a) Overall charge present in Zwitter ions is neutral

b) Zwitter ions are formed only in acidic condition

c) Zwitter ions are formed only in basic condition

d) The overall charge is negative

e) The overall charge is positive

51. Choose the incorrect statement about omega-3 fatty acid.

a) It is an example of a triglyceride.

b) It gives rise to anti-inflammatory mediators such as prostaglandins.

c) It contains a carboxylic acid functional group.

d) It is an unsaturated fatty acid.

52. Which statement about proteins is incorrect?

a) Protein is synthesized by the process of transcription and translation.

b) Proteins are made up of amino acid held together by peptide bonds

c) Hydrogen bonding is the predominant interaction in protein folding.

d) Riboflavin and thiamine are globular proteins.

CONSENT FOR PUBLICATION

Not Applicable.

CONFLICT OF INTEREST

The authors confirm that this chapter contents have no conflict of interest.

ACKNOWLEDGEMENTS

Authors are thankful to Department of Chemistry, North Eastern Regional Institute of Science and Technology,Nirjuli, Itanagar for providing facilities to accomplish the work.

REFERENCES

[1] Silberberg MS. Principles of General Chemistry. 2ndEdition. New York: McGraw Hill Publishing Company 2010.

[2] Pietri J. Hydrogen Bonding. [Updated 30th September 2019]. https://chem.libretexts.org/Textbook_Maps/Physical_and_Theoretical_Chemistry_Textbook_Maps/Supplemental_Modules_(Physical_and_Theoretical_Chemistry)/Physical_Properties_of_Matter/Atomic_and_Molecular_Properties/Intermolecular_Forces/Hydrophobic_Interactions

[3] Smart O. Non-bonded Interactions. [Updated 26th July 1996]. Available from: http://www.cryst.bbk.ac.uk/PPS2/course/section7/os_non.html

[4] Williams LD. Molecular Interactions (Non-covalent Interactions) and the Behaviors of Biological Macromolecules. [updated 31st August 2019]. Available from: https://ww2.chemistry.gatech.edu/~lw26/structure/molecular_interactions/mol_int.htm

[5] Learn Biochemistry (The Chemistry of life). Chapter 2: Water-Non-Covalent Bonds; Van Der Waals Forces. [updated 11th September 2011]. Available from: https://learnbiochemistry.wordpress.com/2011/09/11/chapter-2-water-non-covalent-bonds-van-der-waals-forces

[6] Cliffsnotes. Electrostatic and van der Waals Interactions. Boston: Houghton Mifflin Harcourt 2016. Available from: https://www.cliffsnotes.com/study-guides/biology/biochemistry-i/the-importanc--of-weak-interactions/electrostatic-and-van-der-waals-interactions

[7] Williams LD. Molecular Interactions (Non-covalent Interactions) and the Behaviors of Biological Macromolecules. [updated 31st August 2019] Available from: https://ww2.chemistry.gatech.edu/~lw26/structure/molecular_interactions/mol_int.htm

[8] Than J. Hydrophobic interactions. [updated 6th June 2019] Available from: https://chem.libretexts.org/Bookshelves/Physical_and_Theoretical_Chemistry_Textbook_Maps/Supplemental_Modules_(Physical_and_Theoretical_ Chemistry)/Physical_Properties_of_Matter/Atomic_and_Molecular_Properties/Intermolecular_Forces/Hydrophobic_Interactions

[9] Boyer R. Concept in Biochemistry. 2nd ed., NJ: John Wiley and Sons 2002.

[10] Chemistry Libre Texts. Protein Folding. [updated 26th Ocotober 2019]. Available from: https://chem.libretexts.org/Textbook_Maps/Biological_Chemistry/Proteins/Protein_Structure/Protein_Folding

[11] Berg JM, Tymoczko JL, Stryer L. Biochemistry. 5th ed.., New York: W H Freeman 2002.

[12] Berg JM, Tymoczko JL, Stryer L. Biochemistry. 6th ed.., New York: W.H. Freeman 2007.

[13] Jain JL, Jain S, Jain N. Fundamentals of Biochemistry. New Delhi: S. Chand & Company Ltd 2007.

[14] Shveta A, Arun K S. Spectrometric, Thermodynamic, pH metric and viscometric studies on the binding of teals as surfactant with albumin as biopolymer. Curr Phys Chem 2020; 10(1): 47-64. [http://dx.doi.org/10.2174/1877946809666190913182152]

CHAPTER 4

Introduction To Physical Techniques for Determination of Structure of Biopolymers

Nene Takio[1], **Meera Yadav**[1,*] and **SonamTashi Khom**[1]

[1] *Department of Chemistry, North Eastern Regional Institute of Science And Technology (NERIST), Nirjuli, Itanagar-791109, Arunachal Pradesh, India*

Abstract: Biopolymers perform important cellular processes, which are isolated, purified and studied with the help of a wide range of biophysical, electrochemical and molecular biology techniques such as NMR, X-ray Crystallography, Electron Microscopy, and Electron Tomography. X-ray Crystallography helps us to establish 3D structures of small and large molecules at atomic resolution *via* the X-ray diffraction method, whereas NMR determines the content and purity of a sample. Electron microscopy obtains high-resolution images of biological and non-biological specimens and electron tomography provides three-dimensional density maps of the pleomorphic organism from macromolecular complexes to cells.

All these techniques, discussed in this unit, provide us a greater insight to unravel the mystery of complex organisms and to understand the structure and functional properties of the biomolecules.

Keywords: Fourier transform, Macromolecules, Mosaicity, Pleomorphism, Synchrotons.

INTRODUCTION

Biopolymers such as DNA, RNA and proteins plays key role in the cellular processes like cell differentiation, cell growth, maintenance, repair, recombination, transcription, translation, *etc.* Unlike synthetic polymers, biopolymers have a well-marked structure. These polymers have a uniformly distributed set of molecular mass and appear as a long chain of worms or a curled up string ball under a microscope. This type of polymer is differentiated based on their chemical structure. Biopolymers plays an essential role in nature [1]. Different types of biopolymers extracted from biomolecules has been shown

* **Corresponding author Meera Yadav:** Department of Chemistry, North Eastern Regional Institute of Science And Technology (NERIST), Nirjuli, Itanagar-791109, Arunachal Pradesh, India; E-mail: drmeerayadav@rediffmail.com

below in Fig. (**1**). They are extremely useful in performing functions like storage of energy, preservation and transmittance of genetic information and cellular construction. Some of the uses of biopolymers has been mentioned below:-

- Sugar based polymers, such as polyactides, naturally degenerate in the human body without producing any harmful side effects. This is the reason why they are used for medical purposes. polyactides are commonly used as surgical implants.
- Starch based biopolymers can be used for creating conventional plastic by extruding and injection molding.
- Biopolymers based on synthetic are used to manufacture substrate mats.
- Cellulose based biopolymers, such as cellophane, are used as a packaging material.
- These chemical compounds can be used to make thin wrapping films, food trays and pellets for sending fragile goods by shipping.

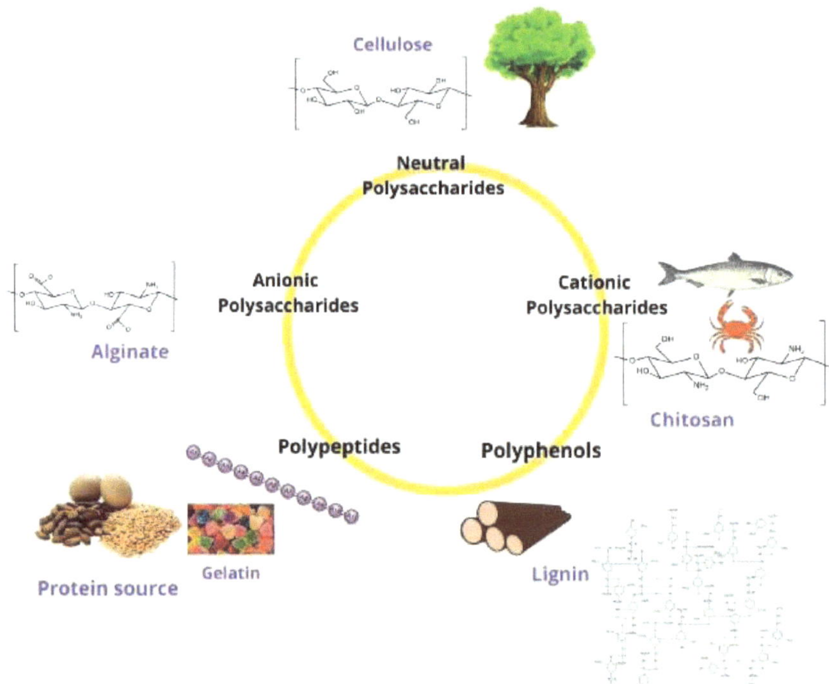

Fig. (1). Biopolymers found in nature.

Immense efforts have been made for their isolation, purification, quantification as well as structural determination and functional characterization. Profound insights into the structural and functional characterization of these biopolymers could help us understand the intricate cellular machinery.A wide range of biochemical, biophysical, electrochemical and molecular biology techniques have really been beneficial in exploring the key and interdependent relationships between the structure and function of these biopolymers. Each instrumentation techniques has its own advantage and disadvantage in terms of their applications, selectivity and sensitivity. In this unit we will discuss some of the physical techniques which provides a greater insight of the structure and functional properties of the biomolecules [2].

BIOPOLYMER TECHNIQUES

X-ray Crystallography

It is a powerful tool to establish 3D structures of molecules at the sub-atomic level. The method works on the principle of X-ray diffraction by the crystal. The electron in the crystal scatters X-rays beam and the 3D structure of molecules is established due to the interference pattern of the scattered X-rays. Unfortunately, crystals required for diffraction are difficult to obtain, and considerable amounts of proteins may be needed, as given below in Figs. (**2** and **3**). Usually, proteins are created as recombinant material from microorganisms and mammalian cells. They require high labour work but are considered a good approach for the detection of even tough membrane proteins.Initially used for the identification of salt crystal structure, this technique helped Linus Pauling to develop his theory of attractive force (structural information combined with quantum mechanics calculations). From the data obtained from salt crystals, Linus Pauling proposed the primary and secondary structure of protein molecules [Pauling and Corey, 1951]. Both atomic coordinates and structural helical configurations of polypeptide chains were detected and confirmed by X-ray crystallographic analysis. Crystals of myoglobin and haemoglobin were first analyzed in the 1960s by Kendrew and co-workers. Since several such materials may be obtained in crystal forms, like salts, minerals, semiconductors, organic, inorganic and biomolecules, x-ray crystallography has been a primary technique for scientific researches and discoveries. In its initial decades of operation, the technique helped to determine the atomic size, length, nature of chemical bonding, and variations in material's atomic scale, particularly minerals and alloys. This method also helps in the identification of structural arrangement and functions of biological molecules like vitamins, proteins, nucleic acids like deoxyribonucleic acid, *etc.* X-ray crystallography continues to be the prime methodology for characterizing the atomic structure of new materials that are rather difficult to characterize by alternative techniques [3].

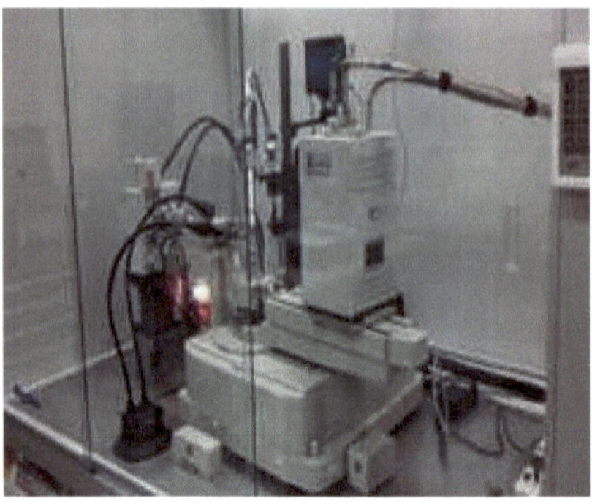

Fig. (2). X-ray crystallography instrument.

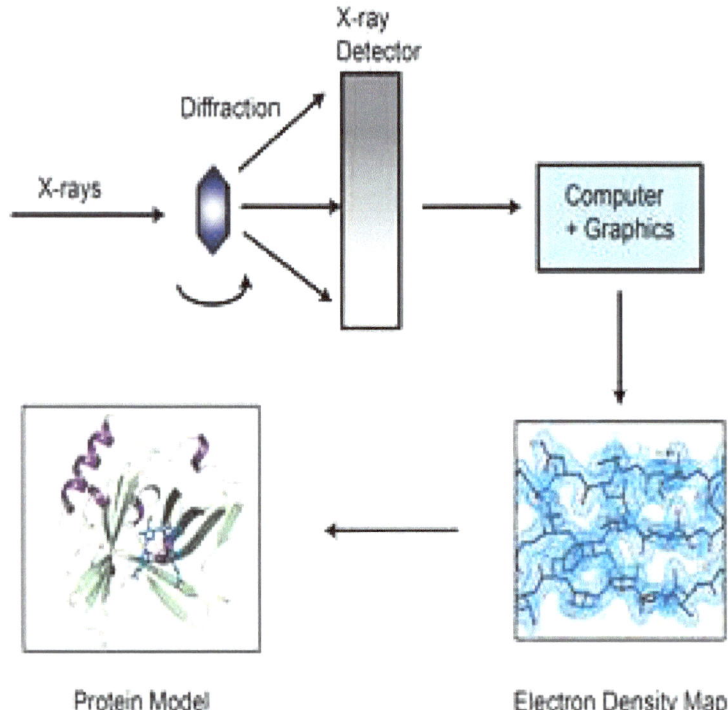

Fig. (3). Schematic process of X-ray crystallography method.

Diffraction in X-ray Crystallography

Crystals are arrays of atoms that scatter x-ray through their electrons. The method requires a crystalline sample and high-intensity monochromatic X-ray beams with constructive interference as a radiation source. These X-rays are produced by an electron beam tube, filtered to supply monochromatic radiation, concentrated and aimed onto the sample. Constructive interference radiation waves are generated *via* the interaction of incident rays with the sample. The understanding of the optical phenomenon in X-ray crystallography can be understood by Bragg's model of diffraction, which explains that the incoming X-rays scatters equally (mirror-like) from every direction and mix constructively when the path difference between scattered X-rays and angle θ is an integer multiple n of the X-ray wavelength λ. When it occurs, the diffracted rays leave the crystal at the same angle as that of incident rays.

Bragg's formula is given by;

$$n\lambda = 2d \sin \theta$$

Where d is that the distance between diffracting planes, θ is the incident angle, n is any number and λ is the wavelength of the X-ray beam. For diffraction to occur, it is needed that photons being shot out towards the crystal have the wavelength equal to the atoms of the crystal.

Crystallisation of Proteins

The X-ray crystallization of proteins aims to obtain high-resolution single crystals with sufficient size. Since it is difficult to conclude the exact state of protein crystallization, broad screening of conditions required for protein crystallization is done to predict the appropriate condition for crystal formation. Precipitant used can be a carbon-based solvent, a mixture of concentrated salt-like $(NH_4)_2SO_4$, or a polythene–glycol mixture. Extra parameters like pH, the concentration of Precipitants and additives like metal ions and detergents need to be assorted. Optimization of conditions is usually done after initial crystallization is obtained, which includes changes in the concentrate of precipitants and additives, their types and pH. For crystallizing of macromolecule-like G-protein coupled receptors, various connected approaches are being developed, that led to achieving dramatic success in the field of biomolecule crystallization. When appropriate crystals are obtained, the diffraction pattern must be measured. In the last twenty years, the development of extremely good synchrotrons has increased its efficiency on data collection of minute crystals of less than 0.1mm in size. The formation of 1µm X-ray beamlines in the synchrotrons has allowed the scanning of existing crystals for easy data assortment. Increased efficiency of routine data collection at synchrotrons is recorded. The crystal is turned around in the axis in small angular, an X-ray beam is passed and a diffraction pattern is recorded in every step. After the initial diffraction pattern is collected, the presence of a well-resolved, high-resolution non-overlapping single spot is inspected instantly, to conclude whether the additional data assortment is required. The finest crystals are those which diffract with high resolution and low mosaicity. Changes in the inner symmetry of crystals affect the diffraction pattern and after the collection of optical diffraction images, the raw pixel intensities of the individual images are merged and the intensities of the successive images are combined to show a full collection of reflection intensities. The raw pixel intensities of individual images are integrated and successive image intensities are combined together to display a complete set of reflection intensities.

The electron-density map of the repeated unit of crystal is calculated by using phases and amplitude of the diffraction data, which helps in the interpretation of raw data. The quality of the protein crystal determines the resolution of diffracted data and also affects data interpretation such as lattice regularity of unit cell of protein and regularity of heavy atom distribution. The data interpretation of the X-ray diffraction depends on data resolution and requires information of protein amino acid sequence since different amino acids may have identical electron densities (*e.g.* Tyr and Phe, or Leu and Ile). There is a need to refine initial models of protein structures due to the limitation in data resolution, which can be achieved by comparing experimental data with structural results from computer modeling. The difference between experimental and hypothetic structure is given as *R-factor*.

Applications

X-ray crystallography has wider applications in physical, chemical, biological and material sciences and conjointly provides information regarding the structure of matter as well its crystalline property, *etc.*

Structure of DNA

Watson & Crick proposed the coiled helical structure of deoxyribonucleic acid using the X-ray Crystallography method and this technique is still being used for structural determination.

Drugs

The technique may be helpful for the identification of drugs and their properties. Every drug has a distinctive XRD pattern that helps in its structural identification.

Textile and Polymer Industries

Textile fibres are made up of crystalline and amorphous molecules; hence using X-ray diffractometry method, the information and characterization of fibres can be achieved by measuring their degree of crystallization.

NMR (Nuclear Magnetic Resonance)

It is a technique utilized in quality control and analysis for the determination of volume, sample purity and its structural properties. In recent years, the NMR technique has grown to become a strong tool to analyse and examine the structures of both liquid and solid biopolymers. It offers a way to research the dynamics of polymers in solution and to analyse the effect of solvents and other factors on the actions of polymers. With the progress of 2D and 3D NMR spectroscopy, it has become possible to study conformations of the small macromolecules such as oligonucleotides and saccharides as shown in Fig. (**4**). The NMR works on the principle that if a nucleus of the sample having torsion is placed in a strong magnetic field irradiated with radio waves, the excitation of the nucleus occurs by producing an NMR signal that is detected and received by sensitive radio receivers. NMR obtains high-resolution data by using a unique technique to calculate distances between atomic nuclei instead of electron density in a molecule. With the help of NMR, a solid, high-frequency force field stimulates the atomic nuclei of the isotopes H-1, D-2, C-13, or N-15 (they have a magnetic spin) and measures the frequency of the field force of the atomic nuclei as they oscillate back to their original state. The point is to identify which resonance comes from which spin. The distance and type of adjacent nuclei determine the resonance frequency of the excited nuclei. This reliance on the neighbouringgroup called the chemical shift (or spin-spin coupling constant), reflects the local electronic environment and the information contained in the 1-DNMR [4].

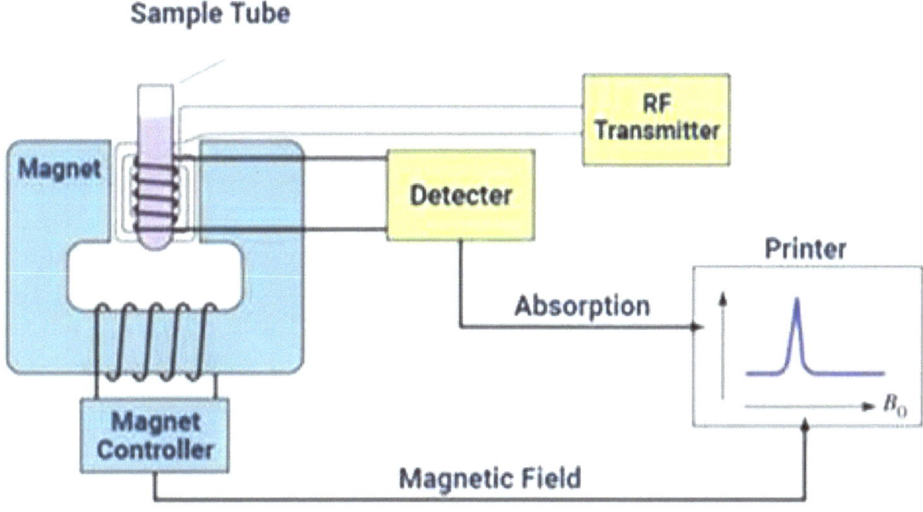

Fig. (4). Schematic diagram of NMR.

For proteins, NMR typically measures the spin of protons.

As shown in Fig. (**5**), the subsequent reasons make the H-1NMR spectroscopic method, the strategy for selection and studying biological macromolecules:- Hydrogen is found in proteins, nucleic acids and polysaccharides at many locations. They have a high abundance for each site and the H nucleus is the most responsive to detect 1-D spectra containing data on all the chemical shifts of the macromolecule. Frequency resolution is usually not sufficient to distinguish between individual chemical shifts. 2-D NMR resolves this drawback by providing details on the relative location of H in molecular structures. 2-D NMR spectra provide information on the interaction between H that is covalently bonded by one or two more atoms (COSY or correlation spectroscopy). Instead, pairs of H may be adjacent to space, even though these are from residues that do not seem to be similar to sequence (NOE spectra, or Nuclear Overhauser Effect). Thus, the entire structure can be determined by sequentially distributing the cross-peak correlations in 2-D spectra.

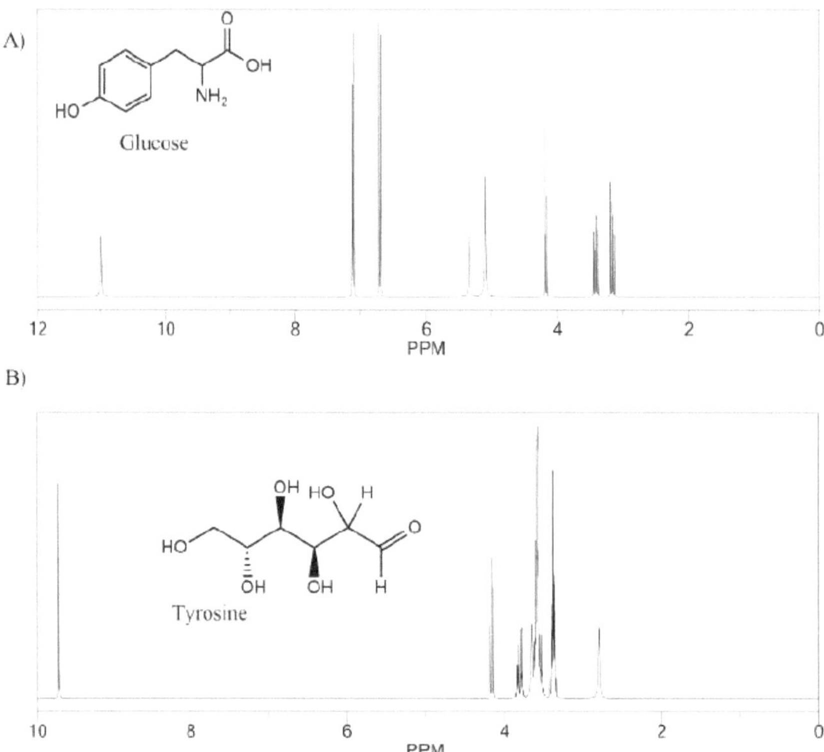

Fig. (5). H-1 NMR spectra of biomolecules, **a)** Glucose **b)** Tyrosine.

Currently, the protein size limit for the NMR structural analysis is approximately

200 amino acids. A significant aspect of the detection of cross-peaks is that normal patterns can be detected from secondary structure sections such as alpha-helical and parallel or antiparallel beta-sheets as they include typical H bonding networks.

NMR also provides the data of the amino acid sequence; however, the protein doesn't have to be in ordered crystal, nevertheless, high concentrations of solubilized macromolecule is obtainable (NMR structures are thus called resolution structures).

In biopolymers, the basic structure (sequence) breaks the molecules into groups of attached spins, after which the residues are divided into one or two groups.This is also applicable to nucleic acids and polysaccharides [5].

Electron Microscopy

Electron microscopy (EM) is a technique that provides high-resolution images of biological and non-biological specimens. This is used for analysis and biomedical research to explore the structure of tissues, cell organelles and macromolecule complexes. As shown in Fig. (**6**), The high-resolution EM image result from the use of electrons which give key information about the cell structures, functions and cellular diseases [6].

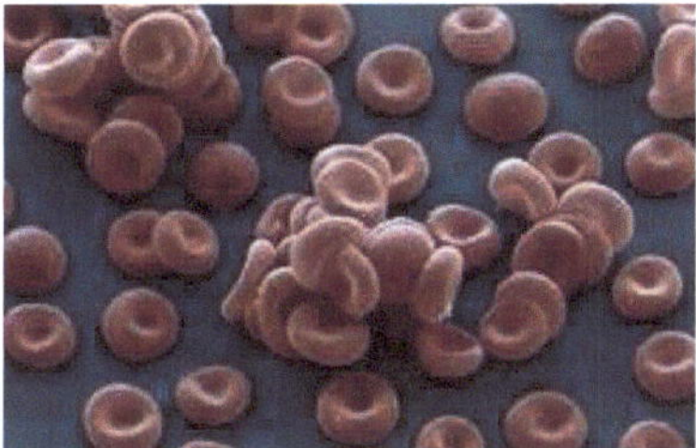

Fig. (6). Image of RBC observed under an electron microscope thatis not visible under the naked eye.

Principle of Electron Microscopy

The electron microscope uses the signals generated by the interaction of the electron beam with the sample to obtain information about the structure, shape and composition. Like photons in lights, electrons are minute particles having a wave nature. The electron beam passes through the sample and then through a series of lenses that magnify the image. The image is obtained from the dispersion of electrons by the atoms in a sample. A heavier atom is normally more successful in scattering and increasing image contrast than an atom with a lower atomic number. All EMs use electromagnetic or electrostatic lenses. These lenses have a coil of wire called a solenoid, wrapped outside the tube [4]. In traditional EM, the electron beam must be in a vacuum because electrons cannot travel a considerable distance in the air under atmospheric pressure. The microscope column is emptied by a vacuum pump and sample, and other necessary devices are evacuated by an airlock. The EM has a varifocal lens that keeps the distance between the sample and the objective lens and the lens division constant. Amplitude or magnification is primarily determined by the current generated by the intermediate lens coil and the projector lens coil. The image is centered by changing the current through the target lens coil. A microgun is used as an electron gun, and an image or electron micrograph is also displayed on the screen instead of the eyepiece [7].

Types of Electron Microscopy

Electron microscopes are useful in analysing the ultrastructure of a broad variety of biological and inorganic materials, including microorganisms, cells, molecules, biopsy laboratory samples, metals and crystals. In industry, EM is usually used for quality control and failure analysis. Modern EM produces electron micrographs using advanced digital cameras and frame grabbers to acquire an image. Some of the techniques of EM have been mentioned as having a huge potential for the structural determination of biomolecules [8].

Cryo-electron-microscopy (Single-particle Analysis)

Cryo-electron-microscopy(cryo-EM) is rising as a significant and widely acceptable technique for deciding the structures of macromolecule assemblies. The widely used term, single particle analysis (SPA), refers to the analysis of sets containing several images of a given particle. In the 1980s, a significant discovery was made that macromolecules could be stored in a near-native state by rapid freezing: this vitrifies the thin buffer film containing the particles that remained water. To reduce radiation damage, a low-dose technique is used. Since the transmission electron micrographs are two-dimensional projections of three-dimensional objects, the geometry of each particle should be determined. Combining this information ends up in a three-dimensional reconstruction of the object.

Electron Crystallography

Electron crystallographic specimens have a two-dimensional arrangement; typically just 1 molecule is thick. Electron diffraction is used to gather Fourier amplitudes (structure factors) while phases are calculated by analysis Fourier transform images. The data is calculated in 3 dimensions by repeating these operations by tilting the specimen across various angles. The resulting three-dimensional Fourier transformation is then inverted to create a three-dimensional density map in real space. The solution is basically anisotropic, being slightly lower within the dimension perpendicular to the plane of the array since only a small range of tilt angles is permitted and information quality tends to deteriorate at higher tilts. This approach is particularly useful for membrane proteins that can be seen in a quasi-native atmosphere with a continuous lipid process.

Transmission Microscopy (TEM)

Transmission is another type of microscope EM that illuminates high voltage electronic radiation towards the sample and creates a violent image of the sample. Electron guns are used to make electron beams. The gun is usually equipped with a metal filament cathode (tungsten filament cathode) which is a source of producing an electron beam. An anode in the gun accelerates the electron beam, and electrostatic and electromagnetic lenses make it easy to focus the electron beam. The schematic diagram of TEM is shown in Fig. (7). As the electron beam moves through the sample samples, the electron disperses and provides an image of the sample's microscope structure, which is further viewed and analyzed under a microscope. This variation is tested by presenting the image on a fluorescent Zn sulphide coated screen. Another process that can be used to record an image is to place a photographic material in an electron beam, which records the image in real-time on a computer. The main limitation of TEM is that it requires a very thin sample, which is less than 100 nm. As a result, most biological samples should be treated with chemicals and dehydrated so that they can be embedded in a polymer resin and viewed under TEM.

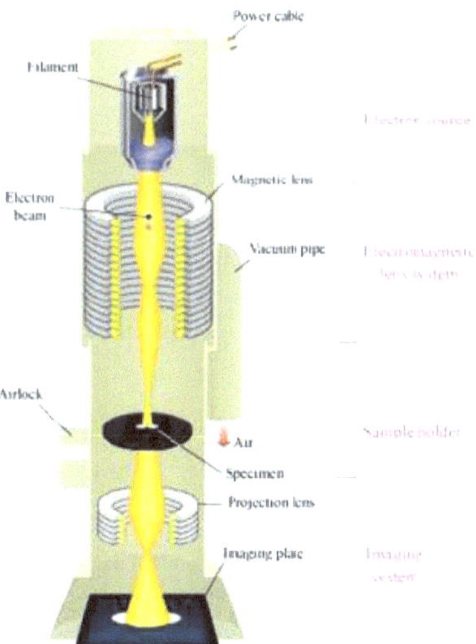

Fig. (7). Outline of Transmission electron microscopy.http://www.hk-phy.org the scanning microscope produces magnified images of the specimen by using a technique known as raster scanning.

Scanning Microscopy (SEM)

It directs the electron beam of interest into the rectangular space of the sample and loses energy as it passes. The energy is converted into alternative energies such as heat, light, secondary electrons, and backscattered electrons. This information is translated to examine the topography and composition of the first specimen. The SEM resolution is slightly lower than the resolution achieved by TEM.

SEM *VS*. TEM

Both SEM and TEM are useful resources in the fields of biological, chemical and physical sciences Summary of difference between SEM vs TEM is given in Table **1**.

By understanding the differences and similarities between SEM and TEM, researchers may find it easier to select the required electron microscope according to the need.

SEM *vs*. TEM Advantages

Compared to TEM, SEM is:

- Cost-effective
- Less time to form the image
- Less amount of sample is required
- Analyse thicker and large samples

Compared to SEMs, TEMs

- Give higher quality images
- Provide crystallographic and atomic data
- Give 2-D images that are easier to interpret than SEM 3-D images
- Enable users to analyse additional properties of a sample

SEM and TEM are similar in many aspects such as both of them are high-resolution microscopes and have an electron source that emits a beam of electrons into a vacuum sample. Both contain lenses and electron aperture to monitor electron beam and record images. Some of the pictorial images are compared, as shown in Figs. (**8** and **9**). Both SEM and TEM have vast functional and mechanical differences. They require different quality of samples and also record images with distinct resolutions.

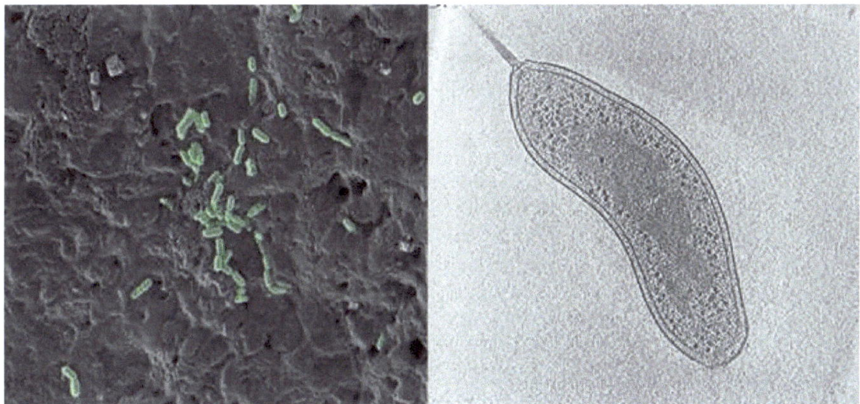

Fig. (8). SEM (left) and TEM (right) views of the bacteria. Whereas SEM shows various bacteria on the surface (green), the TEM picture shows the internal structure of a single bacterium.https://www.thermofisher.com.

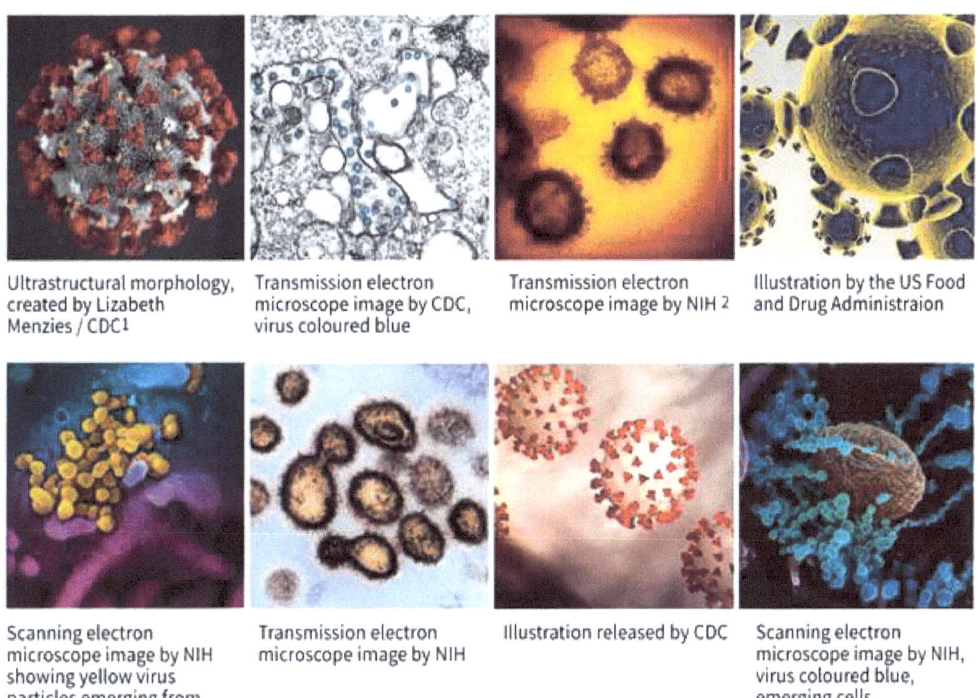

Fig. (9). SEM and TEM images of COVID-19.

Photographic images of Severe Acute Respiratory Syndrome Coronavirus 2 (SARS-CoV-2) that caused Coronavirus Disease 2019 (COVID-19) captured by SEM and TEM are shown below in Fig. (**9**).

Table 1. Summarizes the differences between scanning electron microscopes and transmission electron microscopes.

-	Scanning Electron Microscopes (SEM)	Transmission Electron Microscopes (TEM)
Electron stream	Fine, focused beam	Broad beam
Image taken	Topographical/surface	Internal structure
Resolution	Lower resolution	Higher resolution
Magnification	Up to 2,000,000 times	Up to 50,000,000 times
Image dimension	3-D	2-D
Sample thickness	Thin and thick samples okay	Ultrathin samples only
Penetrates sample	No	Yes
Sample restriction	Less restrictive	More restrictive
Sample preparation	Less preparation required	More preparation required
Cost	Less expensive	More expensive
Speed	Faster	Slower
Operation	Easy to use	More complicated; requires training

Electron Tomography

Electron Tomography is the only imaging method used to display pleomorphic organisms' three-dimensional density maps. Projection pictures are recorded for a tilt series and a tomogram is calculated. In cryo-electron tomography, at very low doses, the individual projections are recorded. Computational procedures are then established to align these noisy images, allowing a three-dimensional map (tomogram) to be computed without data loss. Resolution as in cryo-tomogram is limited to 4–5 nm noise in a plane; however, it can be extended by combining tomograms that reproduce images of a complex of interest in different orientations and by sub-tomogram averages.

SUMMARY

• A wide selection of chemical, biophysical, electrochemical, and biological techniques has very helped explore the key and co-dependent relationships between the structure and functions of these biopolymers. Every instrumentation techniques have its own advantage and disadvantage in terms of their applications, selectivity and sensitivity.

• In the crystallographic method, X-rays are scattered by the electrons within the crystal and the interference pattern of the scattered X-rays is helping in establishing the 3 three-dimensional structures of the molecules in the crystal.

• Nuclear resonance (NMR) technique has evolved to become a robust tool to probe the structures of biopolymers in liquid as well as in solid-state. It provides a way to study the dynamics of polymers in solution and to look at the effects of the substance, solvent, and other factors on polymeric behavior.

• Electron microscopy (EM) is a technique for getting high-resolution images of biological and non-biological specimens. EM images offer key information on cellular structure, functions, and cellular diseases.

• Cryo-electron microscopy (cryo-EM) is an emerging technique for determining the structures of macromolecular assemblies.

KEYWORDS

Mosaicity

It is a measure of crystal plane orientations. A mosaic crystal is an idealised model of imperfect crystal, speculated to encompass varied little excellent crystals that are to some extent randomly disoriented.

Pleomorphism

Variability within the size and form of cells and/or their nuclei.

Fourier Transforms

It's a mathematical formula that relates a signal sampled in time or space to a similar signal sampled in frequency. In the signal process, the Fourier transform can reveal the necessary characteristics of a signal, namely, its frequency components.

Synchrotrons

An accelerator during which charged particles are accelerated around a constant circular path by an electrical field and held together to the path by an increasing magnetic force. It is capable of generating very high energies.

SHORT-ANSWER TYPE QUESTION

1. What is the principle behind X-ray crystallography?
2. Why H-1 NMR spectroscopy is used for the structural determination of macromolecules?
3. Explain STEM microscopy.
4. What is Nuclear Overhauser Effect in NMR?
5. How Optimization of crystalline protein is carried out in X-ray crystallography?
6. What are STEM techniques and also show its schematic diagram?
7. What is cryoelectron microscopy?
8. Draw the schematic diagram of NMR and its principle.
9. What are biopolymers?
10. Explain the principle behind diffraction in x-ray crystallography.
11. What is Bragg's law of diffraction?
12. Explain COSY NMR.
13. What is single-particle analysis?
14. What is constructive Interference?
15. What is Electron Crystallography.

LONG-ANSWER TYPE QUESTIONS

1. Explain the process involved in the crystallization of protein and its structural determination in X-ray crystallography?
2. Explain NMR spectroscopy in detail.
3. What are the different types of electron microscopy?
4. Mention different techniques for the structural determination of biopolymers.
5. What are biopolymers? Explain the uses of physical techniques for the structural determination of biopolymers.
6. Mention the difference between TEM and SEM in detail.
7. Explain Electron Tomography.
8. Define the Schematic process in X-ray Crystallography.
9. What are the various applications of Electron Microscopy?
10. What are 1D NMR and 2D NMR?

MULTIPLE CHOICE QUESTIONS

1. It is an imaging technique to give three-dimensional density maps of pleomorphic specimens from macromolecular complexes to cells

a) Electron crystallography

b) NMR

c) STEM

d) Electron tomography

2. Biopolymers containmonomericunits bonded together by

a) Covalent bond

b) Ionic bond

c) Coordinate bond

d) Polar bond

3. The technique in which crystalline atoms cause a beam of incident X-rays to diffract into many directions is

a) X-ray crystallography

b) Electron Crystallography

c) SEM

d) Cryo electron microscopy

4) The method which gives data plotted in a space defined by two frequency axis is

a) 1D NMR

b) 2D NMR

c) 3D NMR

d) TEM

5) The precipitant used in the crystallisation of proteins is

a) NaCl

b) Ammonia

c) Polyethylene glycol

d) Ethanol

6) Which of the following have non-crystalline structure?

a) Iron

b) Quartz

c) Silica glass

d) Tungsten

7) Which of the following is a characteristic of crystalline structure?

a) High density

b) Low density

c) Range of melting point

d) Varying structure

8) Which of the following factor is not responsible for formation of non-crystalline structure?

a) Atomic packing has open structure

b) Primary bonds are absent

c) Formation of 1-dimensional chain molecule

d) Strong secondary bond

9) In amyloidosis Beta pleated sheet will be seen in

a) X-ray crystallography

b) Electron microscope

c) Spiral electron microscope

d) Congo red stain

10) What is the role of synchrotrons in x-ray crystallography?

a) As a source of crystals

b) As a source of x-ray

c) Provide computational power

d) Provide proton beams.

11) What is the first step in any x-ray crystallography experiment?

a) Build a model of your molecule.

b) Compute an electron density map.

c) Grow a crystal.

d) Measure a diffraction pattern.

12) Which property of X-rays is critical to their use in X-ray crystallography?

a) X-rays are scattered by electrons.

b) X-rays go right through most materials.

c) X-rays can destroy molecules.

d) X-rays are made of photons.

13) Isomorphous crystals...

a) have cell dimensions that differ significantly

b) cannot be of the same space group

c) cannot be used as heavy atom derivatives

d) proportionally equal sides

e) have identical cell dimensions

14) Bragg planes are...

a) positions in space that scatter in phase

b) only relevant if they are also crystal planes

c) Reflection planes.

d) perpendicular to the scattering vector

e) all of the above

15) Mosaicity is...

a) a property of atoms in a crystal.

b) a property of the crystal lattice.

c) a property of the X-ray data.

d) a property of the model.

e) All of the above.

f) None of the above.

16) What is shielding in NMR?

a) Using a curved piece of metal to block an opponent attack

b) Putting metal around an Rf source

c) Blocking part of a molecule from Rf radiation

d) When the magnetic moment of an atom blocks blocks the full induced magnetic field from surrounding nuclei

17) What is used to cool the superconducting coil?

a) Hydrogen

b) Ice

c) Dry ice

d) Liquid helium

18) Coupling causes the peaks in 1H NMR spectra to be split into

a) Two peaks

b) Multiple peaks equal to the number of hydrogen on surrounding atoms

c) Multiple peaks equal to the number of sourrounding carbon atoms

d) Multiple peaks equal to the number of hydrogen on surrounding atoms, plus one

19) When placed in a magnetic field, all the random spins of the nuclei

a) Stop

b) Reverse direction

c) Align with the magnetic field

d) Rotate to 90°C away from the induced field

20) All the hydrogen atoms

a) Have the same resonance frequency

b) Resonate at different frequencies depending on their environment.

c) Are attached to carbon

d) Resonate at about the same frequency as carbon

21) How many absorptions will the following compound have in its carbon NMR spectrum?

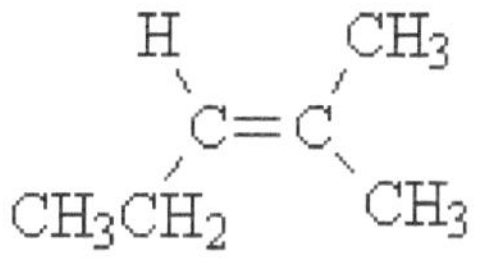

```
   H       CH3
    \     /
     C = C
    /     \
CH3CH2     CH3
```

a) 3

b) 4

c) 5

d) 6

22) What is the multiplicity expected in the hydrogen NMR spectrum for the hydrogen atoms marked by a "star" in the following compound?

```
      O
      ||   *
CH3 - C - CH2 - CH3
```

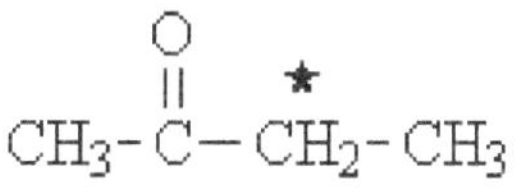

a) Singlet

b) Triplet

c) Quartet

d) Heptet

23)Which technique does the pharmaceutical industry routinely use to screen for polymorphs of a compound?

a) solution NMR spectroscopy

b) solution electronic absorption spectroscopy

c) powder XRD

d) mass spectrometry

24) Which of the following analytical methods would you choose to investigate whether a compound is a monomer, dimer or trimer?

a) Elemental analysis

b) NMR spectroscopy

c) ESI-MS

d) IR spectroscopy

25) Which of the following scientist is associated with the invention of electron microscope and awarded the nobel prize for the same?

a) J.J Thompson

b) Ernst Ruska

c) Louise de Broglie

d) Otto von borris

25) Why are tiny sections of specimen is necessary in transmission electron microscope?

a) Electrons are negatively charged

b) Electrons have a wave nature

c) Electron have no mass

d) Electrons have a poor penetrating power

26) Osmium tetra oxide is used in electron microscope as

a) Precipitator

b) Mordant

c) Staining agent

d) Fixing agent

27) Which instrument is more useful to study the surface details of specimen?

a) Phase contrast microcope

b) SEM

c) Light microscope

d) TEM

28) Which of the following is not true about SEM & TEM?

a) the illuminating source is an electron beam

b) the microscope is focused using electromagnetic

c) can be used to view specimen smaller than 0.2 millimeters

d) the specimen must be sectioned before viewing

29) Kind of electron microscope which is used to study internal structure of cells is

a) scanning electron microscope

b) transmission electron microscope

c) light microscope

d) compound microscope

30) Electrons of Scanning Electron Microscope are reflected through

a) glass funnel

b) specimen

c) metal-coated surfaces

d) vacuum chamber

31) Photograph which is taken from microscope is known as

a) macrograph

b) monograph

c) micrograph

d) pictograp

32) Object can be magnified under electron microscope about

a) 350, 000 times

b) 250, 000 times

c) 300, 000 times

d) 450, 000 times

33) Magnification of light microscope is

a) 1500X

b) 2000X

c) 1000X

d) 2500X

34) As compared to light microscope, the resolving power of electron microscope is

a) 5

b) 10

c) 100

d) 1000

35) Scanning electron microscope is important for their images which are

a) Two dimensional

b) Three dimensional

c) Fluorescent

d) Very large and sharp

36) High magnification of electron microscope is due to

a) High voltage

b) Electron beam

c) Vacuum

d) Electron magnets

37) Which of the following microscope is best for the studying the process of mitosis?

a) Electron microscope

b) Dark field microscope

c) Phase contrast microscope

d) All of the above

38) Which of the following microscopy techniques relies on the specimen interfering with the wavelength of light to produce a high contrast image without the need for dyes or any damage to the sample?

a) Conventional bright field light microscopy

b) Phase contrast microscopy

c) Electron microscopy

d) Fluorescence microscopy

39) The major attractions of the scanning electron microscope (SEM) include all of the following except:

a) Its great depth of focus

b) Its ability to polarise light

c) Its high magnification

d) Its high resolution

40) Which microscope can be used to determine whether or not a suspect has recently fired a gun?

a) A comparison microscope

b) A polarising microscope

c) A scanning electron microscope

d) A compound microscope

41) Which statement regarding the differences between electron microscopy and light microscopy is false?

a) Images produced by electron microscopes are always black and white.

b) The higher resolution of electron microscopes is due to the fact that the wavelength by electrons is about one hundred thousand (1 x 105) times longer than the wavelength of visible light.

c) Viruses and other objects smaller than about 200 nm (0.2 µm) must be examined *via* electron microscopy.

d) Instead of using glass lenses, an electron microscope uses electromagnetic lenses.

42) Which of the following statements regarding resolution is true?

a) In order to achieve high magnification with good resolution, a small objective lens must be used.

b) The white light used in a compound light microscope has a relatively short wavelength in order to resolve structures smaller than 0.002 µm.

c) A general principle of microscopy is that the shorter the wavelength of light used in the instrument, the lower the resolution.

d) Resolution (resolving power) is calculated by multiplying the objective lens power by the ocular lens power.

43) Electron microscope is an invention that uses beam of

a) neuron

b) electron

c) neutron

d) proton

44) EBCT scanners stands for

a) electrical beam computed tomography

b) electric beam computed tomography

c) electronic beam computed tomography

d) electron beam computed tomography

46) What are the tilting methods used in electron tomography?

a) Single axis method

b) Standard free axis method

c) Triple axis method

d) Electron axis method

47) Which of the following is the type of electron tomography?

a) BF-TEM

b) SEM

c) TEM

d) ZRT

CONSENT FOR PUBLICATION

Not Applicable.

CONFLICT OF INTEREST

The authors confirm that this chapter contents have no conflict of interest.

ACKNOWLEDGEMENTS

Authors are thankful to Department of Chemistry, North Eastern Regional Institute of Science and Technology,Nirjuli, Itanagar for providing facilities to accomplish the work.

REFERENCES

[1] Kaushik M, Kumar M, Chaudhary S, *et al.* Advancements in characterization techniques of biopolymers: cyclic voltammetry, gel electrophoresis, circulardichroismand fluorescence spectroscopy. Adv Tech Biol Med 2016; 4(3): 1-8.

[2] Chemistry Learner. https://www.chemistrylearner.com/biopolymer.html

[3] http://www.whatislife.com/reader/techniques/techniques.html

[4] Klebe G. Experimental methods of structure determination. In: Klebe G, Ed. Drug Design. Berlin, Heidelberg: Springer 2013; pp. 265-90. [http://dx.doi.org/10.1007/978-3-642-17907-5_13]

[5] Finley JW, Schmidt SJ, Serianni AS, Eds. NMR Applications in Biopolymers. New York: Springer US 1990; pp. 1-526. [http://dx.doi.org/10.1007/978-1-4684-5868-8]

[6] University of Massachusetts Medical School. What is Electron Microscopy? https://www.umassmed.edu/cemf/whatisem/

[7] https://www.wikilectures.eu/w/Electron_microscopy/principle

[8] Smith Y. https://www.news-medical.net/life-sciences/Types-of-Electron-Microscopes.aspx

CHAPTER 5

Structure and Function of Biological Biomolecules: Carbohydrates, Amino Acids, Proteins, Nucleic Acids, Lipids and Biomembranes

Nagendra Nath Yadav[1,*], **Saroj Yadav**[2] and **Archana Pareek**[1]

[1] *Department of Chemistry, NERIST, Nirjuli, Itanagar-791109 (AP), India*

[2] *Delhi University, Benito Juarez Road, South Campus, South Moti Bagh, New Delhi-110021, India*

Abstract: This chapter deals with the introduction, definition and classification of various biological molecules like carbohydrates, lipids and bio-membranes, amino acids, proteins and nucleic acids. The synthesis of carbohydrates, structure and function of monosaccharides, disaccharides and polysaccharides, absolute and relative configuration of sugar, reducing and non-reducing sugar, Fischer projection formula and Haworth projection formula are also briefly described. A brief knowledge about the type of fatty acids, acid value, soap value, various types of lipids, including phospholipids and sphingolipids, is also included. This unit also discusses the properties of various types of amino acids, proteins and nucleic acids, including structures of nucleoside, nucleotides, DNA and RNA.

Keywords: Carbohydrates, Nucleic acid, Nucleotide, Polypeptides, Proteins, Relative and absolute configuration.

CARBOHYDRATES

Carbohydrates are biomolecules consisting of carbon, hydrogen and oxygen atoms. They perform various important roles in living organisms. For example, some polysaccharides serve as storage of energy; the monosaccharide ribose is the backbone of RNA (genetic material) and is also an important component of coenzymes like ATP, NAD, FAD; the monosaccharide deoxyribose is the backbone of DNA [1]. Carbohydrates also play key roles in fertilization, immune system, blood clotting, *etc.*

In the human diet, the most important carbohydrates are starch and sugars which are found in a variety of foods, both natural and processed, for *e.g.* starch is avail-

* **Corresponding author Nagendra Nath Yadav:** Department of Chemistry, NERIST, Nirjuli, Itanagar-791109 (AP), India; E-mail:nny@nerist.ac.in

Meera Yadav & Prof. Hardeo Singh Yadav (Eds.)

able in cereals, potatoes and even in bread, pizza, pasta, *etc.* (shown in Fig. **1**).

Initially, it was believed that the term carbohydrates refers to the "hydrates of carbon", having general formula $C_x(H_2O)_y$, for *e.g.* glucose ($C_6H_{12}O_6$), cane-sugar ($C_{12}H_{22}O_{11}$), *etc.* But later, it was found that several carbohydrates do not follow the pattern given in the above formula, *e.g.* deoxyribose ($C_5H_{10}O_4$), rhamnose ($C_6H_{12}O_5$), glucosamine ($C_6H_{13}O_5N$). In recent days, carbohydrates are defined as "*Polyhydroxy aldehydes or polyhydroxy ketones or substances that yield such compounds on hydrolysis*", known as carbohydrates [1], for *e.g.* glucose, fructose, galactose, *etc.*

Fig. (1). Food rich in carbohydrates [2 - 4] respectively.

Synthesis of Carbohydrates

Carbohydrates are synthesized in green plants by the process of photosynthesis. Green plants contain chlorophyll, which catalyses the conversion of carbon dioxide into sugar in the presence of water and sunlight [5]. The reaction involved in photosynthesis can be shown as follows:

$$6CO_2 + 6H_2O \xrightarrow[\text{Chlorophyll}]{\text{Sun light}} C_6H_{12}O_6 + 6O_2 \uparrow$$

Carbohydrates are degraded by animals using O_2, providing energy for their survival and releasing an equivalent amount of carbon dioxide.

Function of Carbohydrates

Carbohydrates participate in a wide range of functions. Some of them are:

- They are the most abundant dietary source of energy for all organisms.
- They are the main precursors for most of the organic compounds, *e.g.* fats, amino acids.
- Carbohydrates, especially glycoproteins and glycolipids, participate in the structure of cell membrane and cellular functions like cell growth, *etc.*
- They provide energy and regulate blood glucose, sparing the use of protein for energy.
- They act as important structural units in plants.

Classification

Carbohydrates are also known as Saccharides (Greek: Sakkharon), which refer to the substances which are sweet in taste [1]. The classification of carbohydrates depends on their behaviour, and is given as follows:

Based on their Behavior Upon Hydrolysis

Monosaccharides

These are the simplest unit of carbohydrates and cannot be further hydrolysed into other smaller fragments [1]. The general formula of monosaccharide is $C_n(H_2O)_n$, where n = 3 to 7, for *e.g.* Glucose ($C_6H_{12}O_6$), Fructose ($C_6H_{12}O_6$), Galactose ($C_6H_{12}O_6$), Ribose ($C_5H_{10}O_5$), *etc.*

Oligosaccharides

The carbohydrates which provide a definite number (2-9) of molecules of monosaccharide on hydrolysis are called oligosaccharides [1]. Based on the number of monosaccharide units obtained after hydrolysis, oligosaccharides can be further classified into the following:

Disaccharides

The carbohydrates which on hydrolysis give two units of the same or different monosaccharides are called disaccharides [1]. The general formula of monosaccharide is $C_n(H_2O)_{n-2}$, for *e.g.* Sucrose ($C_{12}H_{22}O_{11}$), Maltose ($C_{12}H_{22}O_{11}$), Lactose ($C_{12}H_{22}O_{11}$), *etc.*

$$\underset{\text{Sucrose}}{C_{12}H_{22}O_{11}} + H_2O \longrightarrow \underset{\text{Glucose}}{C_6H_{12}O_6} + \underset{\text{Fructose}}{C_6H_{12}O_6}$$

$$\underset{\text{Maltose}}{C_{12}H_{22}O_{11}} + H_2O \longrightarrow \underset{\text{Glucose}}{C_6H_{12}O_6} + \underset{\text{Glucose}}{C_6H_{12}O_6}$$

$$\underset{\text{Lactose}}{C_{12}H_{22}O_{11}} + H_2O \longrightarrow \underset{\text{Glucose}}{C_6H_{12}O_6} + \underset{\text{Galactose}}{C_6H_{12}O_6}$$

- **Trisaccharides**: The carbohydrates which upon hydrolysis give three units of the same or different monosaccharides are called trisaccharides [1]. The general formula of trisaccharides is $C_n(H_2O)_{n-2}$, for *e.g.* Rhamninose, Raffinose, *etc.*

$$\underset{\text{Raffinose}}{C_{18}H_{32}O_{16}} + 2H_2O \longrightarrow \underset{\text{Galactose}}{C_6H_{12}O_6} + \underset{\text{Glucose}}{C_6H_{12}O_6} + \underset{\text{Fructose}}{C_6H_{12}O_6}$$

- **Tetrasaccharides**: They produce four units of the same or different monosaccharides on hydrolysis. The general formula of tetrasaccharides is $C_n(H_2O)_{n-3}$, for *e.g.* Stachyose ($C_{24}H_{42}O_{21}$).

$$\underset{\text{Stachyose}}{C_{24}H_{42}O_{21}} + 3\,H_2O \longrightarrow \underset{\text{Galactose}}{2\,C_6H_{12}O_6} + \underset{\text{Glucose}}{C_6H_{12}O_6} + \underset{\text{Fructose}}{C_6H_{12}O_6}$$

- **Polysaccharides**: Carbohydrates that give more than ten molecules of the same or different monosaccharides on hydrolysis are called polysaccharides [1]. The general formula of polysaccharides is $(C_6H_{10}O_5)_n$, for *e.g.* starch, cellulose, pectin, glycogen.

$$\underset{\text{Starch}}{(C_6H_{10}O_5)n} + nH_2O \xrightarrow{H^+} \underset{\text{Glucose}}{n\,C_6H_{12}O_6}$$

Polysaccharides that exhibit only a single type of monosaccharides are called homopolysaccharides and those that provide a mixture of monosaccharides and their derivatives are called heteropolysaccharides.

Based on their Taste

Sugars

Carbohydrates that are sweet in taste are called sugars. Sugars are generally monosaccharides or disaccharides soluble in water [6], for *e.g.* Glucose, Fructose, Sucrose, Lactose, *etc.*

Non-sugars

Carbohydrates that are not sweet in taste are called non-sugars. Non-sugars are generally polysaccharides that are less insoluble in water, for *e.g.* Starch and Cellulose [6].

Based on their Reaction (Oxidation)

Reducing Sugars

Carbohydrates that can react with Fehling's solution and Tollen's reagent are called reducing sugars. They contain either a free α-hydroxy aldehyde or α-hydroxy ketone in their hemiacetal forms, for *e.g.* D-glucose, D-fructose, maltose, lactose, *etc* . [6].

Non-reducing Sugars

Carbohydrates that are unable to react with Fehling's solution and Tollen's reagent are called non-reducing sugars. They do not have free aldehyde or ketonic group; instead, these have a stable acetal and ketal group, for *e.g.* sucrose, starch, cellulose, *etc.* [6].

CLASSIFICATION OF MONOSACCHARIDES

Based on the functional group present, monosaccharides can be classified into two categories [6]:

1. **Aldoses**: Monosaccharides having an aldehyde (-CHO) group are termed as aldoses. The structures of some D-aldoses are shown below.

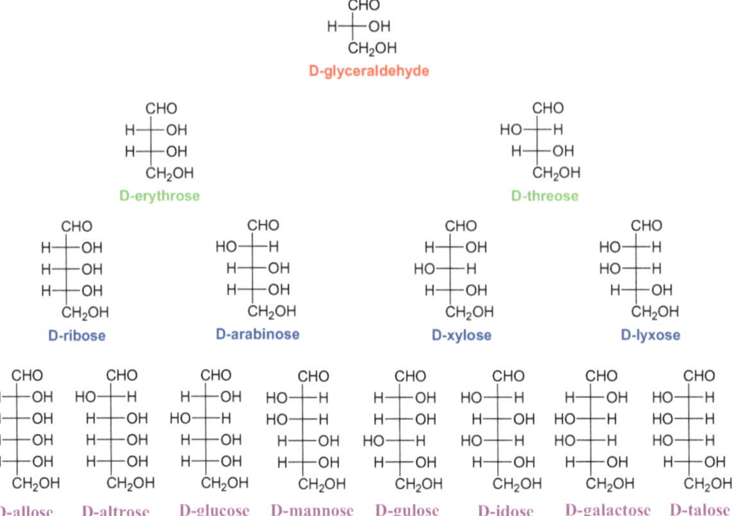

2. **Ketoses**: Monosaccharides having ketone (>C=O) group are termed as ketoses. Structures of some D-ketoses are shown below.

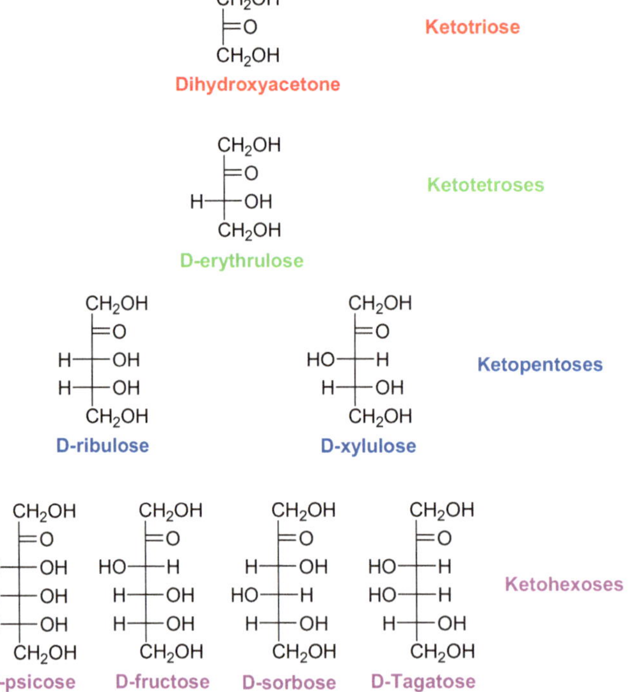

Based on the number of carbon atoms present, monosaccharides may further be classified as [6]-

- 3 carbon atoms = trioses. Aldotriose (if aldehyde group is present) and ketotriose (if ketone group is present).
- 4 carbon atoms = tetroses. Aldotetrose (if aldehyde group is present) and ketotetrose (if ketone group is present).
- 5 carbon atoms = pentoses. Aldopentose (if aldehyde group is present) and ketopentose (if ketone group is present).
- 6 carbon atoms = hexoses. Aldohexose (if aldehyde group is present) and ketohexose (if ketone group is present).

```
                      CH2OH
                        |
  CHO                  C=O
   |                    |
 (CHOH)n              (CHOH)n
   |                    |
  CH2OH                CH2OH
(n = 1,2,3,4,5)      (n = 0,1,2,3,4)
   Aldoses              Ketoses
```

STRUCTURE OF MONOSACCHARIDES

Monosaccharides being the simplest carbohydrates cannot be further hydrolyzed to smaller carbohydrates. Their general chemical formula is $(C \cdot H_2O)_n$. These are the building blocks of nucleic acid [6].

A. **Open Chain Structures**: Trioses are the simplest carbohydrates (monosaccharides) and have the general formula $C_3H_6O_3$. The structure of aldotriose and ketotriose is shown below.

```
    CHO                 CH2OH
     |                    |
    CHOH                 C=O
     |                    |
    CH2OH               CH2OH

Glyceraldehyde       Dihydroxyacetone
(Aldotriose)         (Ketotriose)
```

Glyceraldehydes have one chiral centre (asymmetric carbon) and therefore exist in two optically active forms, called D-form and L-form [No. of possible isomer is 2^n].

Mirror

H–C(=O)–C(H)(OH)–CH_2OH | O=C(H)–C(HO)(H)–CH_2OH

D(+)-glyceraldehyde L(-)-glyceraldehyde

D & L CONFIGURATION (RELATIVE CONFIGURATION)

The prefixes D & L in the name of monosaccharide refer to their configuration and depend upon the configuration of glyceraldehydes [7], for *e.g.* structure of D & L Glucose.

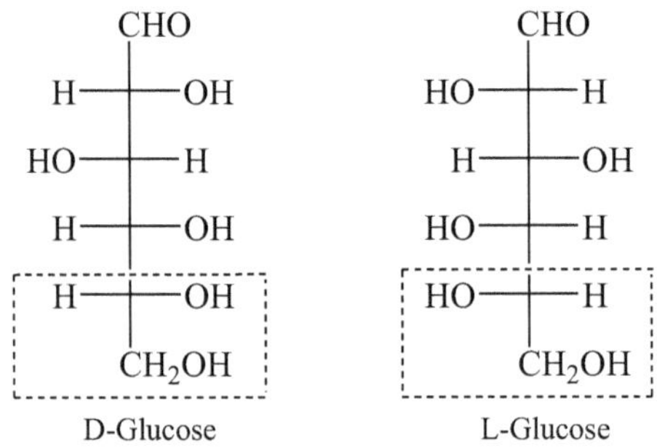

D-Glucose L-Glucose

Other examples of D-SUGAR

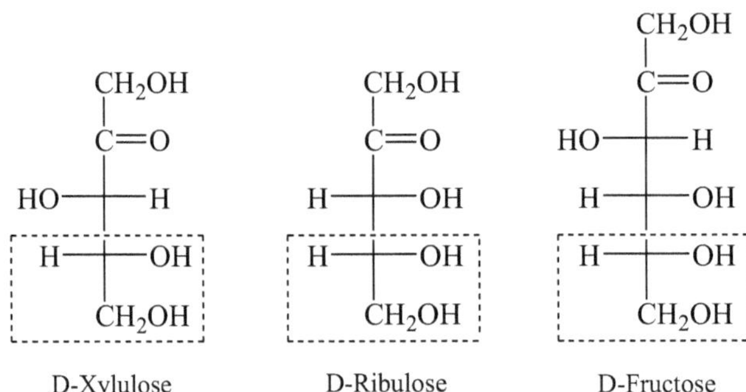

D-Xylulose D-Ribulose D-Fructose

ENANTIOMER

Stereoisomers that are non-superposable mirror images of each other are called enantiomers. Each member of the enantiomeric pair is known as an enantiomorph and the phenomenon is termed as enantiomerism, for *e.g.* D-Glucose and L-

Glucose, D-glyceraldehyde and L- Glyceraldehyde, *etc.* The chemical and physical properties of enantiomers are identical in nature except that they rotate the plane of polarised light in an equal but opposite direction. Therefore, they are sometimes also known as optical isomers. If the compound rotates the plane of polarised light towards the right, then it is known as dextrorotatory (+ or d), whereas if it rotates the plane of polarised light towards the left, then it is known as levorotatory (- or l) [7].

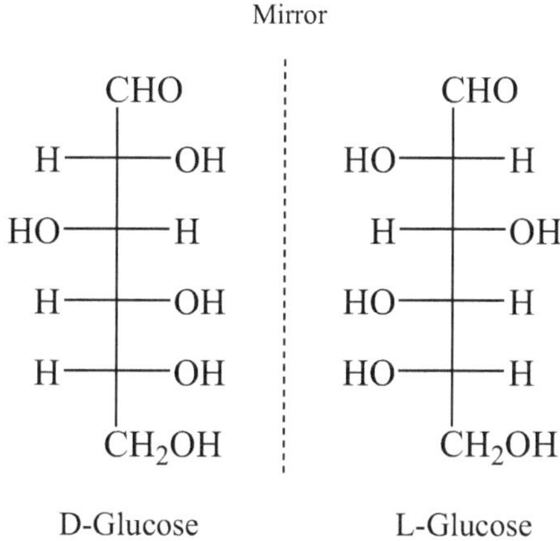

D-Glucose L-Glucose

DIASTEREOMER

Stereoisomers that differ from each other in their configuration around one or more (but not all) carbon atoms are called diastereomers. These are not mirror images of each other. If they differ from each other at only one stereo centre then the diastereomers are known as epimers. Diastereomers have different physical and chemical properties [7], for *e.g.* D-Threose and D-Erythrose, D-Glucose and D-Mannose, *etc.*

CHO
H—OH
HO—H
H—OH
H—OH
CH_2OH

D-Glucose

CHO
HO—H
HO—H
H—OH
H—OH
CH_2OH

D-Mannose

CHO
HO—H
H—OH
CH_2OH

D-Threose

CHO
H—OH
H—OH
CH_2OH

D-Erythrose

Optical activity of sugars: (d+) or (l-) form

When a plane of polarized light is passed through a solution of optically active compounds, it will be rotated either to the right or left direction.

Isomer, which rotates the plane of polarized light in the right direction, is called dextrorotatory (d+), and the isomer which rotates the plane of polarized light in the left direction is called levorotatory (l-) [7].

Racemic mixture: (±) or (*d*, *l*) form

Mixture of *d*- and *l*- isomer in equal concentration is known as racemic mixture or *dl*- mixture. The net rotation of plane-polarized light is zero because though the two enantiomers rotate the plane of polarised light to an equal extent but in the opposite direction, which cancels out each other, as they are present in equal amounts. Therefore racemic mixtures are optically inactive [7].

ABSOLUTE CONFIGURATION (R, S - FORM)

Spatial arrangement of the atoms and/or groups in a chiral molecule along with its stereochemical description is known as absolute configuration, for *e.g.* R- and S-form, where R stands for Rectus and S stands for Sinister. The R- and S-configurations of a chiral molecule are often obtained from X-ray crystallography [7].

e.g., Glyceraldehyde

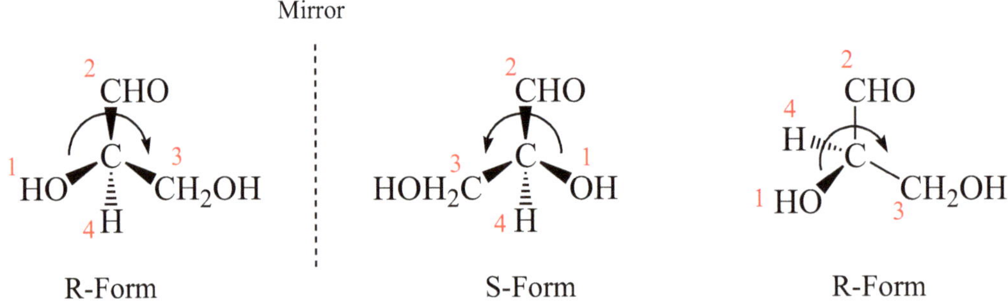

CIP Rule (Cahn-Ingold-Prelog Priority rules)

R= Rectus and S= Sinister

If the lowest priority group as determined by the CIP rule is away from observer and

- Priority decreases in a clockwise direction, then it should be (*R*)- form
- Priority decreases in anticlockwise direction, then it is (*S*)- form

FISCHER PROJECTION FORMULA OF CARBOHYDRATES

It is a convenient way to depict three-dimensional structures to two-dimensional structures by projection. In Fischer projections of carbohydrates, the longest chain is drawn vertically with the most highly oxidised carbon (*i.e.*, carbonyl group) on the top.

D-glyceraldehyde or R-(+)-Glyceraldehyde (CHOCHOHCH$_2$OH)

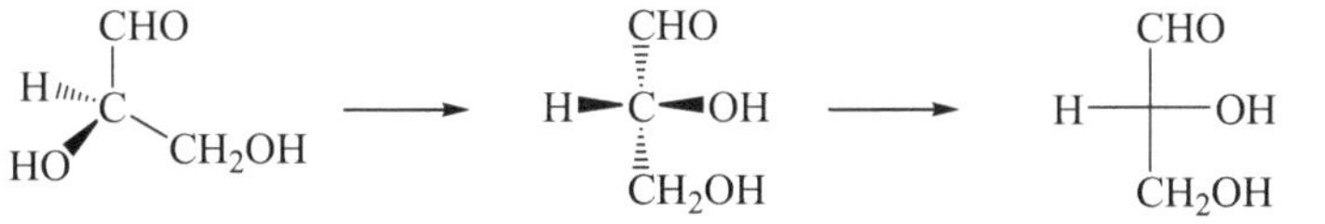

3-D structure 2-D structure

The D- and L- notation of the compounds is based on these structures. According to definition, (R)-(+)-glyceraldehyde is D-glyceraldehyde and (S)-(-)-glyceraldehyde is L-glyceraldehyde [7].

D-Glucose and D-Fructose

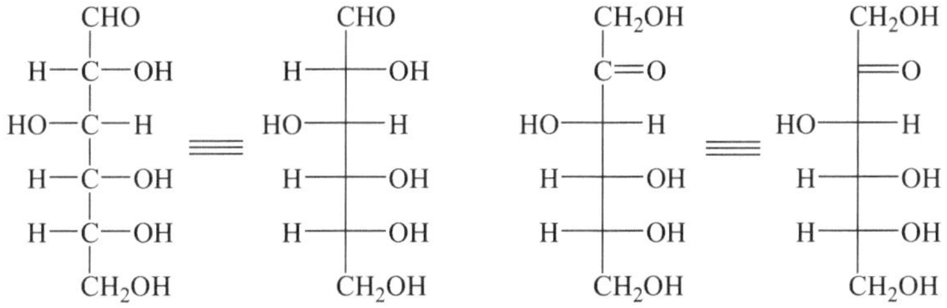

D-Glucose D-Fructose

Other carbohydrates are assigned as D- or L- by comparing the stereochemistry of the chiral center next to the -CH$_2$OH group, *i.e.* the chirality centre furthest from the carbonyl group [7].

CYCLIC STRUCTURE OF MONOSACCHARIDES

A monosaccharide undergoes a nucleophilic addition reaction between the carbonyl group and one of the hydroxyl groups of the same molecule and converts acyclic form to a cyclic form. This reaction forms a ring of carbon atoms with a bridging O-atom. The resulting molecule will be consisting of a hemiacetal or a hemiketal group based on whether the linear form has an aldehyde or ketone group [7].

$$R{-}\ddot{O}H + R^1{-}\overset{\overset{O}{\|}}{C}{-}H \rightleftharpoons R^1{-}\underset{OR}{\overset{OH}{C}}{-}H \xrightarrow[-H_2O]{ROH} R^1{-}\underset{OR}{\overset{OR}{C}}{-}H$$

Hemiacetal Acetal

$$R{-}\ddot{O}H + R^1{-}\overset{\overset{O}{\|}}{C}{-}R^{11} \rightleftharpoons R^1{-}\underset{OR}{\overset{OH}{C}}{-}R^{11} \xrightarrow[-H_2O]{ROH} R^1{-}\underset{OR}{\overset{OR}{C}}{-}R^{11}$$

Ketal

The formation of an intramolecular hemiacetal or a hemiketal in a monosaccharide results in a five-membered (furanose) or a six-membered (pyranose) cyclic structure, as shown below.

Furan (5-membered) Pyran (6-membered)

PYRANOSE AND FURANOSE STRUCTURE

The aldehyde group of glucose at C_1 reacts with alcohol group either at C_4 carbon to generate two types of cyclic hemiacetal (α- and β-) or at C5 carbon to generate two types of cyclic hemiacetal named α-pyranose or β-pyranose, as shown below.

β-D-Glucopyranose (62%) $[\alpha]_D = +18.7°$

C1 + C5 Pyranose

D-Glucose

C1 + C5 Pyranose

α-D-Glucopyranose (37%) $[\alpha]_D = +122.2°$

C1 + C4 Furanose

C1 + C4 Furanose

β-D-Glucofuranose (0.5%)

α-D-Glucofuranose (0.5%)

Structure of D-Fructose

HO–C–CH_2OH		CH_2OH		HOH_2C–C–OH
HO—C—H	⇌	C=O	⇌	HO—C—H
H—C—OH		HO—C—H		H—C—OH
H—C—OH		H—C—OH		H—C—OH
HOHC		H—C—OH		HOHC
		CH_2OH		
β–D-Fructose or β–D-Fructopyranose		D-Fructose		α–D-Fructose or α–D-Fructopyranose

HO–C–CH_2OH		CH_2OH		HOH_2C–C–OH
HO—C—H	⇌	C=O	⇌	HO—C—H
H—C—OH		HO—C—H		H—C—OH
H—C—O		H—C—OH		H—C—O
CH_2OH		H—C—OH		CH_2OH
		CH_2OH		
β–D-fructofuranose		D-Fructose		α–D-fructofuranose

HAWORTH PROJECTION FORMULA

It is the common form of writing the structural formula which is used to represent the cyclic structures of monosaccharides in the 3D form [1, 7]. Its characteristics are-

- In the figure the carbon atoms in the structure are given a specific number (1-6); these are the implicit type of atoms. Carbon 1 is an anomeric carbon.
- The bold line indicates that the atoms are closer to the observer. Here carbon-2 and -3 are closer to the observer and carbon-5 and -6 are at the farthest.
- The groups on the right-hand side of the Fischer projection corresponds to the groups below the plane of the ring in the Haworth projection formula. But this rule is not followed by the groups on the two ring carbon atoms which are bonded to an endocyclic oxygen atom. Haworth projection formula of some carbohydrates are given below-

α–D-Glucopyranose

Hawarth projection formula of α–D-Glucopyranose

β–D-Glucopyranose

Hawarth projection formula of β–D-Glucopyranose

α–D-Glucofuranose

Hawarth projection formula of α–D-Glucofuranose

β–D-Glucofuranose

Hawarth projection formula of β–D-Glucofuranose

ANOMERS AND MUTAROTATION

Anomers: Monosaccharides that differ in configuration from each other, specifically at the hemiacetal carbon, are called anomers. *e.g.* α-D-Glucose and β-D-Glucose are the anomeric forms of D-Glucose. The hemiacetal carbon (C1) of glucose is called anomeric carbon or glycosidic carbon. The process of conversion of one anomer to the other form is known as anomerization. The two anomers of a compound are named as α- and β- form of that compound. The different anomers will differ in their physical properties like specific rotations and melting points [1, 7].

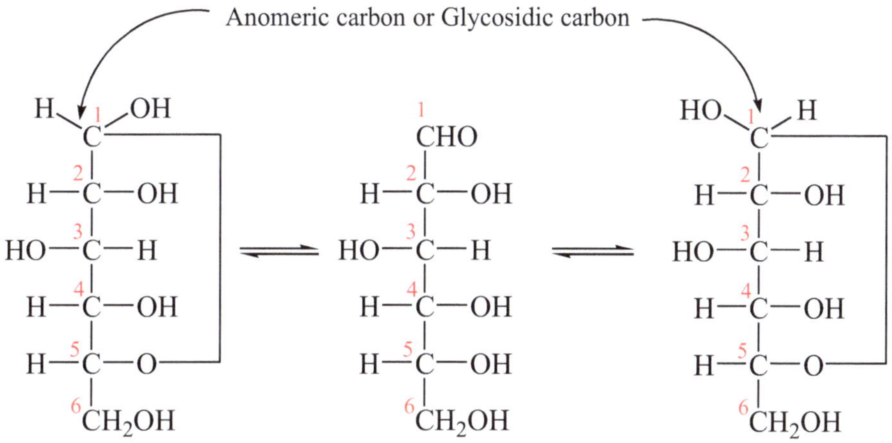

Anomeric carbon or Glycosidic carbon

α–D-Glucose		D-Glucose		β–D-Glucose
H–C1–OH (ring O)		1 CHO		HO–C1–H (ring O)
H–C2–OH		H–C2–OH		H–C2–OH
HO–C3–H	⇌	HO–C3–H	⇌	HO–C3–H
H–C4–OH		H–C4–OH		H–C4–OH
H–C5–O		H–C5–OH		H–C5–O
6 CH_2OH		6 CH_2OH		6 CH_2OH
α–D-Glucose $[\alpha]_D = +122.2^o$		D-Glucose		β–D-Glucose $[\alpha]_D = +18.7^o$

Mutarotation: The change in specific optical rotation of a compound due to a change in the equilibrium between its two anomers is called mutarotation. *e.g.* the specific optical rotation of aqueous solution of D-Glucose changes with time.

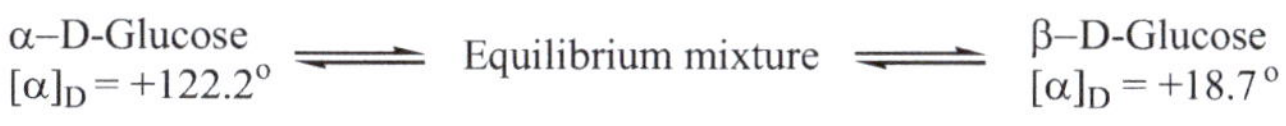

α–D-Glucose $[\alpha]_D = +122.2^o$ ⇌ Equilibrium mixture ⇌ β–D-Glucose $[\alpha]_D = +18.7^o$

The equilibrium mixture contains approximately 37% of α-anomer, 62% of β-anomer and very little (<1%) of open chair form. In an aqueous solution, β-anomer is more predominant due to its stable conformation [7].

The mechanism of mutarotation or interconversion of α and β anomers is shown below-

Structure of few monosaccharides

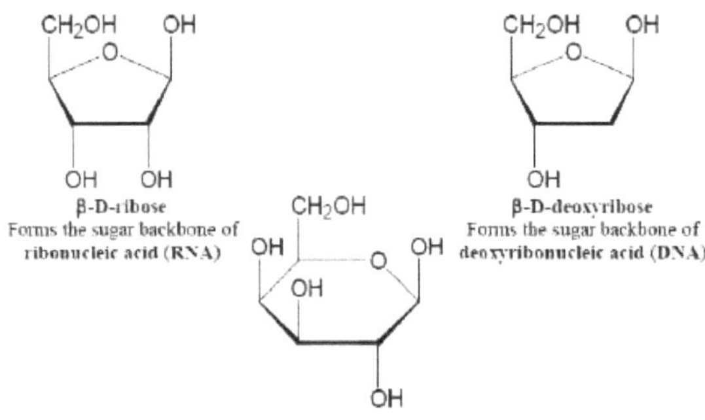

β-D-galactose (Milk Sugar)

β-D-glucose β-D-fructose

GLUCOSE (Dextrose or Grape Sugar or Blood Sugar)

Glucose is the most important member of the monosaccharide family. It is a simple sugar and its molecular formula is $C_6H_{12}O_6$. It is a hexose as it has 6-carbon atoms. It occurs in a free state as well as a combined state in nature and is quite abundant [6]. [Free state = honey, most of the fruits such as grapes, human blood, urine and Combined state = sucrose, maltose and lactose]

Chemical Properties of Glucose (Tautomerization):

When glucose is kept in an alkaline solution then it undergoes isomerization to form D-fructose and D-mannose and vice-versa. This reaction is known as the Lobry de Bruyn-van Ekenstein rearrangement reaction. Tautomerism can also take place in acidic or neutral conditions [6].

D-Glucose & D-Fructose & D-mannose ⇌ Enediol (common intermediae)

Because of the above rearrangement, even D-fructose does not have an aldehyde group but still reduces Fehling's solution and Tollen's reagent. The mechanism of an aldose to ketose transformation is shown below-

Aldose ⇌ Endiol (deprotonierte Form) ⇌ ⇌ Ketose

Formation of Glycosides

The chemical bond formed between the anomeric carbon (hemiacetal or hemiketal) and the hydroxyl group of another molecule (carbohydrate or non-carbohydrate) is called a glycosidic bond and the product formed is known as glycoside [6]. The non-carbohydrate moiety is termed an aglycone. *e.g.* glucose reacts with methyl alcohol in the presence of dry HCl and produces a mixture of α-Methylglucoside and β-Methylglucoside. *i.e.*

$$C_6H_{12}O_6 + CH_3OH \xrightarrow{HCl} C_7H_{14}O_6$$

or

$$C_6H_{11}O_5OH + HOCH_3 \xrightarrow[-H_2O]{HCl} C_6H_{11}O_5OCH_3$$

Methylglucoside

α-D-glucose + CH_3OH $\xrightarrow{H^+}$ methyl α-D-glycopyranoside + methyl β-D-glycopyranoside

Oxidation Reaction

a) Glucose on oxidation with a mild oxidizing agent such as bromine water, give gluconic acid [6].

$$\underset{\text{Glucose}}{CHO-(CHOH)_4-CH_2OH} \xrightarrow{Br_2/H_2O} \underset{\text{Gluconic acid}}{COOH-(CHOH)_4-CH_2OH}$$

Since glucose is easily oxidized therefore it acts as a reducing agent. It

- Decolourises the bromine water.
- Reduces Tollen's reagent to form a silver mirror.
- Reduces Fehling's solution and Benedict's solution and form red ppt. of Cu_2O.

b) Glucose on oxidation with a strong oxidizing agent gives saccharic acid [6].

$$CHO-(CHOH)_4-CH_2OH \xrightarrow{\text{Conc. } HNO_3} \underset{\text{Saccharic acid}}{COOH-(CHOH)_4-COOH}$$

While fructose under similar reaction conditions gives glycolic acid and tartaric acid.

$$\underset{\text{Fructose}}{\begin{array}{c} CH_2OH \\ | \\ CO \\ | \\ (CHOH)_3 \\ | \\ CH_2OH \end{array}} \xrightarrow{HNO_3} \underset{\text{Glycolic acid}}{\begin{array}{c} CH_2OH \\ | \\ COOH \end{array}} + \underset{\text{Tartaric acid}}{\begin{array}{c} COOH \\ | \\ (CHOH)_2 \\ | \\ COOH \end{array}}$$

Some other reactions of glucose are-

- When reacts with hydrogen iodide glucose reduces to n-hexane [6].

$$\underset{\text{Glucose}}{\begin{array}{c} CHO \\ | \\ (CHOH)_4 \\ | \\ CH_2OH \end{array}} \xrightarrow{HI/P} \underset{\text{n-Hexane}}{\begin{array}{c} CH_3 \\ | \\ (CH_2)_4 \\ | \\ CH_3 \end{array}}$$

- It gives sorbitol if reacts with sodium amalgam [6].

$$\underset{\text{Glucose}}{\begin{array}{c} CHO \\ | \\ (CHOH)_4 \\ | \\ CH_2OH \end{array}} \xrightarrow{Na/Hg} \underset{\text{Sorbitol}}{\begin{array}{c} CH_2OH \\ | \\ (CHOH)_4 \\ | \\ CH_2OH \end{array}}$$

- With H_2O, it forms a neutral solution.
- With hydroxylamine, it forms oximes [6].

$$\underset{\text{Glucose}}{\begin{array}{c}CHO\\|\\(CHOH)_4\\|\\CH_2OH\end{array}} \xrightarrow{NH_2OH} \underset{\text{Glucose oxime}}{\begin{array}{c}H\\|\\C{=}NOH\\|\\(CHOH)_4\\|\\CH_2OH\end{array}}$$

- It forms cyanohydrins when reacts with HCN [6]

$$\underset{\text{Glucose}}{\begin{array}{c}CHO\\|\\(CHOH)_4\\|\\CH_2OH\end{array}} \xrightarrow{HCN} \underset{\text{Glucose cyanohydrin}}{\begin{array}{c}CN\\|\\(CHOH)_5\\|\\CH_2OH\end{array}}$$

- With acetic anhydride it forms pentaacetate: This reaction indicates the presence of five hydroxyl groups [6].

$$\underset{\text{Glucose}}{\begin{array}{c}CHO\\|\\(CHOH)_4\\|\\CH_2OH\end{array}} \xrightarrow[C_5H_5N]{5(CH_3CO)_2O} \underset{\text{Glucose pentaacetate}}{\begin{array}{c}CHO\\|\\(CHOCOCH_3)_4\\|\\CH_2OCOCH_3\end{array}}$$

- With phenyl hydrazine it forms glucosazone [6]:

$$\underset{\mathbf{1}}{\begin{array}{c}CHO\\|\\CHOH\\|\\(CHOH)_3\\|\\CH_2OH\end{array}} \quad 3\ C_6H_5.NH.NH_2 \quad \underset{\mathbf{2}}{\begin{array}{c}CH_2OH\\|\\C{=}O\\|\\(CHOH)_3\\|\\CH_2OH\end{array}}$$

$$\longrightarrow \underset{\text{Glucosazone}}{\begin{array}{c}HC{=}N.NH.C_6H_5\\|\\C{=}N.NH.C_6H_5\\|\\(CHOH)_3\\|\\CH_2OH\end{array}} + C_6H_5NH_2 + NH_3 + 2H_2O$$

DISACCHARIDES (SUCROSE, MALTOSE AND LACTOSE)

Disaccharides consist of two same or different monosaccharide units held together by a glycosidic bond. They are crystalline, water-soluble and sweet in taste. Disaccharides are of two types- reducing disaccharides and non-reducing disaccharides. Reducing disaccharides are those which have a free α-hydroxy aldehyde or α-hydroxy ketone or are found in unstable hemiacetal or hemiketal form, *e.g.* maltose, lactose, while non-reducing disaccharides are with no free α-hydroxy aldehyde or ketone and generally have stable acetal or ketal form, *e.g.* sucrose, maltose and lactose. The general formula for disaccharides is $C_{12}H_{22}O_{11}$ [6].

Sucrose (table sugar, cane sugar, beet sugar):

The common table sugar is known as sucrose. It is composed of glucose and fructose and is naturally produced by plants. For human consumption, it is extracted from plants like sugarcane and sugar beet and is refined (shown in Fig. (**2**) [8].

Fig. (2). Sugar [8] (image courtesy: pexels.com).

Occurrence: Sugar cane (Saccharum Officinarum, 16-20%), sugar beets (10-15%), other sources include honey, mango, banana, *etc.*

Biological Importance

- An important source of dietary carbohydrate.
- Applied as a sweetening agent in food industries.
- The intestinal enzyme sucrase hydrolyses sucroseand converts it to glucose and fructose, which are then absorbed in the blood circulation.

Inversion of Cane Sugar

Sucrose (dextrorotatory, +66.5°), on hydrolysis, produces a mixture of glucose and fructose (levorotatory, -28.2°). This process of change of optical rotation is called inversion and the hydrolysed mixture of sucrose-containing glucose and fructose is known as invert sugar. It is made by hydrolysis of sucrose *i.e.* it can be made simply by heating table sugar with water. Since the optical rotation of the mixture and the original sugar is in opposite direction hence the mixture is known as invert sugar [6]. The equation of hydrolysis of sucrose into glucose and fructose is shown below-

$C_{12}H_{22}O_{11}$ (Sucrose) + H_2O (Water) → $C_6H_{12}O_6$ (Glucose) + $C_6H_{12}O_6$ (Fructose)

Structure of Sucrose

Sucrose is made up of *α*-D-Glucose and *β*-D-Fructose and joined together by a glycosidic bond ($\alpha_1 \rightarrow \beta_2$) between C_1 of α-D-Glucose and C_2 of *β*-D-Fructose.

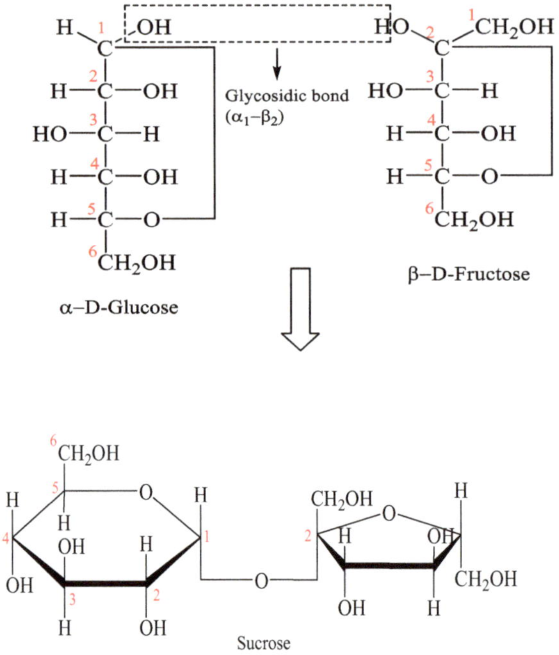

Glycosidic bond (α_1–β_2)

α–D-Glucose

β–D-Fructose

Sucrose

Maltose (malt sugar):

Maltose is produced when β-amylase breaks down the starch molecules removing two glucose units at a time. It is a reducing sugar because one of the two glucose units can open presenting a free aldehyde group while the other can't, due to glycosidic linkage.

Maltose (image shown in Fig. **3**) can be hydrolysed by maltase enzyme yielding two units of a glucose molecule. It is sweet in taste but only about 30 to 35% as that of sugar [6].

Fig. (3). Maltose [9] (image courtesy: pexels.com).

Occurrence: Product of starch hydrolysis, germinating seeds.

Biological Importance

It acts as an important intermediate in the digestion of starch.

Structure

Maltose is a disaccharide. It is made by the combination of two α-D-Glucose units *via* α(1 → 4) glycosidic bond between C_1 of one molecule with C_4 of other glucose molecules. The two glucose units are in pyranose form.

Glycosidic bond (α_1–α_4)

α–D-Glucose α–D-Glucose

Maltose

The structure of maltose can also be written as:

α-D-glucose + α-D-glucose → Maltose + H_2O

cyclic hemiacetal

α(1→4) glycosidic linkage

Maltose

Lactose

It is a disaccharide composed of galactose and glucose. Lactose is white in colour, non-hygroscopic, water-soluble solid which is mildly sweet in taste [6].

Fig. (4). Lactose [10] (image courtesy:pexel.com).

Occurrence

It is a disaccharide and is found in milk (image shown in Fig. **4**) of mammals. It is available in milk, cheese, yogurt, milk solids or milk powder, buttermilk, curds, whey, cream, *etc.*

Biological Importance

Lactose of milk is the most important carbohydrate in the nutrition of young mammals. It plays a role in the absorption of calcium and other minerals like zinc and copper. It is hydrolysed by intestinal enzyme lactase to glucose and galactose. Lactose deficiency causes diarrhea.

Structure: The structure of lactose consists of β-D-Galactose and β-D-Glucose held together by $\beta(1 \rightarrow 4)$ glycosidic bond.

β–D-Galactose

Glycosidic bond (β1–β4)

β–D-Glucose

Lactose

POLYSACCHARIDES (Starch, glycogen, and cellulose)

Polymers are the carbohydrate molecules in which the monosaccharides are linked together through glycosidic linkages. When hydrolysed these polysaccharides gives monosaccharides or oligosaccharides. Their structures vary from linear chains to branch [6].

Fig. (5A). Starch [11] (image courtesy: pexels.com).

Starch (Homopolysaccharides from D-Glucose)

Starch (shown in Fig. **5A**) is a polymer of glucose *i.e.* a large number of glucose units are joined together by glycosidic linkage in a starch molecule. It is produced mostly by green plants.

Occurrence

It occurs as reserve food material in higher plants and is found in cereals, legumes, potatoes and other vegetables. Pure starch is powder white in colour, tasteless, odourless and insoluble in cold water and alcohol.

Biological Importance

- Used as a constituent of food such as rice, potato, cornflour.
- In the human diet, starch is one of the major nutrients. Enzymes present in our body break down the starch into glucose which is used to produce energy.
- Used in the manufacture of dextrin and adhesive.
- Starch is the primary form in which plants store food.

Structure

Starch consists of two polysaccharides components-

- Amylose (15-20%), water soluble
- Amylopectin (80-85%), water insoluble

Amylose

Amylose is a long unbranched chain of 200-1000 D-Glucose units held together by $\alpha(1 \rightarrow 4)$ glycosidic linkage. It has a helical structure with six glucose units per turn. Due to its structure amylose is resistant to digestion and hence is an important form of resistant starch. As amylase is linear in shape and therefore takes less space, hence it is the preferred form of starch for storage in plants [6, 7]. The structure of α-amylose is given below:

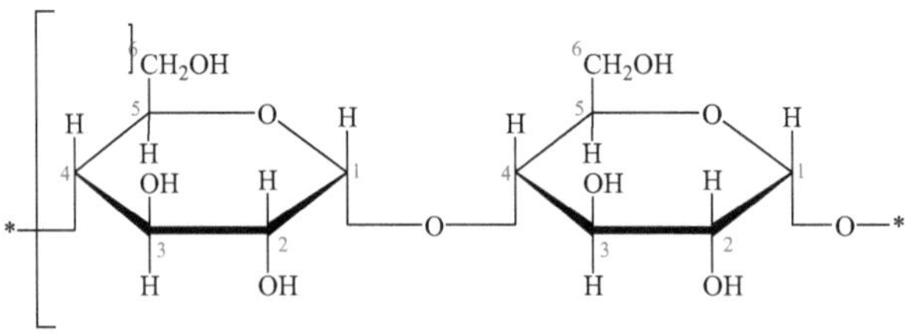

CH_2OH H O H H OH H O H OH H H OH CH_2OH H O H H OH H O H OH

α–Amylose

Amylopectin

It is a branched-chain with α(1 → 6) glycosidic bond at branching point and α(1 → 4) glycosidic bond at the main chain. Amylopectin consists of few thousand of glucose units. After every 25-30 glucose units of amylopectin, a branched-chain is present. These branched chains prevent the formation of a helical structure in amylopectin and allow the formation of a spongy reticular structure [6, 7].

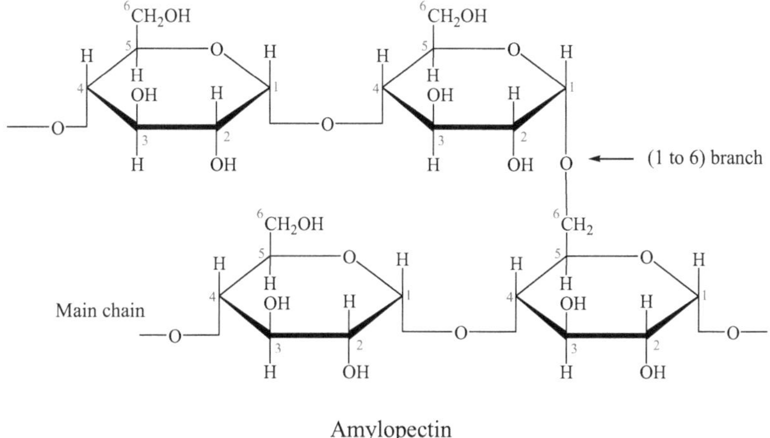

Amylopectin

Glycogen

It is a multi-branched polysaccharide of glucose and is the major food reserve in animals, called animal starch. It is present in high concentration in the liver followed by muscle and brain *etc.* [6, 7].

Structure

The structure is similar to that of amylopectin *i.e.* α(1→4) glycosidic bond in the main chain and α(1→6) glycosidic bond in the branched-chain but is more extensively branched than starch (Fig. **5B**).

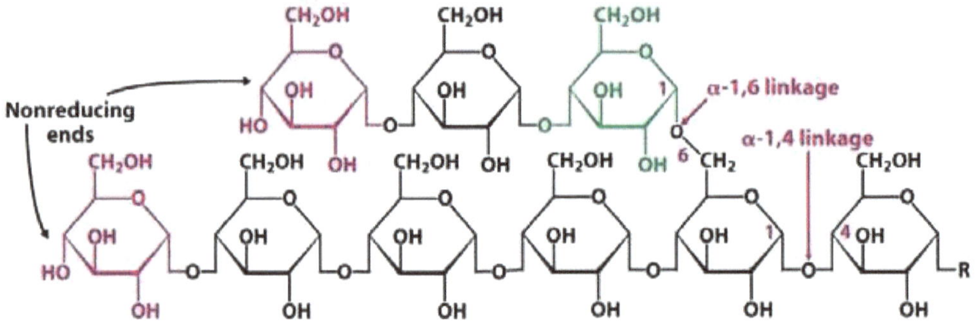

Glycogen

Fig. (5B). Structure of Glycogen [6] (figure courtesy https://www.bankofbiology.com).

Cellulose

Cellulose occurs in plants and is the most abundant organic substance in the plant kingdom. It is totally absent in the animal body. It cannot be digested by mammals including humans due to the lack of the enzyme that cleaves the β-glycosidic bond. Its general formula is $(C_6H_{10}O_5)_n$. Cellulose is tasteless, odourless, insoluble in water, chiral and biodegradable in nature [6].

Structure

Cellulose is composed of β-D-Glucose units held together by β (1→4) glycosidic linkage. It is a straight-chain polymer with no branching and coiling and therefore it adopts a long, stiff rod-like conformation. The hydroxyl groups form a hydrogen bond with the same or neighbouring chain, thus holding the side of the chain by side forming microfibrils with high tensile strength [6]. If at high-temperature cellulose is treated with a concentrated mineral acid, it breaks down into its constituents *i.e.* glucose.

The structures of cellulose, starch, and glycogen are compared and shown in Fig. (6).

Structure of Cellulose (n = >1000)

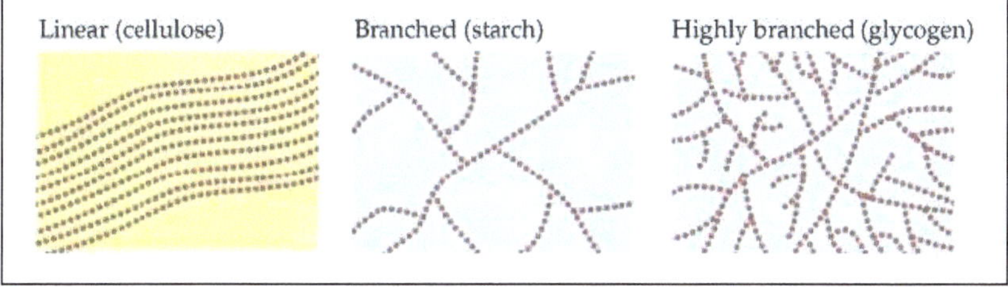

Fig. (6). Comparison of structures of Cellulose, Starch and Glycogen [12] (figure courtesy https://www.bankofbiology.com)

AMINO ACIDS

The carboxylic acids in which α-carbon is replaced with the amino group are called amino acids, *e.g.*

There are several amino acids that occur in nature, but only twenty amino acids are known as standard amino acids because they are repeatedly found in the structure of proteins, isolated from different forms of life *i.e.* animal, plant and microbial [7]. The key elements of amino acids are carbon (C), nitrogen (N), oxygen (O) and hydrogen (H).

$$R-\overset{\overset{\displaystyle H}{|}}{\underset{\underset{\displaystyle NH_2}{|}}{C_\alpha}}-\overset{\overset{\displaystyle O}{\|}}{C}-OH$$

Most amino acids exist in an ionic form known as Zwitter ion or dipolar ion, *e.g.*

$$R-CH(NH_2)-C(=O)-OH \rightleftharpoons R-CH(NH_3^{\oplus})-C(=O)-O^{\ominus}$$

Optical Isomerism in Amino Acid

The common form of amino acids found in nature is α-amino acids, with the α-carbon atom being chiral in nature. All the amino acid (except glycine) consists of four different groups at α-carbon and therefore shows optical isomerism (D & L form) [7] *e.g.*

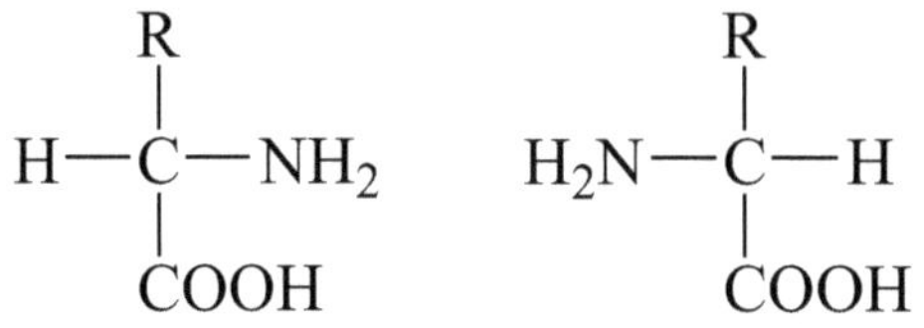

$$\begin{array}{cc} H-C(R)(COOH)-NH_2 & H_2N-C(R)(COOH)-H \\ \text{D-Amino acid} & \text{L-Amino acid} \end{array}$$

The structures of D- and L- amino acids are written, based on D- and L- Glyceraldehyde *i.e.*

$$\begin{array}{cc} H-C(CHO)(CH_2OH)-OH & HO-C(CHO)(CH_2OH)-H \\ \text{D-Glyceraldehyde} & \text{L-Glyceraldehyde} \end{array}$$

The natural amino acids are L-α-amino acid and highly soluble in water. The structure of a few D- and L- amino acids can be seen as follows in Fig. (7) [1]:

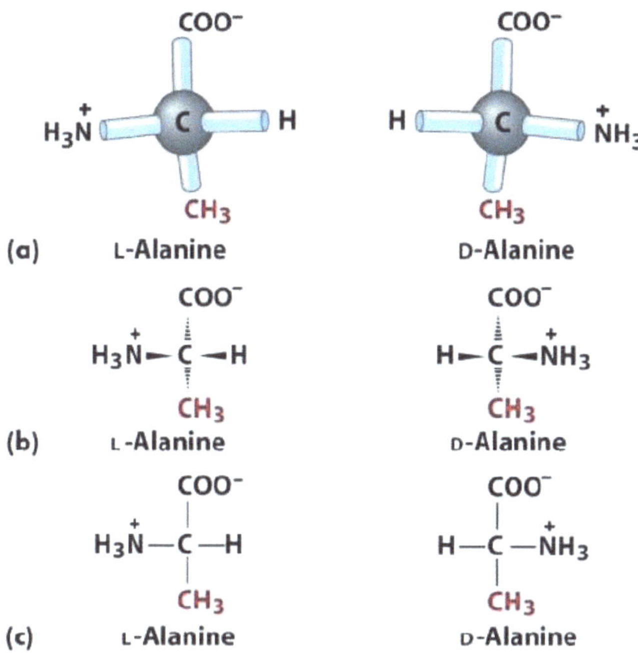

Fig. (7). Structure of different amino acids.

Classification of Amino Acids

There are different ways of classification of amino acids [7]. Based on different R-group, amino acids are classified into the following groups-

A. Non-polar amino acids having hydrophobic R-groups.
B. Polar amino acids with no change on R-group.
C. Polar amino acids with positive R-group.
D. Polar amino acids with negative R-group.

Each amino acid is assigned with a three-letter or one-letter symbol, *e.g.* Gly for glycine, Ala for alanine, *etc.* In each group of amino acids, there is a considerable difference in size and nature of the R-group.

Non-polar Amino Acids having Hydrophobic R-groups

This group of amino acids contains 8 amino acids in which 5 amino acids have alkyl side chain (Alanine, Valine, Leucine, Isoleucine, and Proline), two have aromatic R-group (Phenylalanine, Tryptophan) and one amino acid has sulphur containing alkyl group (Methionine).

• **Alanine (Ala):** It is the least hydrophobic among all 8 non-polar amino acids because of the small methyl side chain. It is synthesized in the body, hence Alanine (structure shown in Fig. **8**) is a non-essential amino acid and need not be obtained through diet. However, it is available in a variety of foods, particularly in meat. Its L-isomer is used in the synthesis of proteins in the body [1].

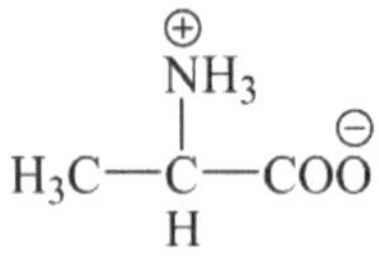

$$H_3C-\overset{\overset{\oplus}{NH_3}}{\underset{H}{C}}-COO^{\ominus}$$

Fig. (8). Alanine.

• **Valine (Val):** It is a branched-chain amino acid with an isopropyl group in its side chain. Valine (structure shown in Fig. **9**) is used in the synthesis of proteins. It is an essential amino acid and therefore needs to be obtained through diet. Valine is available in foods that consist of proteins like meat, soy products, dairy products, beans, *etc.* [1].

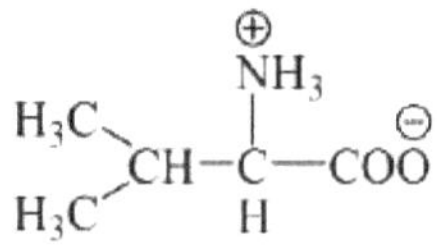

$$(H_3C)_2CH-\overset{\overset{\oplus}{NH_3}}{\underset{H}{C}}-COO^{\ominus}$$

Fig. (9). Valine.

• **Leucine (Leu):** It is also a branched-chain amino acid with an isobutyl group in its side chain (structure shown in Fig. **10**). Leucine too is used in the synthesis of proteins. Like valine, it is an essential amino acid, *i.e.* it cannot be synthesized in our body and therefore need to be obtained through diet. Leucine is available in milk, fish, chicken, beef, oats, lentils, eggs, almonds, peanuts, soy proteins, *etc.* [1].

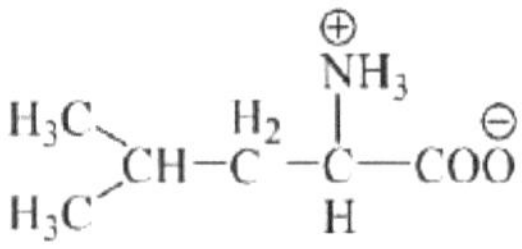

$$(H_3C)_2CH-\overset{H_2}{C}-\overset{\overset{\oplus}{NH_3}}{\underset{H}{C}}-COO^{\ominus}$$

Fig. (10). Leucine.

• **Isoleucine (Ile):** It is an isomer of leucine and has two asymmetric carbons (shown in Fig. **11**). Like valine and leucine, it is also an essential amino acid and must be obtained from the diet. Food rich in isoleucine is fish, cheese, lamb, chicken, turkey, seaweed, eggs, soy protein, *etc.* Isoleucine is also used in the synthesis of proteins [1].

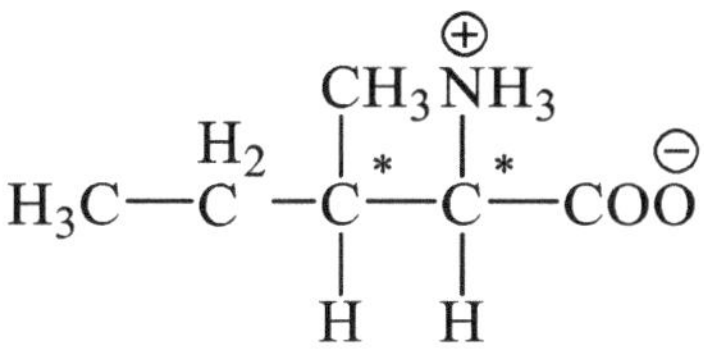

Fig. (11). Isoleucine.

• **Proline (Pro):** It is a pyrrolidine containing amino acid and has an imino group (>NH) instead of an amine group (-NH_2) (shown in Fig. **12**). It is the only amino acid with a secondary amine group. It is a non-essential amino acid *i.e.* it can be synthesised in our body and need not be supplied in diet [1].

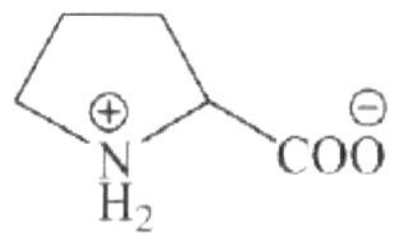

Fig. (12). Proline.

• **Phenylalanine (Phe):** It is also known as an aromatic amino acid. It can be said that the methyl group in alanine is replaced by a benzyl group in phenylalanine (shown in Fig. **13**). It is an essential amino acid and is obtained from the diet. The breast milk of mammals is rich in phenylalanine [1].

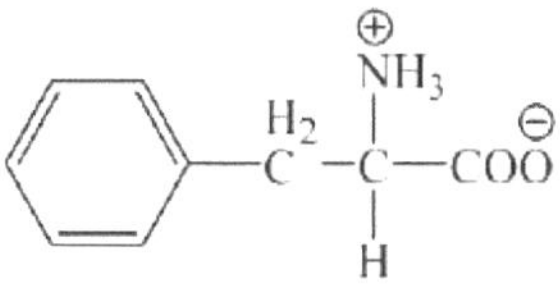

Fig. (13). Phenylalanine.

• **Tryptophan (Try):** It has indole as a side chain (shown in Fig. **14**). It is an essential amino acid and is obtained from food like egg, chicken, beef, soybeans, cheese, sunflower seeds, *etc*. [1].

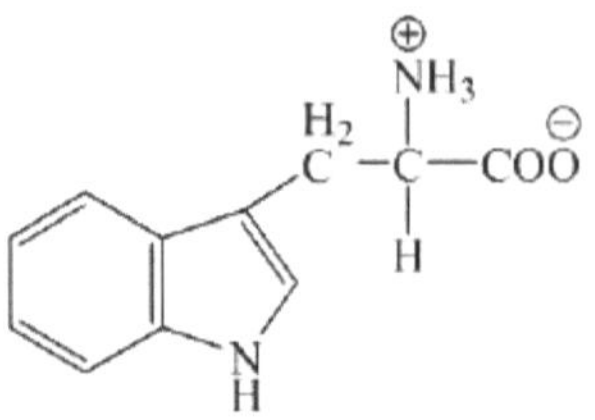

Fig. (14). Tryptophan.

• **Methionine (Met):** It has a thio-ether group in the side chain (shown in Fig. **15**). It is an essential amino acid and obtained from the diet. Egg, meat, fish, sesame seeds, cereal, Brazil nuts, *etc.* are some food sources rich in methionine. Methionine also plays an important role in the metabolism of humans and other species [1].

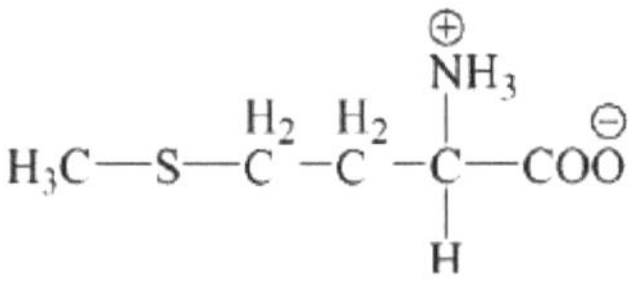

Fig. (15). Methionine.

Polar Amino Acids with no Change on R-group

These amino acids are relatively more soluble in water because they have a functional group that can form a hydrogen bond with water [7]. This category includes 7 amino acids *i.e.*

• **Glycine (Gly):** It is the simplest amino acid and doesn't have asymmetric carbon (shown in Fig. **16**). It has a hydrogen atom at its side chain. It is a colourless crystalline solid and sweet in taste. Glycine is present in a wide variety of foods *i.e.* fish, egg, meat, seeds, beans, lentils, nuts, cheese, *etc.*

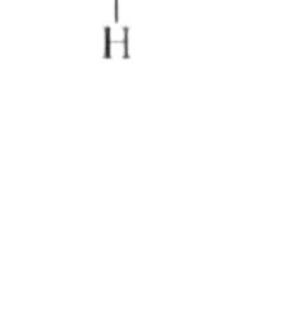

Fig. (16). Glycine.

• **Serine (Ser):** It has a hydroxymethyl group at its side chain (shown in Fig. **17**). The hydroxyl group makes H-bonding and makes it more hydrophilic. It is a non-essential amino acid.

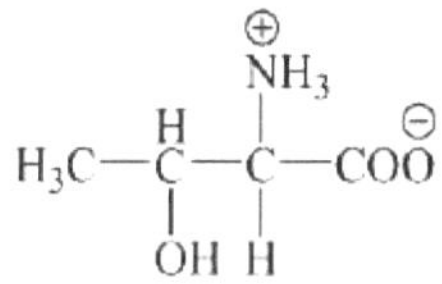

Fig. (17). Serine.

• **Threonine (Thr):** It is used in the biosynthesis of proteins. It consists of a hydroxyl group in its side chain that makes it polar in nature (shown in Fig. **18**). It is an essential amino acid *i.e.* cannot be synthesised in our body and must be obtained from a diet like fish, meat, poultry, cheese, sesame seeds, lentils, nuts, *etc.*

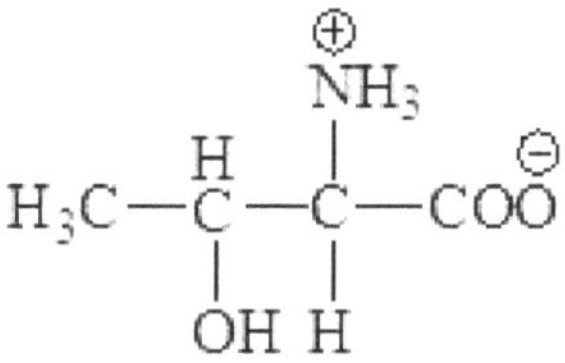

Fig. (18). Threonine.

• **Tyrosine (Tyr):** It is an aromatic amino acid and contains a phenolic group, *i.e.* it has a polar side group (shown in Fig. **19**). It is generally a hydrophobic amino acid, but it is more hydrophilic than phenylalanine. It is a non-essential amino acid however available in high protein food sources like meat, fish, milk, cheese, peanuts, almonds, sesame seeds, soy products, pumpkin seeds, avocados, *etc.*

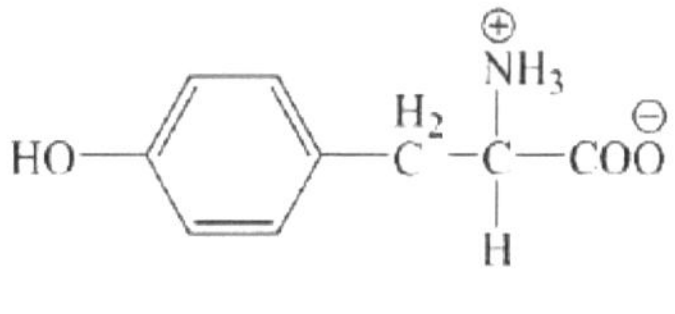

Fig. (19). Tyrosine.

• **Cysteine (Cys):** It is a sulfur-containing semi-essential amino acid. Its structure is similar to serine with one of the oxygen atoms replaced by sulphur (shown in Fig. **20**). It is available in meat, eggs, dairy products, garlic, onion, broccoli, oats, lentils, *etc.*

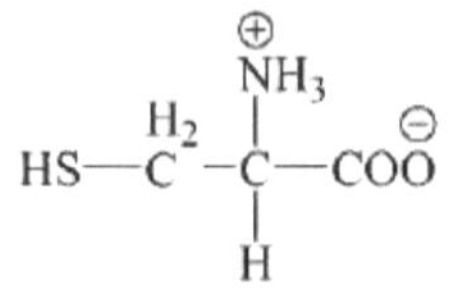

$$HS-\overset{H_2}{C}-\underset{H}{\overset{\overset{\oplus}{NH_3}}{C}}-COO^{\ominus}$$

Fig. (20). Cysteine.

• **Asparagine (Asp):** In this amino acid, an amide group is present in the side chain that makes it polar in nature (shown in Fig. **21**). It is a non-essential amino acid however available in food sources like dairy products, seafood, whey, meat, fish, potatoes, nuts, seeds, soy, whole grains, *etc.*

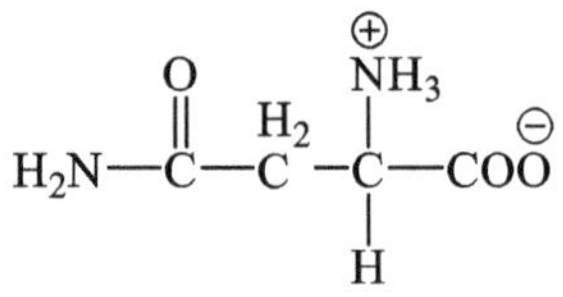

$$H_2N-\overset{O}{\overset{\|}{C}}-\overset{H_2}{C}-\underset{H}{\overset{\overset{\oplus}{NH_3}}{C}}-COO^{\ominus}$$

Fig. (21). Asparagine.

• **Glutamine (Glu):** Its side chain consists of an amide group (shown in Fig. **22**). It is a non-essential and also a conditionally essential amino acid for humans, *i.e.* the body can synthesise it but at some conditions, the body requires a large amount of glutamine and during that time it becomes an essential amino acid and should be obtained from the diet. Glutamine is obtained through the catabolism of proteins present in protein-rich foods.

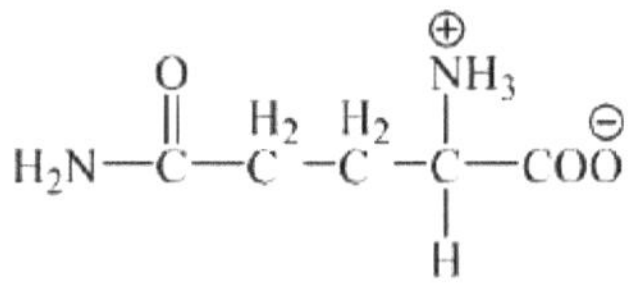

$$H_2N-\overset{O}{\overset{\|}{C}}-\overset{H_2}{C}-\overset{H_2}{C}-\underset{H}{\overset{\overset{\oplus}{NH_3}}{C}}-COO^{\ominus}$$

Fig. (22). Glutamine.

Polar Amino Acids with a Positive Charge on R-group

These are basic amino acids with an additional amino group in R-group and show a positive charge at $P^H = 7$ [1, 7]. There are three amino acids present in this group-

• **Lysine (Lys):** It consists of a side chain 'lysyl', $(CH_2)_4NH_2$, due to which it is basic in nature (shown in Fig. **23**). It cannot be synthesised in our body hence it is an essential amino acid and obtained from the diet. Lysine is mostly available in meat sources. A limited amount of lysine is also found in cereal crops and vegetables.

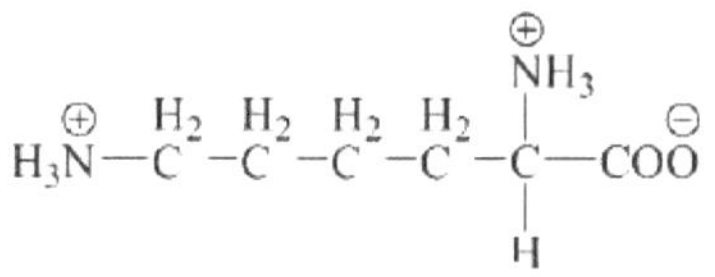

$$H_3\overset{\oplus}{N}-\overset{H_2}{C}-\overset{H_2}{C}-\overset{H_2}{C}-\overset{H_2}{C}-\underset{H}{\overset{\overset{\oplus}{NH_3}}{C}}-CO\overset{\ominus}{O}$$

Fig. (23). Lysine.

• **Arginine (Arg):** It has a positively charged guanidinium ion inside the chain (shown in Fig. **24**). For a human being, it is a semi-essential or conditionally essential amino acid. Food sources for arginine are meat, eggs, dairy products, cereals, nuts, beans, *etc.*

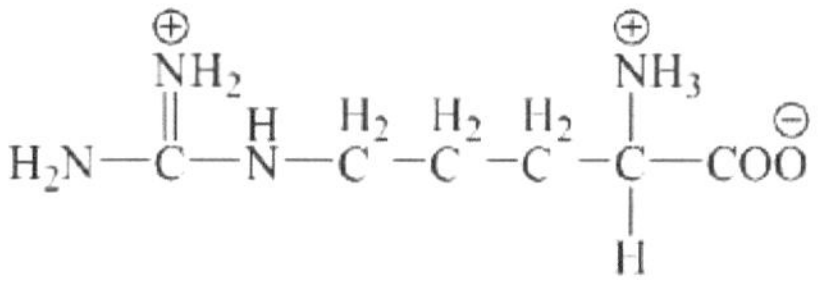

$$H_2N-\overset{\overset{\oplus}{NH_2}}{\overset{\|}{C}}-\overset{H}{N}-\overset{H_2}{C}-\overset{H_2}{C}-\overset{H_2}{C}-\underset{H}{\overset{\overset{\oplus}{NH_3}}{C}}-CO\overset{\ominus}{O}$$

Fig. (24). Arginine.

• **Histidine (His):** It has an imidazolium group as its side chain (shown in Fig. **25**). It is an essential amino acid hence must be obtained from the diet. Food sources rich in histidine are protein-rich foods like meat, poultry products, fish, dairy products, rice, wheat, *etc.*

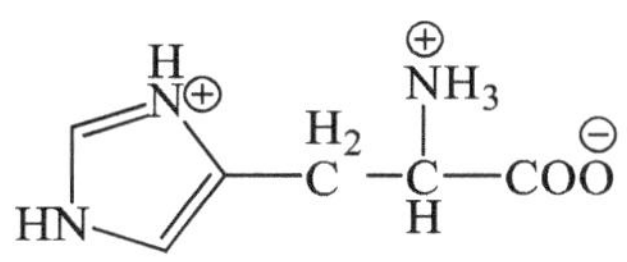

H N⊕ / HN (imidazolium ring) $-\overset{H_2}{C}-\underset{H}{\overset{\overset{\oplus}{NH_3}}{C}}-CO\overset{\ominus}{O}$

Fig. (25). Histidine.

Polar Amino Acids with Negative R-group

This group contains two amino acids having an additional –COOH group [1, 7].

• **Aspartic acid (Asp):** The side chain of aspartic acid is the CH_2COOH group shown in Fig. (**26**). It is synthesised in our body hence is a non-essential amino acid. However, there are some food sources where aspartic acid is found. These are oysters, meat, avocado, sugarcane, oats, sprouting seeds, *etc.*

$$^{\ominus}OOC-\overset{H_2}{C}-\underset{H}{\overset{\overset{\oplus}{N}H_3}{C}}-COO^{\ominus}$$

Fig. (26). Aspartic acid.

$$^{\ominus}OOC-\overset{H_2}{C}-\overset{H_2}{C}-\underset{H}{\overset{\overset{\oplus}{N}H_3}{C}}-COO^{\ominus}$$

Fig. (27). Glutamic acid.

• **Glutamic acid (Glu):** It is a non-essential amino acid for human beings as the body itself can synthesise it (shown in Fig. **27**). It is the most abundant amino acid in the vertebrate nervous system. It is also available in food sources like poultry products, fish, eggs, meat and dairy products. Some protein-rich plant foods are also a great source of glutamic acid. Sickle cell disease is characterized by a point mutation involving the substitution of glutamic acid at position 6 to valine [13].

Classification of Amino Acids Based on Nutritional Requirement

Based on the nutritional requirement, amino acids are classified into two groups, essential amino acid and non-essential amino acid [7].

• **Essential amino acid:**

The amino acid which cannot be synthesized by the body and is required for proper growth, therefore, must be supplied through diet, are called essential amino acids, *e.g.* Arginine, Valine, Histidine, Isoleucine, Leucine, Lysine, Methionine, Phenylalanine, Threonine and Tryptophan.

• **Non-essential amino acids:**

The amino acid which can be synthesized by the body and therefore these need not be consumed through diet is called a non-essential amino acid, *e.g.* Glycine, Alanine, Serine, *etc.*

Properties of amino acids:

Physical Properties

▪ Solubility: Most of the amino acids are usually soluble in water and insoluble in organic solvents.

▪ Melting points: Amino acids possess higher melting points *i.e.* >200°C because of ionic interactions.

▪ Taste: Amino acids may be sweet (Gly, Ala, Val), tasteless (Leu), or bitter (Arg, Ile) in taste.

▪ Optical properties: Except glycine, all amino acids have asymmetric carbon atom and therefore shows optical isomers. Few amino acids have two asymmetric carbon *i.e.* isoleucine, threonine.

▪ Amino acids as ampholytes: Substances that can donate a proton or accept a proton are called ampholytes. Since amino acids contain both acidic group (COOH) and basic group (NH_2), therefore they are termed as ampholytes.

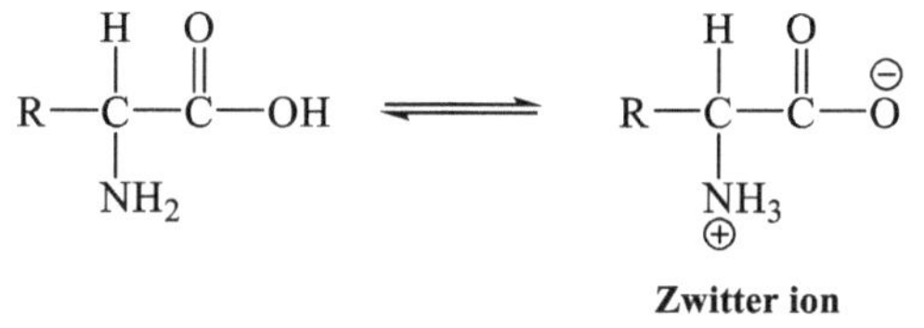

$$R-CH(NH_2)-COOH \rightleftharpoons R-CH(\overset{\oplus}{N}H_3)-COO^{\ominus}$$

Zwitter ion

Fig. (28). Zwitter ion formation.

A hybrid molecule containing both positive and negative ionic groups is called Zwitter ion (Fig. **28**). Most amino acids exist in ionic form (Zwitter ion) [1].

In a strongly acidic pH (low pH), amino acids are positively charged (cation), while in a strongly alkaline pH (high pH), it is negatively charged (anion) (Fig. **29**). For each amino acid, there is a characteristic pH at which it carries both positive and negative charge and exists as a zwitter ion (*e.g.* Leucine, pH 6.0).

A particular pH at which a molecule exists as a Zwitter ion or dipolar ion and carries no net charge is known as isoelectric pH [1].

$$\underset{\text{Cation}}{R-\overset{H}{\underset{NH_3^{\oplus}}{C}}-\overset{O}{\overset{\|}{C}}-OH} \xleftarrow[+H^{\oplus}]{\text{at low }P^H} R-\overset{H}{\underset{NH_2}{C}}-\overset{O}{\overset{\|}{C}}-OH \xrightarrow[-H^{\oplus}]{\text{at high }P^H} \underset{\text{Anion}}{R-\overset{H}{\underset{NH_2}{C}}-\overset{O}{\overset{\|}{C}}-O^{\ominus}}$$

Fig. (29). Positively and negatively charged amino acids.

Chemical Properties

Reaction Due to –COOH Group

- **Salt formation**: Amino acids reacts with bases to form a salt (-COONa).

$$R-\overset{H}{\underset{NH_2}{C}}-COOH + NaOH \longrightarrow R-\overset{H}{\underset{NH_2}{C}}-COONa + H_2O$$

- **Ester formation**: Amino acids reacts with alcohol to form ester *i.e.*

$$R-\overset{H}{\underset{NH_3^{\oplus}}{C}}-\overset{O}{\overset{\|}{C}}-O^{\ominus} \xrightarrow{R^1OH} \underset{\text{Ester}}{R-\overset{H}{\underset{NH_2}{C}}-\overset{O}{\overset{\|}{C}}-OR^1}$$

- **Decarboxylation**: Amino acid undergo decarboxylation reaction to form amines *i.e.*

$$R-\overset{H}{\underset{NH_3^{\oplus}}{C}}-\overset{O}{\overset{\|}{C}}-O^{\ominus} \longrightarrow R-\overset{H}{\underset{NH_3^{\oplus}}{C}}-H + CO_2\uparrow$$

Decarboxylation reaction in the living cells is responsible for the formation of many biologically important amines *i.e.* histamine from histidine, γ-amino butyric acid (GABA) from glutamic acid, tyramine from tyrosine, tryptamine from tryptophan, cadaverine from lysine, *etc.*

▪ **Reaction with ammonia**: Dicarboxylic amino acids react with ammonia to form amide *e.g.* Aspartic acid + NH_3 → Asparagine

Glutamic acid + NH_3 → Glutamine

$$R{-}CH(NH_2){-}COOH + H{-}NH{-}R_1 \longrightarrow R{-}CH(NH_2){-}C(=O){-}NH{-}R_1 + H_2O$$

Amino acid Amine Amide Water

Reaction due to –NH_2 group

▪ **Salt formation**:

Amino acid reacts with acids and forms salt.

$$R{-}CH(\overset{\oplus}{N}H_3){-}C(=O){-}O^{\ominus} + HCl \longrightarrow R{-}CH(\overset{\oplus}{N}H_3\ Cl^{\ominus}){-}COOH$$

Hydrochloride salt

▪ **Reaction with ninhydrin**:

α-amino acids react with ninhydrin (strong oxidizing agent) to form a purple-blue or pink colour complex (Ruhemann's purple).

Amino acid + Ninhydrin ⟶ Aldehyde + NH_3 + CO_2 + Hydrindantin

Hydrindantin + NH_3 + Ninhydrin ⟶ Ruhemann's purple

$$\text{Ninhydrin} + R{-}CH(\overset{\oplus}{N}H_3){-}C(=O){-}O^{\ominus} \longrightarrow \text{Hydridantin} + CO_2 + RCHO + NH_3$$

Hydridantin (Partially reduced form)

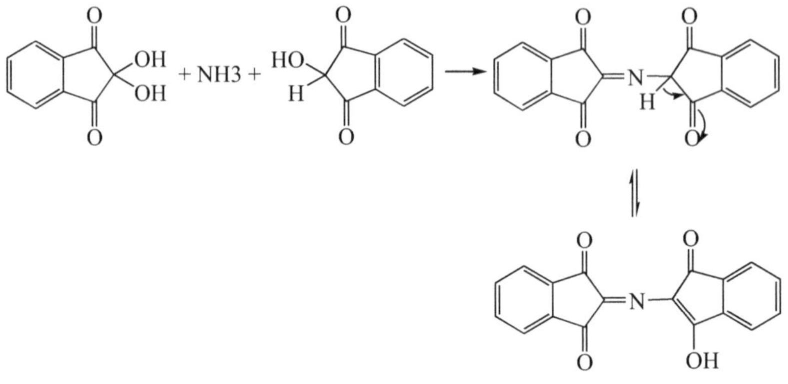

+ NH3 +

Ruhemann's purple

- **Reaction with benzaldehyde**:

$$R-\overset{H}{\underset{NH_2}{C}}-COOH + O{=}\overset{C_6H_5}{CH} \longrightarrow R-\overset{H}{\underset{N=\underset{H}{C}-C_6H_5}{C}}-COOH + H_2O$$

Amino acid Benzaldehyde Schiff's base Water

Amino acid reacts with benzaldehyde to give Schiff's base.

- **Colour reaction of amino acids**:

Amino acids can be identified by specific colour reaction with various reagents *i.e.*

- Biuret reaction is used to detect amino acids with two peptide linkage.
- Ninhydrin reaction can be used to detect α-Amino acids.
- Xanthoproteic reaction is useful in the detection of aromatic amino acids (Phe, Tyr, Trp).
- Millions reaction helps to detect amino acids with a phenolic group (Tyr).
- Nitroprusside reaction is for amino acids with sulfhydryl groups (Cys).

▪ **Transamination**:

Transfer of an amino group from an amino acid to a ketoacid to form a new amino acid is called transamination (Fig. **30**). It is a very important reaction in amino acid metabolism [1].

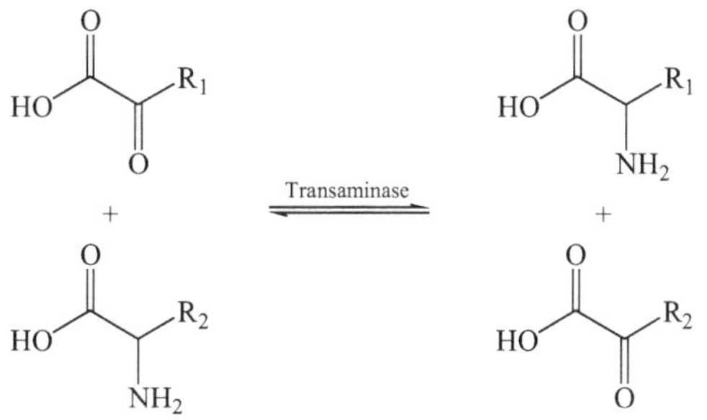

Fig. (30). Transamination reaction.

▪ **Oxidative Deamination**:

Oxidative deamination of amino acids results in the generation of NH_3 and keto acids and mostly takes place in the liver and the kidney. The liberated NH_3 is used for urea synthesis while keto acids are used for a variety of reactions including energy generation [1].

PROTEINS

Mulder (Dutch chemist) used the term protein for the high molecular weight nitrogen-rich and a most abundant substance present in animals and plants. Proteins are the polymer of amino acids held together by a peptide bond. A peptide bond is formed between two amino acids by joining the α-carbonyl group of one amino acid with an α-amino group of other amino acids with loss of water molecule [7]. *e.g.*

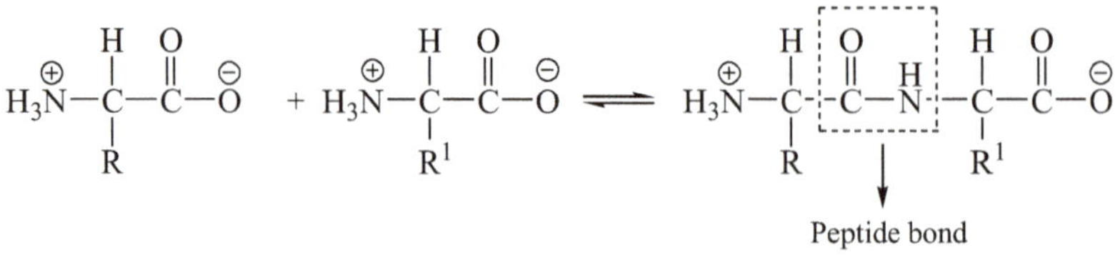

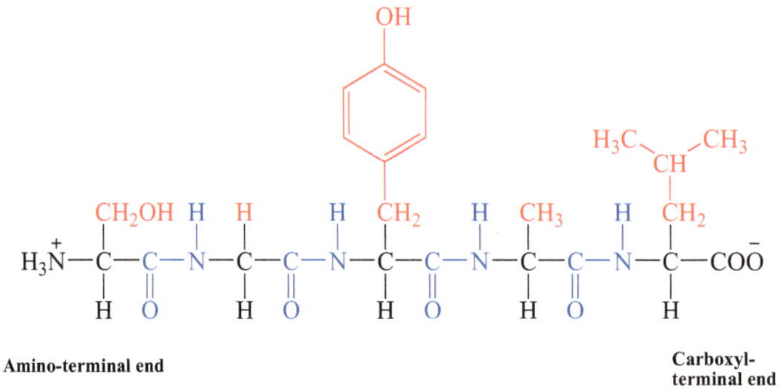

Fig. (31). Structure of a simple peptide (Ser-Gly-Tyr-Ala-Leu).

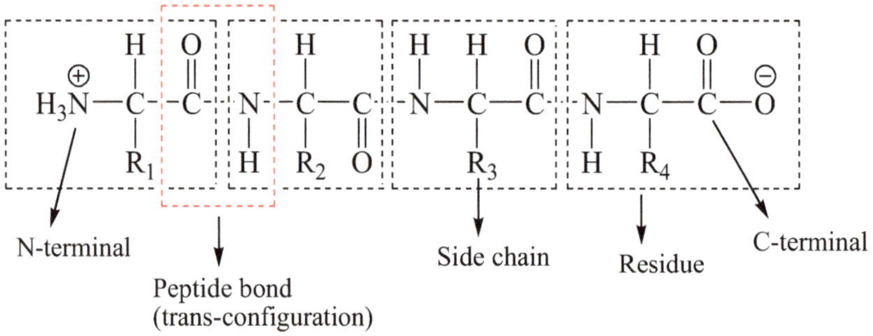

Fig. (32). General structure of a polypeptide.

The chain formed by linking many amino acids together by a peptide bond is called a peptide chain (shown in Fig. **31**). Many amino acids, usually more than 100 amino acids join together by a peptide bond to form a polypeptide chain. Amino acid unit in a polypeptide is known as residue and the two ends of the peptide chain are named amino-terminal (N-terminal) and carbonyl terminal (C-terminal) (shown in Fig. **32**). The terminal amino acid with a free amino group is called the N-terminal amino acid and the amino acid with a free carbonyl group at the other end is called the C-terminal amino acid.

Polypeptide chain consists of a regular repeating unit, called the main chain and the variable parts are known as side chains. The main chain is also termed as the backbone. The peptide bond is rigid and planar with a partial double bond character. Both >C=O and >NH groups are *trans-* to each other and involve in hydrogen bond formation.

Structure of proteins

Proteins are polymers of L-α-amino acids and the structure of proteins are rather complex, which can be divided into four types [7]-

Primary structure

The primary structure of a protein refers to the number, nature and sequence of amino acids, *e.g.* Lys-Ala-His-Gly-Lys-Lys-Val-Leu-Gly. These are held together by peptide bonds. One end of a polypeptide chain is known as the carboxyl terminus if the carboxylic group is present and another end is called the amino terminus if the amino group is present. Every protein has a unique sequence that defines its structure and function.

The protein may consist of either one or more peptide chains. *e.g.* Insulin, it consists of two chains, in one chain there are 31 amino acids and another chain consists of 20 amino acid *i.e.* a total of 51 amino acids is present. The determination of the primary structure of a protein involves three stages-

- Determination of amino acid composition.
- Degradation of proteins or polypeptide into the smaller fragment.
- Determination of amino acid sequence.

2. **Secondary structure of protein**:

The conformation that gives polypeptide backbone in a coil is known as secondary structure. Two types of secondary structure *i.e.* α-helix and β-sheet, are mainly identified.

α-Helix structure:

Pauling and Corey proposed the secondary structure of proteins as α-helix and β-sheet. The salient feature of α-helix can be seen as follows-

- The α-helix structure is the common motif form in the secondary structure of proteins.
- The α-helix structure is also known as Pauling-Corey-Branson α-helix.

• The α-helix is a tightly packed coiled structure with amino acids side chain extending outward from the central axis.

• The α-helix is stabilized by extensive hydrogen bonding between the >NH and >C=O groups of peptide bonds. The >C=O group of each amino acid is hydrogen-bonded to the >NH group of amino acid that is situated four residues ahead in the linear sequence *i.e.* (Fig. **33**).

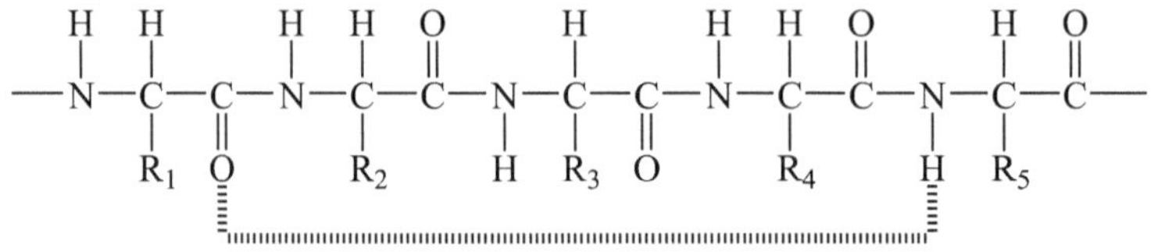

```
    H   H        H   H   O        H        H   H   O       H   O
    |   |        |   |   ||       |        |   |   ||      |   ||
 —N—C—C—N—C—C—N—C—C—N—C—C—N—C—C—
       |   ||       |        |    |   ||       |        |    |
       R1  O        R2       H   R3   O        R4       H   R5
           :..............................................:
```

Fig. (33). Hydrogen bonding in a polypeptide chain.

The oxygen of R_1 amino acid is hydrogen-bonded with>NH group of R_5 amino acid.

• All peptide bonds except first and last in the polypeptide chain, participate in hydrogen bonding.

• The H-bonding occurs spontaneously, which results in a rod-like structure with well-defined dimensions.

• Each turn of α-helix contains 3.6 amino acids and travels a distance of 5.4 Å. The spacing of each amino acid is 1.5 Å.

• The α-helix may be right or left-handed. The right-handed α-helix (shown in Fig. **34**) is more stable than the left-handed α-helix.

• Experimentally, it has been found that the oxygens of all the backbone carbonyl point downward and slightly outwards, towards the carbon terminus, and the hydrogen bonds formed are approximately parallel to the axis of the helix.

• When two or more coiled α-helices wrap around each other they form a 'supercoil' structure which is a highly stable form.

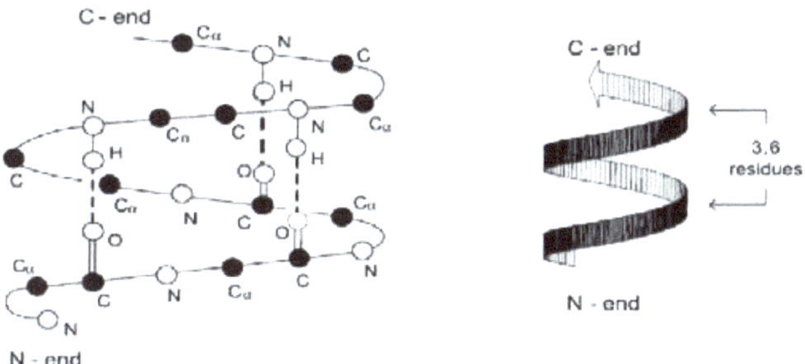

Fig. (34). Right-handed α-helix structure [14].

β-pleated Sheet

These are composed of two or more segments of fully extended peptide chains. In the β-sheets, the hydrogen bonds are formed between the neighbouring segments of polypeptide chains forming twisted pleated sheets (as shown in Fig. **35**) [7].

Fig. (35). Right-handed α-helix structure [14].

There are two types of pleated sheet arrangement-

• Parallel β-sheets: In this, all the chains run in the same direction *i.e.* all the N-terminal strands of the amino acids are oriented in the same direction (Fig. **36**). This type of arrangement is less stable because here the hydrogen bonds within the strands occur at an angle which makes them longer and weaker [7].

Fig. (36). Bonding in parallel β-sheets.

• Anti-parallel β-sheet: In this type of arrangement the polypeptide chain runs alternatively in opposite directions *i.e.* the N-terminus of one strand of amino acid will be opposite to the C-terminus of the other stand (Fig. **37**). These are more stable as the hydrogen bonds are aligned directly opposite each other which makes them stronger [7].

Parallel β–Sheet hydrogen bonding pattern

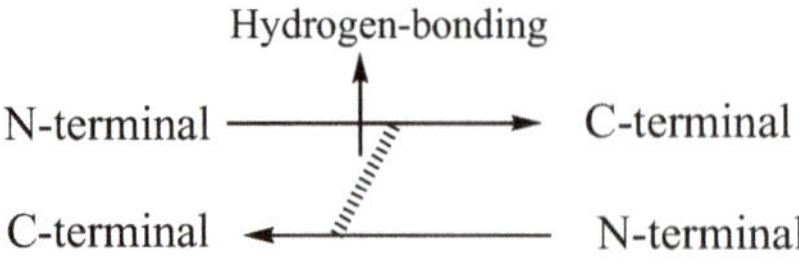

Fig. (37). Bonding in anti-parallel β-sheets.

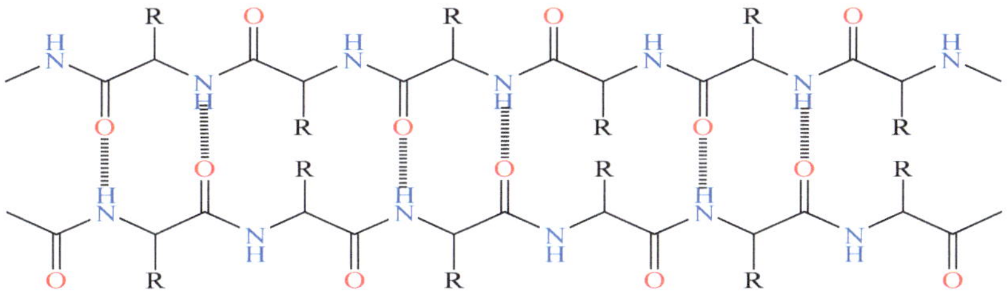

Antiparallel β–Sheet hydrogen bonding pattern

Tertiary Structure of Protein

The three-dimensional arrangement of the polypeptide chain is called the tertiary structure of the protein (shown in Fig. **38**). In the tertiary structure of proteins, the backbone has regular folding which is maintained by H-bonding while the weak interaction between different regions of the chain holds it to produce tertiary structure [7]. The following type of weak interaction is present in the tertiary structure of proteins-

- H-bonding
- Disulfide bonds (S-S)
- Ionic bonding (Electrostatic bonds)
- Hydrophobic interaction
- Vander Waals forces

Fig. (38). Tertiary structure of the protein [15].

Quaternary Structure of Protein

A majority of the proteins are composed of single polypeptide chains. However, some of the proteins consist of two or more identical or different polypeptides. Such proteins are termed as oligomers and possess a quaternary structure. Haemoglobin, DNA polymerase, ion channels are some of the proteins with quaternary structures [7].

In quaternary structure, protein organization involves the association of two or more individual protein units each with its own 3D structure, into a complex and a functional unit. Such association of various protein units is known as quaternary structure (shown in Fig. **39**). The weak interaction (bond) responsible for tertiary structure (except disulfide bond) also operates in quaternary structure.

The individual polypeptide chains are known as monomers. The dimer consists of two polypeptides while a tetramer has four polypeptide units and so on [7].

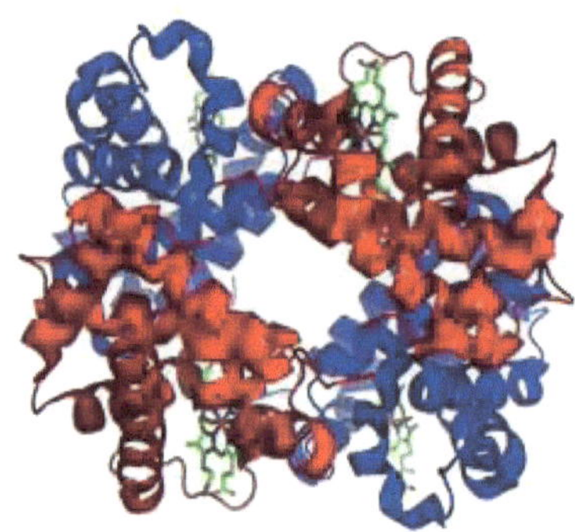

Fig. (39). Quaternary structure of protein [16].

Properties of Proteins

- **Solubility:** Proteins form colloidal solutions instead of true solutions in water because of the huge size of protein molecules [7].
- **Molecular weight:** Proteins are high molecular weight compounds composed of 40 to 4000 amino acids with a molecular weight ranging from 4000 to 440000g, *e.g.* Hemoglobin- 64450g [1, 7].
- **Isoelectric pH:** For each protein, there is a particular pH at which the sum of the positive charges equals to the sum of the negative charges and therefore the net charge on the protein is zero. This pH value at which the uncharged protein molecule does not migrate in an electric field is known as isoelectric pH. It is denoted by pI [7].

- **Precipitation at pI:** Proteins are least soluble at isoelectric pH. Few proteins (*e.g.* casein) get easily precipitated when the pH is adjusted to pI (4.6 for casein) *e.g.* formation of curd from milk. This occurs due to the lactic acid produced by bacteria which lowers the pH to the pI of casein.
- **Hydrolysis:** Proteins can be hydrolyzed into intermediate peptides or amino acids depending upon the hydrolysis conditions. Hydrolysis can be performed by using dil. HCl, dil. H_2SO_4, alkali or enzymes.

$$\text{Proteins} \rightarrow \text{Polypeptide} \rightarrow \text{Peptides} \rightarrow \text{Amino acids}$$

- **Optical activity and precipitation:** All proteins are optically active. Proteins can be coagulated and precipitated from an aqueous solution by heat, the addition of alkali, acids, salts, organic solvent, *etc.*
- **Denaturation:** "The phenomenon of disorganization of native protein structure is known as denaturation". Denaturation (shown in Fig. **40**) results in the loss of the secondary, tertiary and quaternary structure of proteins and causes a change in the physical, chemical and biological properties of protein molecules [7].

Proteins on treatment with heat or strong acids, alkalis or various other reagents, undergo remarkable changes in their behaviour. These changes may be reversible or irreversible and referred to as denaturation. The reversal of denaturation is known as renaturation or refolding. If any protein present in a cell is denatured then the cell activity disrupts, finally causing cell death.

Cooked food contains some denatured proteins. *e.g.* egg whites, which are actually egg albumins in water, are transparent liquids but when exposed to heat, these are converted to an opaque, white coloured solid mass, this is due to the denaturation of egg albumin. Similarly, meat when cooked becomes firm due to the denaturation of proteins present in it.

Fig. (40). Denaturation and Renaturation of Protein (image courtesy: www.socratic.org) [17].

Colour Reactions

Proteins give characteristics colour reaction, on treatment with some specific reagent. However, all proteins do not give all the tests [7]. Some important colour reaction of protein are as follows:

Xanthoprotic Test

Protein usually gives a yellow colour with conc. HNO_3 and the colour becomes orange when the solution is made alkaline. This test is characteristics of those proteins that have at least one molecule of aromatic amino acid. This reaction is due to the nitration of the aromatic ring.

Biuret Test

Protein gives purple/violet colour on the addition of dil. solution of $CuSO_4$ in alkaline medium, this reaction is due to the presence of peptide bonds.

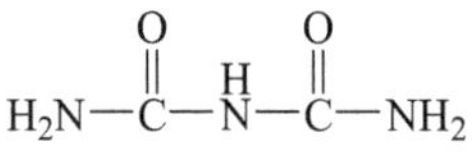

Biuret

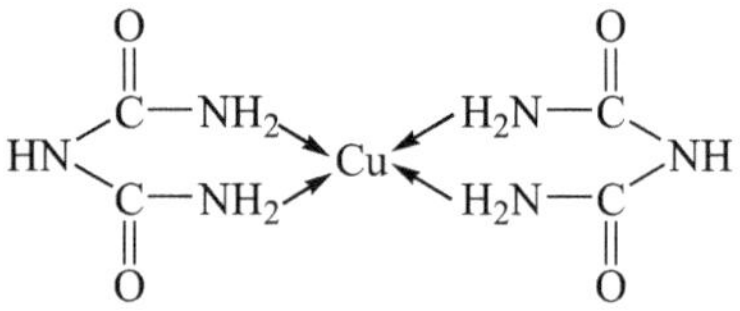

Complex forming in biuret test

Millon's Reaction

When proteins are heated with Millon's reagent ($HgNO_3$ + HNO_2) a red colour develops. This reaction is characteristic of phenol and is shown by the protein which gives tyrosin on hydrolysis.

Ninhydrin Reaction

Protein-containing free amino acids give purple colour with ninhydrin.

Classification of Protein

There are several ways of classification of proteins and the major type of classification is based on their function, chemical nature, solubility, and nutritional importance.

Based on Biological Function

- Structural proteins: Keratin of hair and nails, collagen of bones.

- Enzymes or catalytic proteins: Hexokinase, pepsin.
- Transport proteins: Haemoglobin, Lipoproteins.
- Hormonal proteins or regulatory proteins: Insulin, growth hormone.
- Contractile proteins: Actin, myosin.
- Storage and nutrient proteins: Ovalbumin, casein.
- Genetic proteins: Nucleoproteins.
- Defense proteins: Snake venoms, antibodies.

B. Based on chemical nature and solubility: Based on amino acid composition, structure, shape, and solubility, proteins are classified into three major groups [7]-

Simple Proteins

Proteins that yield only amino acids on hydrolysis are called simple proteins. Simple proteins have been further divided into the following-

a. Globular Proteins: These are spherical or oval in shape, soluble in water or other solvents, and digestible, *e.g.* Albumins, globulins, prolamines, histones, *etc.* (shown in Fig. **41**).

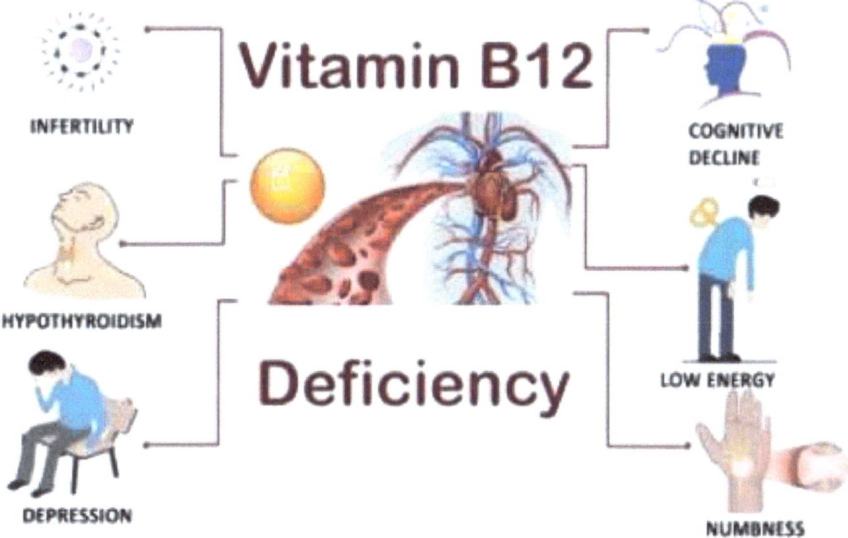

Fig. (41). Structure of simple proteins [18].

b. Fibrous proteins: These are fiber-like in shape, insoluble in water and common solvents, and resistant to digestion. They serve as the chief structural material of animal tissues, *e.g.* nails, hair, *etc.* (shown in Fig. **41**).

• **Conjugated proteins**: Besides the amino acids, these proteins contain a non-protein moiety known as the prosthetic group. The nature and type of prosthetic group is responsible for the further classification of conjugated proteins into the following-

a. Nucleoproteins: Prosthetic group is a nucleic acid (DNA and RNA).

b. Glycoproteins: Prosthetic group is a carbohydrate or its derivative.

c. Lipoproteins: Prosthetic group is lipids.

d. Chromoproteins: Prosthetic group is color material, *e.g.* Hemoglobin cytochrome.

e. Phosphoproteins: Prosthetic group is phosphoric acid.

• **Derived proteins**:

These proteins are derived by the degradation of natural protein by the action of heat, chemicals or enzymes. These are of two types-

a. Primary derived proteins: These are the denatured or coagulated or first hydrolyzed products of protein, *e.g.* coagulated proteins, proteins, metaproteins.

b. Secondary derived proteins: These are the degraded (due to the breakdown of peptide bonds) products of proteins, *e.g.* Proteoses, peptones, polypeptides.

Biological function of proteins:

Proteins are very important biomolecules present in living organisms and perform various functions [1]. Some important functions are as follows:

• Many proteins act as a catalyst. *e.g.* Hexokinase, pepsin.

• The fibrous proteins serve as components of the tissues holding the skeletal elements together. *e.g.* collagen.

• The nucleoproteins serve as carriers of genetic characters.

• Protein also performs transport functions. *e.g.* haemoglobin.

• Blood plasma, a solution of protein in water, is used for the treatment of shock produced by serious injuries and operations.

NUCLEIC ACIDS

"Nucleic acids are the chemical basis of life and heredity and a biopolymer composed of mononucleotides as their repeating units". These are composed of nucleotides. Nucleotides are monomers that are composed of a 5-carbon sugar, a phosphate group and a nitrogenous base. Nucleic acids are abundant in all living beings. Their function is to create, encode and store information in the nucleus of living cells in living beings. These also transmit and express the functions created, inside and outside the cell nucleus.

There are two types of nucleic acids-

- **Deoxyribonucleic acid** (DNA)
- **Ribonucleic acid** (RNA)

Structures of nucleic acids: Nucleic acids are polymers of nucleotides.

n-Nucleotides → Nucleic acids (Polynucleic acid)

The structure of nucleic acids is divided into four different levels: primary structure, secondary structure, tertiary structure and quaternary structure.

Primary Structure

The linear sequence of nucleotides in which they are linked through phosphodiester bonds is called the primary structure of the nucleic acids. The monomer nucleotide consists of three components: a pentose sugar, a nitrogenous base and a phosphate group [1].

Nucleotides = a nitrogenous base + a pentose sugar (C_5) + a phosphate

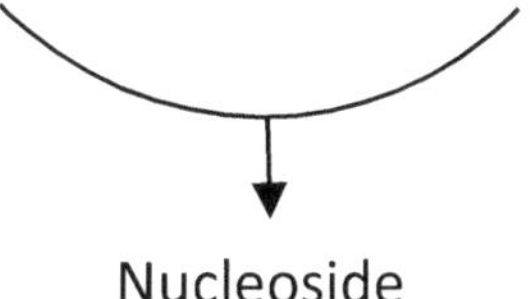

Nucleotide = Nucleoside + Phosphate

Therefore nucleotide consists of a nitrogenous base (Nucleobase), sugar and phosphate and nucleosides refer to a base + sugar.

1. Nitrogenous bases (Nucleobase): There are two types of nucleobase present in nucleic acids *i.e.* purines and pyrimidines [1].

Structure of purine and Structure of pyrimidine

There are two purine bases and three pyrimidine bases present in nucleic acids (shown below)-

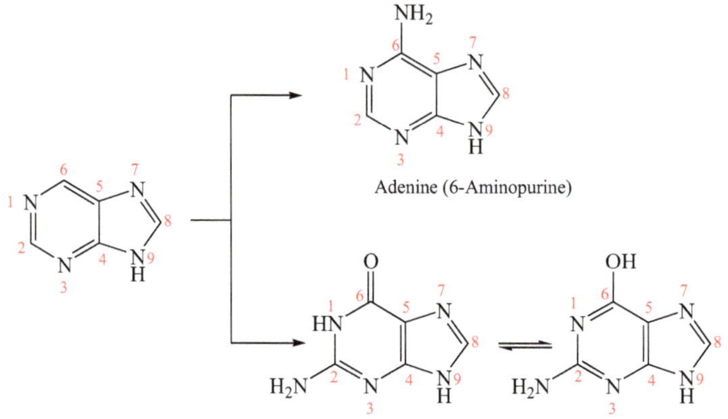

Adenine (6-Aminopurine)

Guanine (2-amino-6-hydroxypurine)

Cytosine (4-amino-2-hydroxypyrimidine)

Thymine (only found in DNA)

Uracil or 2,4-dihydroxypyrimidine
(only found in RNA)

Therefore, nucleobases present in nucleic acids are-

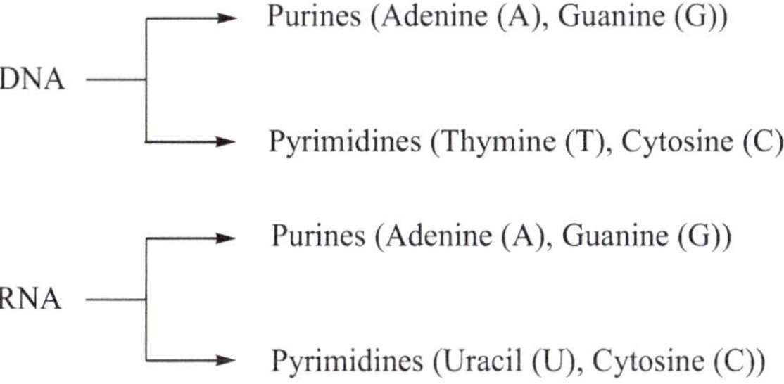

2. Pentose sugar:

The five-carbon monosaccharides (pentose) are present in nucleic acids. DNA contains D-deoxyribose while RNA contains D-ribose sugar [1] *i.e.*

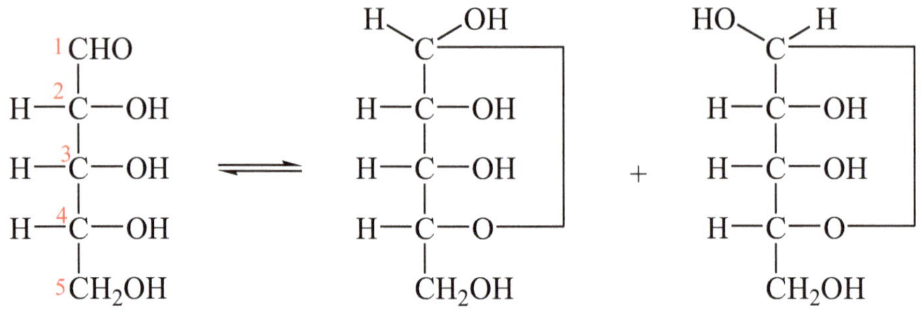

D-Ribose ⇌ α-D-Ribose + β-D-Ribose

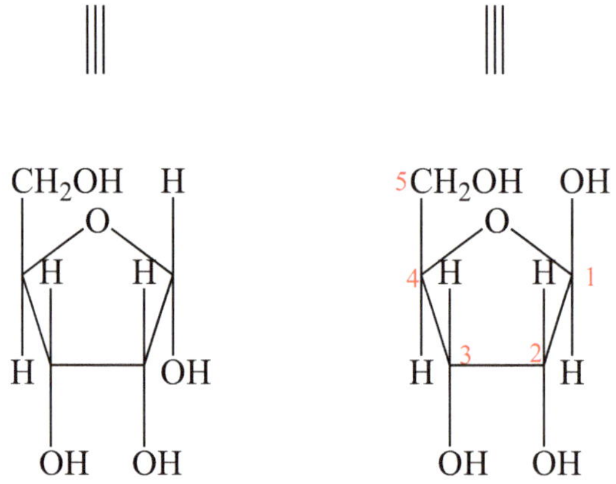

β-D-Ribofuranose
(present in RNA)

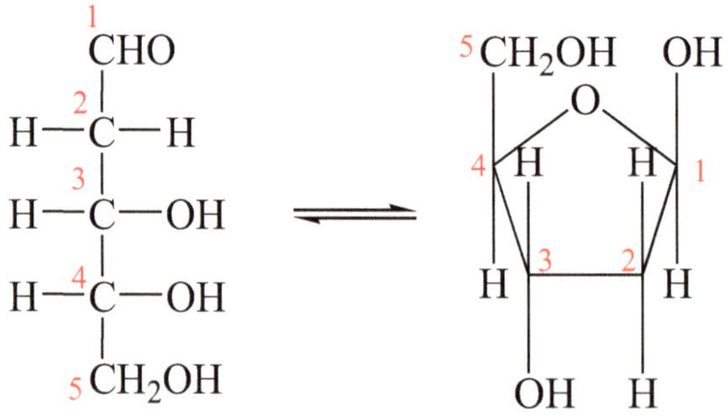

D-2-deoxyribose ⇌ D-2-deoxyribofuranose
(present in DNA)

3. Phosphate (PO_4^{3-})

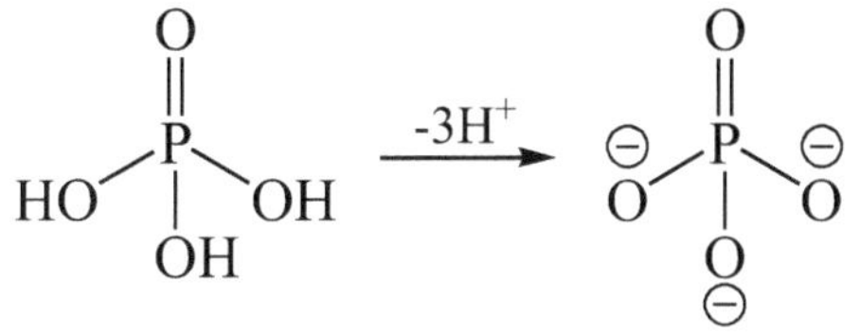
-3H+

Phosphoric acid Phosphate

Bonding and Nomenclature of Nucleotides and Nucleosides

The pentoses are bonded to nitrogenous bases by *β*-N-glycosidic bonds. The nomenclature of nucleotides and nucleosides is given in Table **1**. The glycosidic linkage is formed between the N-9 of purines and N-1 of pyrimidines with C-1 of pentose. The phosphate group linked with C-5 or C-3 of pentose (as shown below) [1, 7].

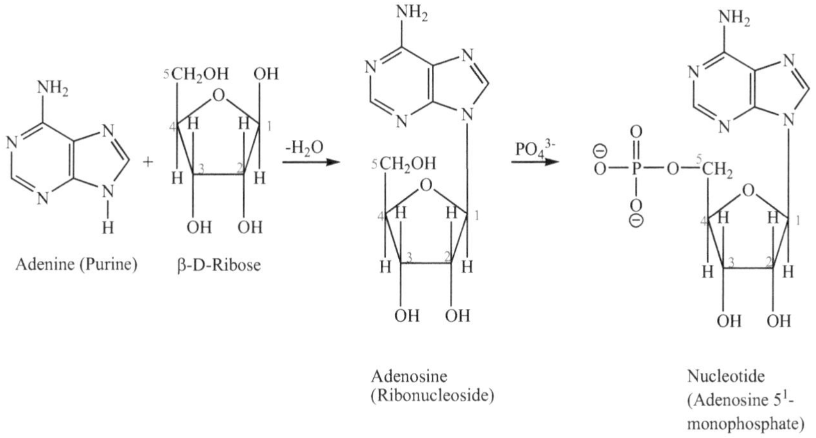
-H2O
PO4 3-
Adenine (Purine)
β-D-Ribose
Adenosine
(Ribonucleoside)
Nucleotide
(Adenosine 5¹-
monophosphate)

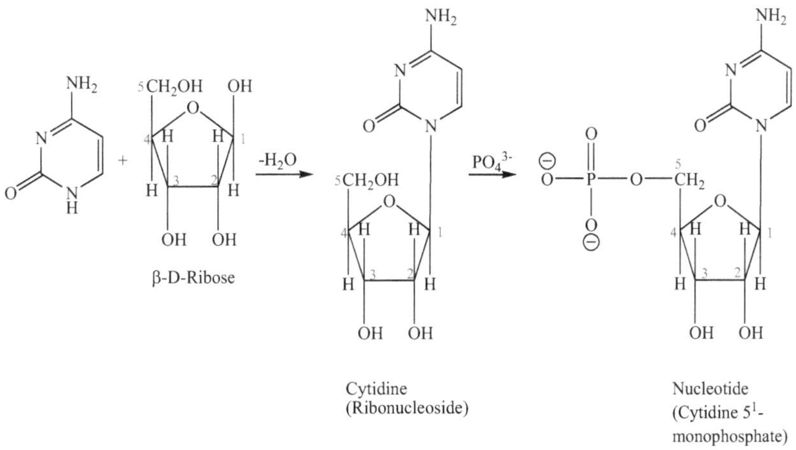
-H2O
PO4 3-
β-D-Ribose
Cytidine
(Ribonucleoside)
Nucleotide
(Cytidine 5¹-
monophosphate)

Similarly, if pentose sugar is 2-Deoxyribose then the nucleoside will be Deoxyribonucleoside and the nucleotide will be Deoxyribonucleotide *i.e.*

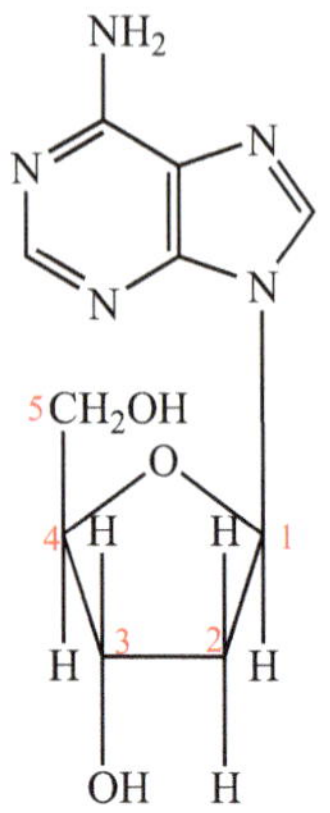

Deoxyadenosine
(Deoxyribonucleoside)

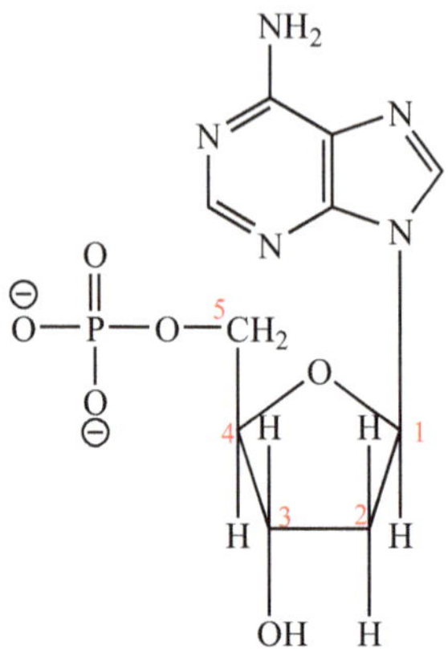

Nucleotide
(Deoxyadenosine 5[1]-monophosphate)

Few other examples of nucleotide can be seen as follows:

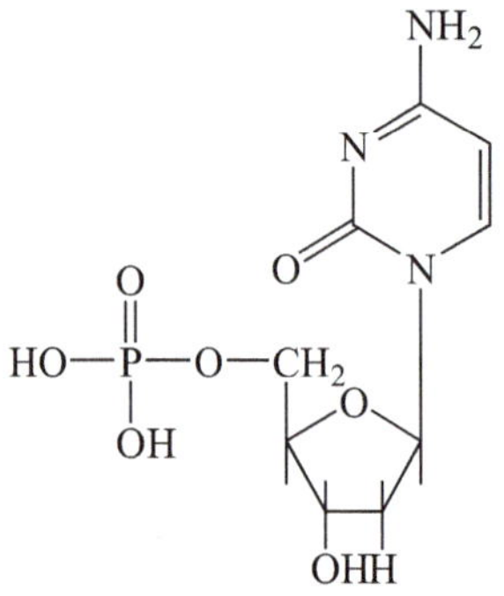

Deoxycytidine 5'-monophosphate

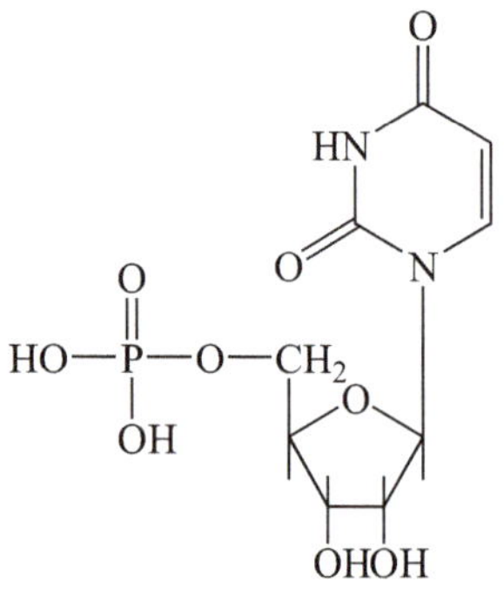

Uridine 5'-monophosphate

Table 1. Nomenclature of nucleotides and nucleosides.

Base	**Ribonucleoside**	**Ribonucleotide (5[1]-monophosphate)**
Adenine (A)	Adenosine	Adenosine 5[1]-monophosphate (AMP)
Guanine (G)	Guanosine	Guanosine 5[1]-monophosphate (GMP)
Cytosine(C)	Cytidine	Cytidine 5[1]-monophosphate (CMP)
Uracil (U)	Uridine	Uridine 5[1]-monophosphate (UMP)
Base	**Deoxyribonucleoside**	**Deoxyribonucleotide (5[1]-monophosphate)**
Adenine (A)	Deoxyadenosine	Deoxyadenosine 5[1]-monophosphate (dAMP)

Base	**Ribonucleoside**	**Ribonucleotide (5[1]-monophosphate)**
Guanine (G)	Deoxyguanosine	Deoxyguanosine 5[1]-monophosphate (dGMP)
Cytosine (C)	Deoxycytidine	Deoxycytidine 5[1]-monophosphate (dCMP)
Thymine (T)	Deoxythymidine	Deoxythymidine 5[1]-monophosphate (dTMP)

Nucleoside di- and tri-phosphate

The term mononucleotide is used when a single phosphate moiety is added to a nucleoside *i.e.*

Adenoine monophosphate (AMP) = Adenine + Ribose + Phosphate

The addition of second or third phosphates to the nucleoside results in nucleosidediphosphate (*e.g.* ADP) or tri-phosphate (*e.g.* ATP), respectively [7].

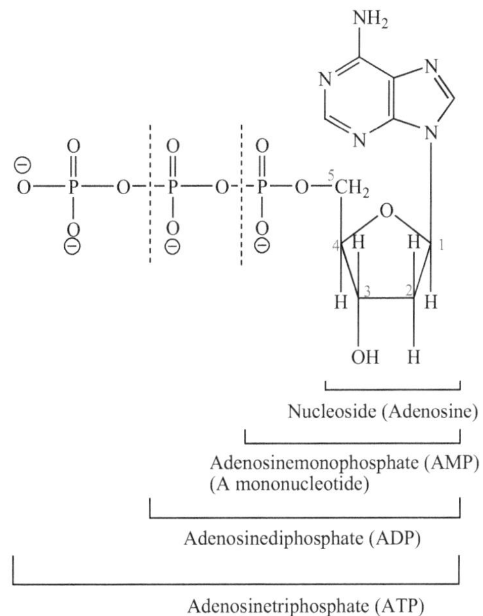

i.e.

$$AMP + H_3PO_4 \rightleftharpoons ADP + H_2O$$

$$ADP + H_3PO_4 \rightleftharpoons ATP + H_2O$$

$$ATP + H_2O \xrightleftharpoons{\text{Hydrolysis}} ADP + P_i + \text{Energy}$$

$$ADP + H_2O \xrightleftharpoons{\text{Hydrolysis}} AMP + P_i + \text{Energy}$$

P_i = Inorganic phosphate

Structure of polynucleotide (nucleic acid)

A polynucleotide (nucleic acid) is a polymer of Nucleotides (Deoxyribonucleotides in case of DNA and Ribonucleotides in case of RNA). The structure of polynucleotides can be seen as follows (Fig. **42**) [7, 19].

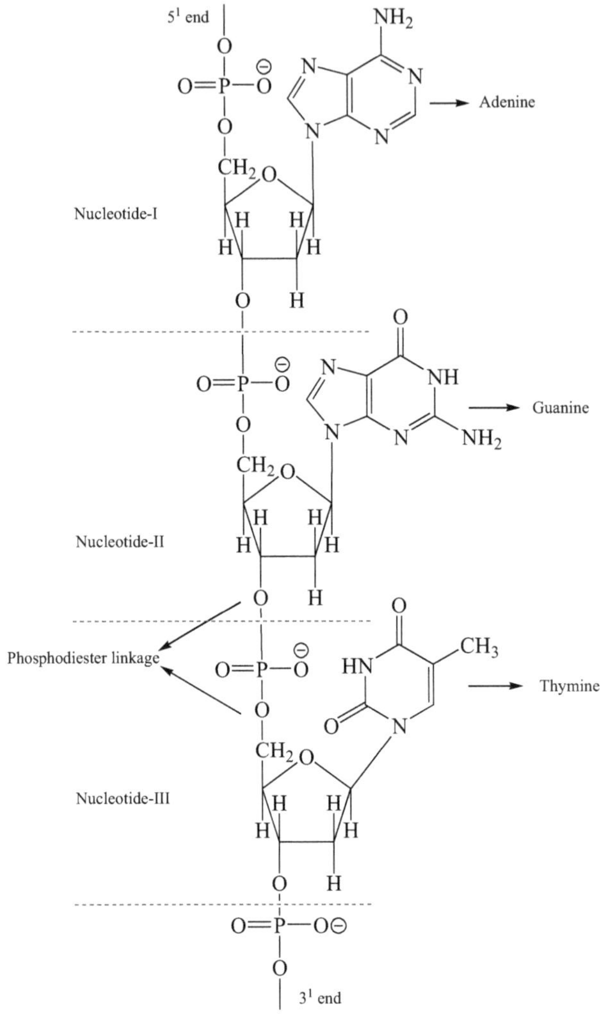

Fig. (42). Structure of polydeoxyribonucleotide.

Secondary Structure

Secondary structure shows the interaction between the bases, *i.e.* it shows which base of a nucleotide on one strand pairs with which base of other nucleotides on another strand (shown in Fig. **42**). These bases are paired up through hydrogen bonds. It is the secondary structure that is responsible for the assumed shapes of nucleic acids. In nucleosides, a purine base always pairs with a pyrimidine base, *i.e.* guanine (G) always pairs with cytosine (C) and adenine (A) pairs with either thymine (T) or uracil (U) [20].

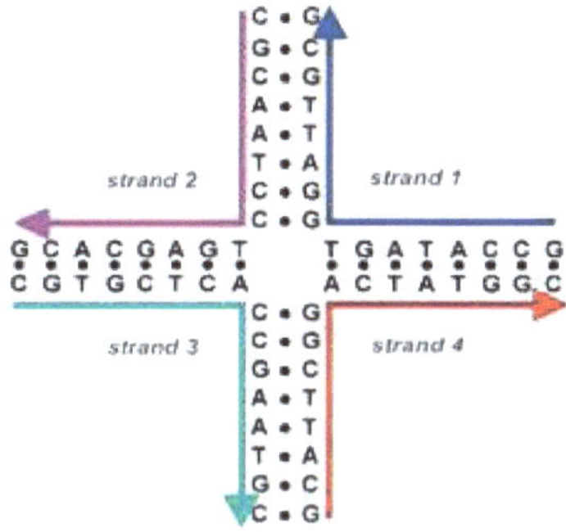

Fig. (43). Structure of polydeoxyribonucleotide' to Secondary Structure of polydeoxyribonucleotide [21].

Tertiary Structure

The 3-Dimensional structures of nucleic acids taking into consideration the positions of atoms, their geometrical and steric constraints are known as tertiary structures of nucleic acids. In this structure folding of linear polymer chain occurs which entirely folds into a specific three-dimensional structure. The DNA helical structure is an example of a tertiary structure (shown in Fig. **44**) in which the two strands fold around each other and forms a helix-like structure [1].

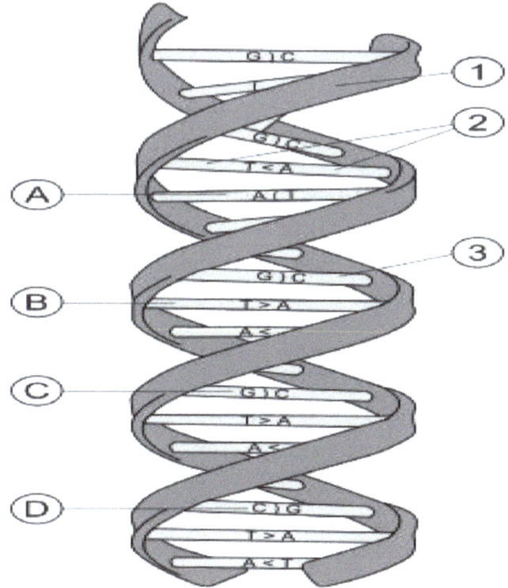

Fig. (44). Tertiary structure of nucleic acid (image courtesy: www.jing.fm) [22].

Quaternary Structure

This structure is similar to the quaternary structure of proteins, however, all the concepts are not exactly the same. Quaternary structure refers to the interaction between nucleic acids and other molecules and interaction between separate RNAs in the ribosomes [19].

Structure of DNA: DNA is a polymer of deoxyribonucleotide and held together by 5′, 3′-Phosphodiester bridges.

Chargaff's rule of DNA composition: According to Chargaff's rule, DNA contains an equal number of Adenine and Thymine (A=T) and an equal number of Guanine and Cytosine residue (G=C).

Double Helix structure of DNA: The double helix structure of DNA was proposed by James Watson and Francis Crick in 1953 and shared the 1962 Noble Prize [7]. The salient feature of Watson - Crick Model of DNA can be seen as follows-

1. The DNA is a right-handed double helix and consists of two poly-deox--ribonucleotide chains twisted around each other on a common axis. (shown in Fig. **45**)

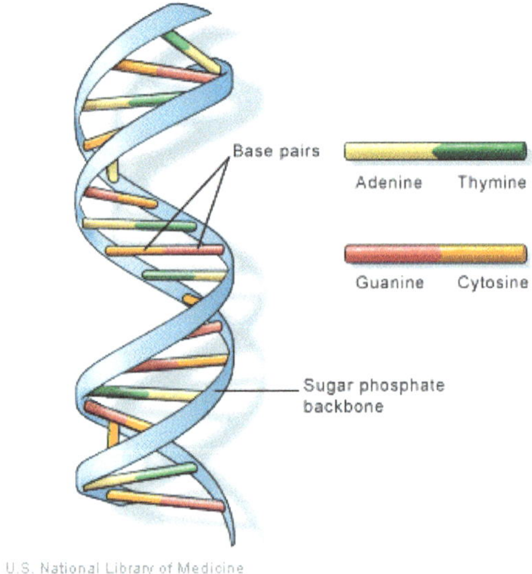

Fig. (45). Double helix structure of DNA [23].

2. The two strands are anti-parallel, *i.e.* one strand run in the 5′- 3′ direction while the other in 3′- 5′ direction.

3. The diameter of the double helix is 20 Å.

4. Each turn of helix is 34 Å and contains 10 pairs of nucleotides.

5. The distance between each pair is about 3.4 Å.

6. Each strand of DNA has a hydrophilic deoxyribose phosphate backbone (3′- 5′ phosphodiester bond) on the outside of the molecule while the hydrophobic bases are present inside.

7. The two polynucleotide chains are not identical but complementary to each other because of base pairing.

8. The two strands are held together by hydrogen bonding formed between complementary base pairs. The A-T pair has two hydrogen bonds while the G-C pair has three hydrogen bonds (shown in Fig. **46**) *i.e.*

A=T and G≡C

9. The base pairing in DNA is according to Chargaff's rule, *i.e.* the content of Adenine is equal to Thymine while Guanine is equal to Cytosine.

10. The double helix has major grooves and minor grooves along the phosphodiester backbone.

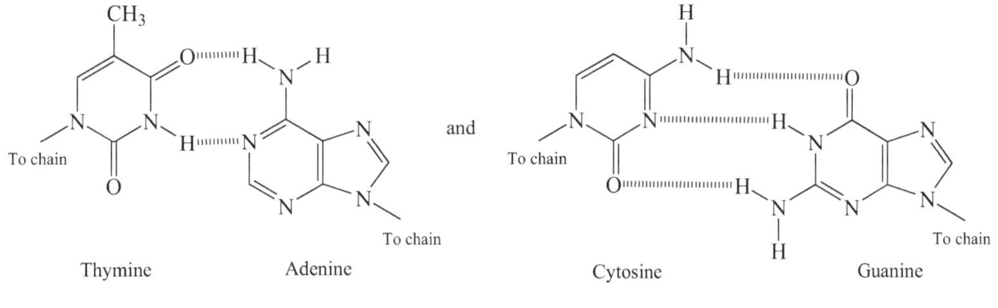

Fig. (46). Hydrogen bonding in A-T and G-C.

Structure of RNA

RNA is a polymer of ribonucleotides held together by 3′, 5′-phosphodiester bridges. The RNA molecule is single-stranded and contains ribose sugar. The four bases present in RNA are Adenine (A), Guanine (G), Cytosine (C) and Uracil (U)

[24]. Due to a single strand in RNA, there is no specific relation between purine and pyrimidine. There are three major types of RNAs present in cells [7]:

i. Messenger RNA (mRNA)

ii. Transfer RNA (tRNA)

iii. Ribosomal RNA (rRNA)

Besides the above three RNAs, another RNA is present in the cells. These include heterogeneous nuclear RNA (hn-RNA), small nuclear RNA (sn-RNA), small nucleolar RNA (snoRNA) and small cytoplasmic RNA (sc-RNA).

i. **Messenger RNA (mRNA)**:

The mRNA is synthesized in the nucleus (in eukaryotes) and present in only 2% of the total RNA. Its main function is to transfer genetic information from genes to ribosomes to synthesize proteins.

ii. **Transfer RNA (tRNA)**:

Transfer RNA, also known as soluble RNA, contains 71-80 nucleotides with a molecular weight of about 25000. Its main function is to transfer amino acids to mRNA for protein synthesis.

Cloverleaf structure of t-RNA: t-RNA contains mainly four arms, each with a base-paired stem. (Shown in Fig. **47**).

• The acceptor arm: This arm is capped with a sequence CCA (5′-3′). The amino acid is attached to the acceptor's arm.

• The anticodon arm: This arm, with the three specific nucleotide bases (anticodon), is responsible for the recognition of the triplet codon of mRNA. The codon and anticodon are complementary to each other.

• The D arm: It is so named due to the presence of dihydrouridine.

• The TψC arm: This arm contains a sequence of T, pseudouridine (psi, ψ), and C.

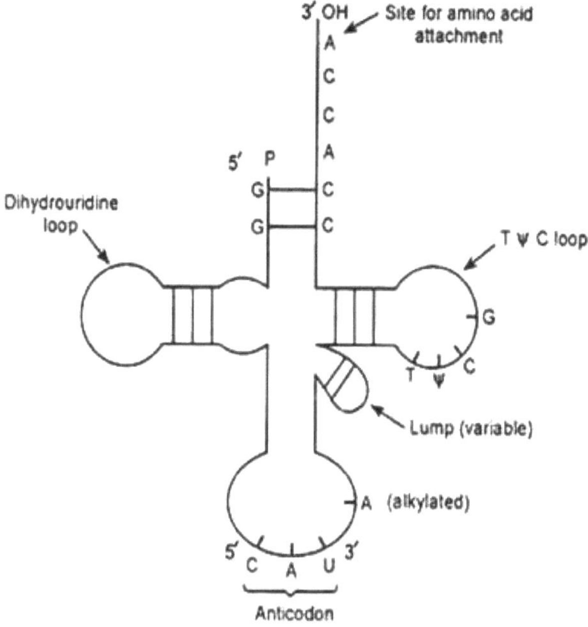

Fig. (47). Diagrammatic representation of the cloverleaf model of tRNA [25].

iii. **Ribosomal RNA (rRNA)**:

It is the most abundant RNA in cells and contributes about 50-80% of the total RNA. It provides a structural framework for ribosomes. It is also known as the factory of protein synthesis. The eukaryotic ribosomes are composed of two major nucleoprotein complexes- 60S subunit and 40S subunit.

LIPIDS

"Lipids are organic substances that are relatively insoluble in water and soluble in organic solvents (alcohol, ether, *etc.*)". Lipids are structural components of cell membranes and their function is to store energy. Some lipids can be synthesized in our body while some are not. These essential lipids should be obtained from the diet [1].

Unlike polysaccharides, proteins, and nucleic acids, lipids are not a polymer.

Classification of Lipids

Based on chemical composition, lipids are classified into the following-

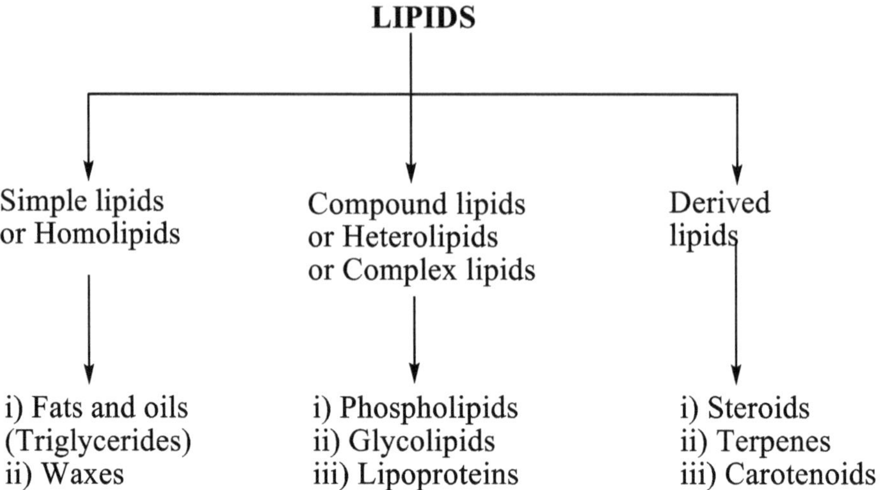

Lipids produce energy more than twice as compared as carbohydrates and therefore serve as the storage form of energy.

1. **Simple lipids**:

Simple lipids are the esters of fatty acids with alcohols. These are mainly of two types-

• **Fats and Oils (Triglycerides)**: These are the esters of fatty acids with glycerol. The basic difference between fats and oil is their physical state *i.e.* oil is liquid while fat is solid at room temperature. Since all the three hydroxyl groups of glycerol are esterified, fats are also known as triglycerides. Fats prevent the loss of body heat. They are found in nature in large quantities and are the best-reserved food material in the body. They protect the internal organs like padding material.

Fats are insoluble in water but are soluble in organic solvents like ether, chloroform, benzene, *etc.* and they themselves are a very good solvent for other fats and fatty acids. They are colourless, tasteless, and odourless and behave neutrally towards reactions. They have very low melting points [26].

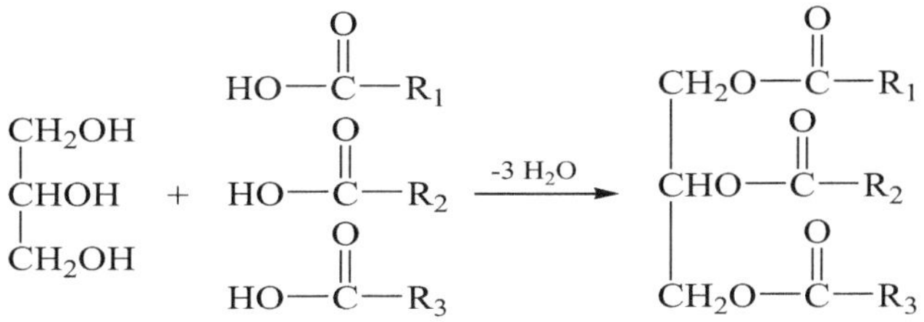

$$\begin{array}{l} CH_2OH \\ | \\ CHOH \\ | \\ CH_2OH \end{array} + \begin{array}{l} HO-\overset{O}{\overset{\|}{C}}-R_1 \\ HO-\overset{O}{\overset{\|}{C}}-R_2 \\ HO-\overset{O}{\overset{\|}{C}}-R_3 \end{array} \xrightarrow{-3\,H_2O} \begin{array}{l} CH_2O-\overset{O}{\overset{\|}{C}}-R_1 \\ | \\ CHO-\overset{O}{\overset{\|}{C}}-R_2 \\ | \\ CH_2O-\overset{O}{\overset{\|}{C}}-R_3 \end{array}$$

Glycerol Triglycerides (oils and fats)

e.g.

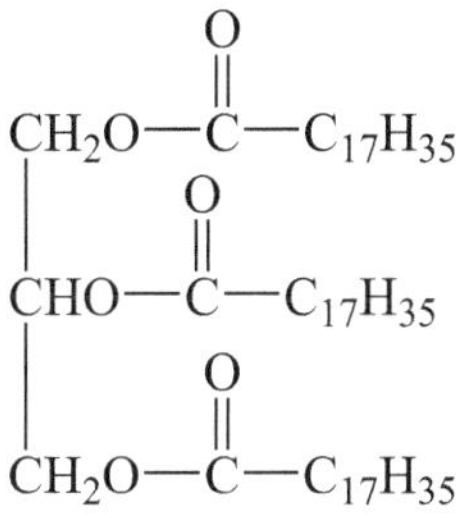

$$\begin{array}{l} CH_2O-\overset{O}{\overset{\|}{C}}-C_{17}H_{35} \\ | \\ CHO-\overset{O}{\overset{\|}{C}}-C_{17}H_{35} \\ | \\ CH_2O-\overset{O}{\overset{\|}{C}}-C_{17}H_{35} \end{array} \qquad \begin{array}{l} CH_2O-\overset{O}{\overset{\|}{C}}-C_{17}H_{33} \\ | \\ CHO-\overset{O}{\overset{\|}{C}}-C_{17}H_{33} \\ | \\ CH_2O-\overset{O}{\overset{\|}{C}}-C_{17}H_{33} \end{array}$$

Tristearin (Tristearylglycerol) (Fat, solid at room temp.) and Triolein (Trioleylglycerol) (Oil, liquid at room temp.)

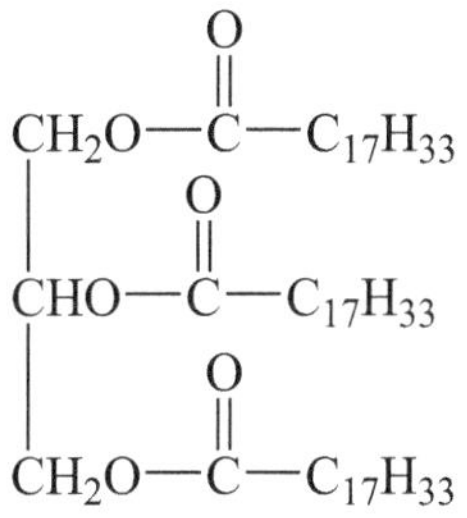

- **Waxes**:

These are the esters of fatty acids with alcohols other than glycerol. The alcohols may be aliphatic or alicyclic. The most common alcohol present in waxes is Cetyl alcohol ($C_{16}H_{33}OH$). *e.g.* beeswax (ester of palmitic acid with myricyl alcohol or Tricontanol-1) CH_3-$(CH_2)_{14}$-$COOCH_2$-$(CH_2)_{28}$-CH_3. The common waxes in the human body are esters of cholesterol [1].

2. **Compound lipids**:

These are the ester of fatty acid with alcohols containing additional groups such as phosphates, carbohydrates, protein, *etc.* Based on the nature of the additional group they are further divided into the following-

- **Phospholipids**:

In addition to alcohol and fatty acids, they contain phosphoric acid and a nitrogenous base. They are mainly present in the brain, liver, kidney, nerve tissue, heart and pancreas. Phospholipids carry the inorganic ions across the membranes, helps in blood clotting, speed up fatty acid oxidation and also acts as a prosthetic group for certain enzymes. These are further classified into the following-

- **Glycerophospholipids or Phosphoglycerides**: These phospholipids contain glycerol as alcohol, *e.g.* Cephalin.

- **Sphingophospholipids or Phosphosphingosides**: Sphingosine is the alcohol part of this group of phospholipids, *e.g.* Sphingomyelin.

- **Glycolipids**: These lipids contain fatty acids, carbohydrates and a nitrogenous base. Glycerol and Phosphate are absent, *e.g.* Gangliosides.

- **Lipoproteins**: These are macromolecular complexes of lipids with proteins. With the increase in protein content the density of lipoprotein increases. Their function is to transport and deliver lipids to tissues.

Derived Lipids

These are the derivatives obtained from the hydrolysis of simple and compound lipids. These include glycerol, other alcohol, fatty acids (obtained by hydrolysis of fatty acids), mono- and di-acylglycerols, steroid hormones, hydrocarbons, *etc.* The most common derived lipids are terpenes, carotenoids and steroids. Terpenes are available in plants in the form of essential oils. Carotenoids are tetraterpenoids produced in plants and are also found in animal bodies if carotenoids producing plants are eaten. Steroids are available in almost all species of animals. They do not contain fatty acids.

Fatty acids: Fatty acids are carboxylic acids with a long hydrocarbon chain.

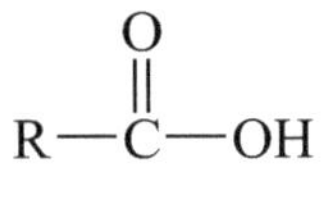

$$R-\overset{\overset{O}{\|}}{C}-OH \qquad \text{Fatty acids}$$

$$R = H \qquad H-\overset{\overset{O}{\|}}{C}-OH \qquad \text{Formic acids}$$

$$R = CH_3 \qquad H_3C-\overset{\overset{O}{\|}}{C}-OH \qquad \text{Acetic acids}$$

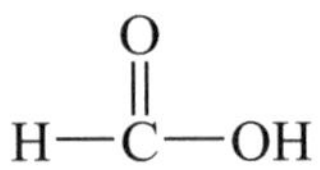

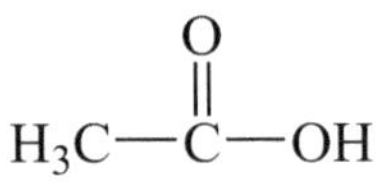

$$R = C_2H_5 \qquad H_3C-\overset{H_2}{C}-\overset{\overset{O}{\|}}{C}-OH \qquad \text{Propionic acids}$$

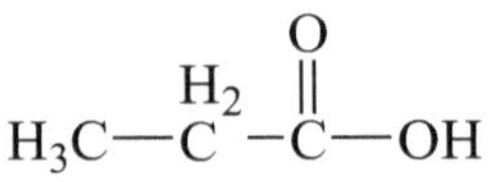

If R having 4 to 30 carbon atoms, then it is known as fatty acids *i.e.* fatty acids are carboxylic acids with R = 4-30 carbon atoms in their hydrocarbon chain. Most of the naturally occurring fatty acids contain 14-20 carbon atoms (even number of carbon atoms). These are a very important dietary source of fuels and structural components of cells in animals [1].

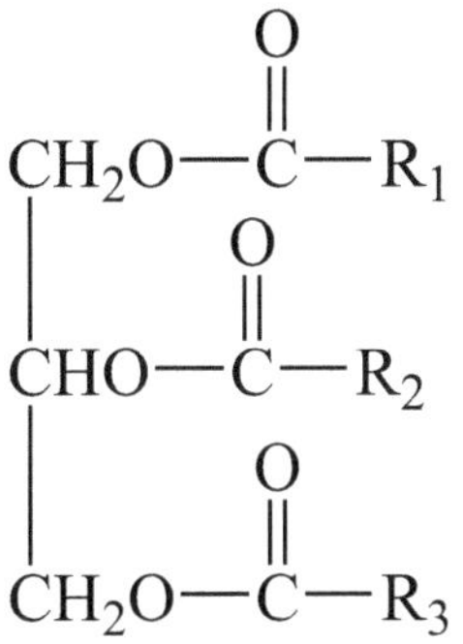

$$\begin{array}{l} CH_2O-\overset{\overset{\displaystyle O}{\|}}{C}-R_1 \\ | \\ CHO-\overset{\overset{\displaystyle O}{\|}}{C}-R_2 \\ | \\ CH_2O-\overset{\overset{\displaystyle O}{\|}}{C}-R_3 \end{array}$$

Triglycerides (oils and fats)
R_1, R_2, R_3 may be same or different

Fatty acids are of two types- Saturated fatty acids and unsaturated fatty acids.

- **Saturated fatty acids**:

These contain only carbon-carbon single bonds, *e.g.* butter. General formula: $CH_3(CH_2)_mCOOH$ or $C_nH_{(2n+1)}COOH$, *e.g.* Palmitic acid (m=14), Stearic acid (m=16).

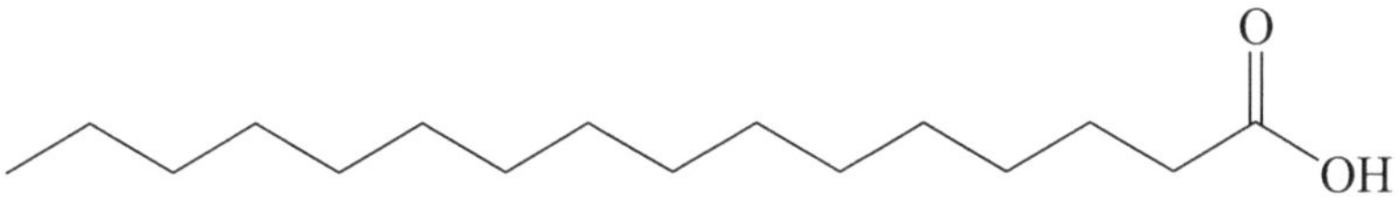

O
OH

Palmitic acid, $C_{15}H_{31}COOH$

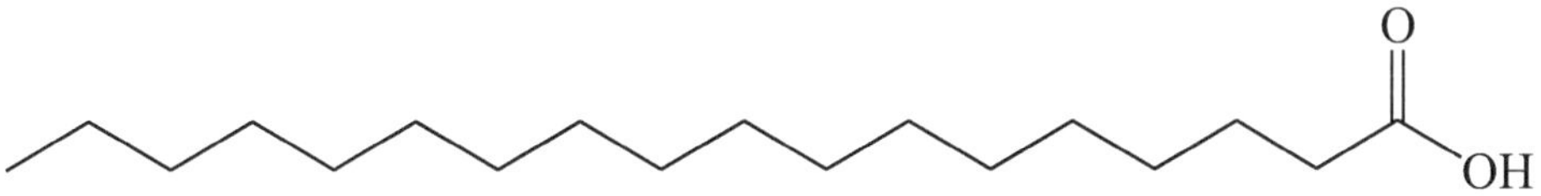

O
OH

Stearic acid, $C_{17}H_{35}COOH$

Saturated fatty acids have tightly packed crystal lattice. Most fatty acids are solid at room temperature and show a high melting point.

• **Unsaturated fatty acids**:

These fatty acids contain one or more double bonds in the alkyl chain, *e.g.* olive oil (one double bond), sunflower oil (72 double bonds). Unsaturated fatty acids are generally liquid at room temperature and double bonds have the cis-configuration, *e.g.* oleic acid.

General formula: Unsaturated fatty acids having one double bond *i.e.* monoethenoid acids = $C_nH_{2n-1}COOH$, two double bonds (Diethenoid acids) = $C_nH_{2n-3}COOH$, three double bonds (Triethenoid acids) = $C_nH_{2n-5}COOH$, and so on.

Linoleic acid

Structure of *Cis*- and *Trans*- Fatty acids (shown in Fig. **48**):

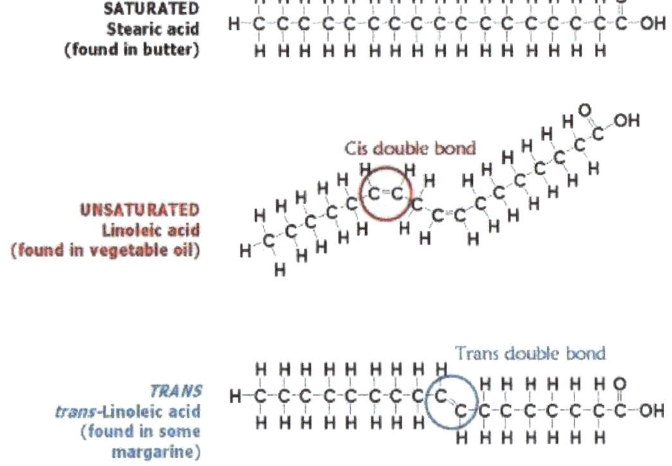

Fig. (48). Saturated and unsaturated (cis- and trans-) fatty acid [27].

Nomenclature of Fatty Acids

In fatty acids, the position of carbon atoms can be indicated either from the carboxyl-end (C-1, C-2, C-3.....) or from the methyl-end (ω-1, ω-2, ω-3...).

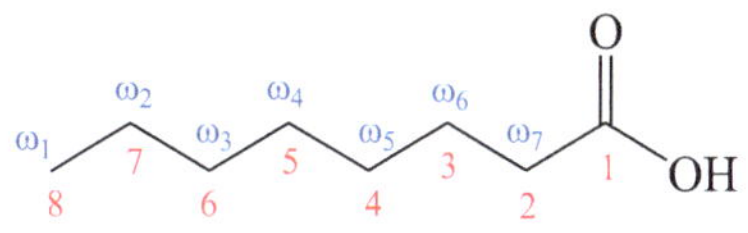

n-octanoic acid

Omega number refers to the position of double bond nearest to the methyl end of the carbon chain [28].

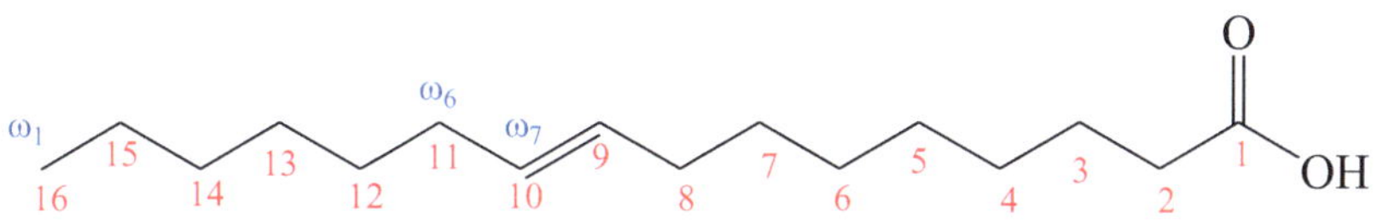

Palmitoleic acid, an omega-7 fatty acid
(9-hexadecenoic acid)

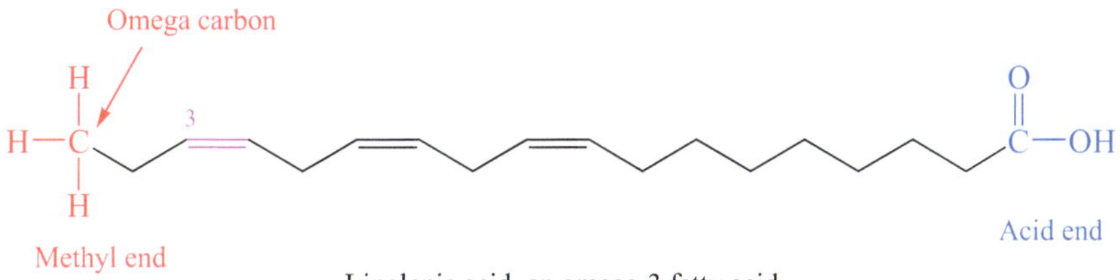

Linolenic acid, an omega-3 fatty acid

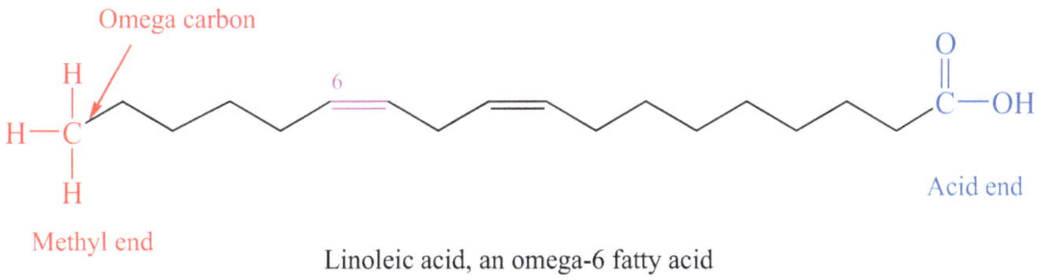

Linoleic acid, an omega-6 fatty acid

Abbreviation of Nomenclature of Fatty Acids

Total number of carbon: Number of the double bond; Position of the double bond

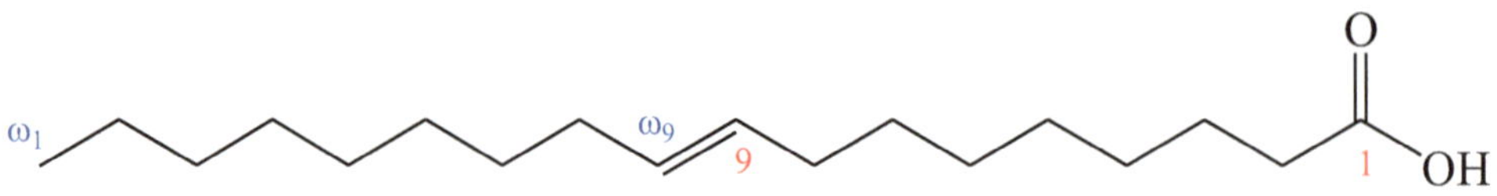

Oleic acid, an omega-9 fatty acid
9-octadecenoic acid, 18:1;9

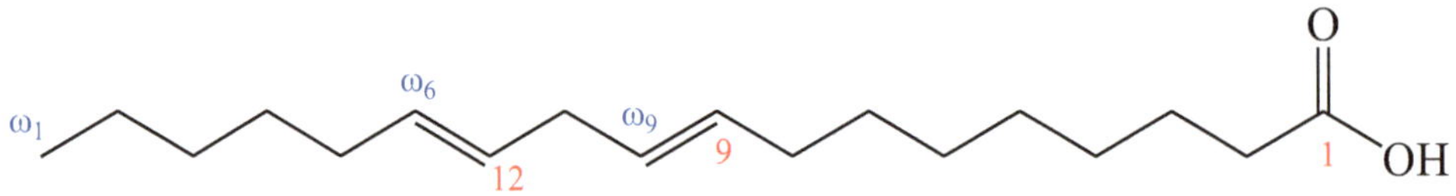

Linoleic acid, an omega-6 fatty acid
cis,cis-9,12-octadeca-dienoic acid, 18:02;9,12

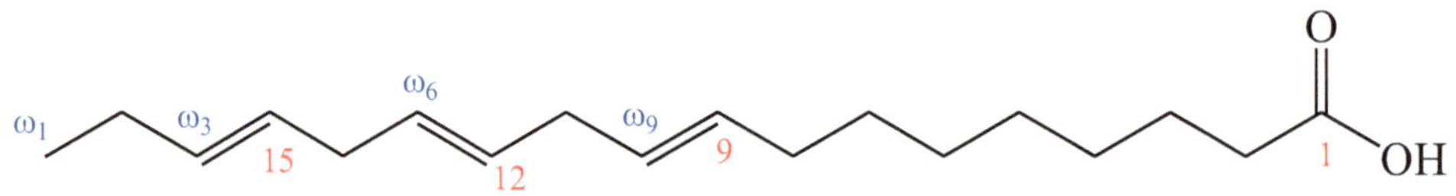

Linolenic acid, an omega-3 fatty acid
cis,cis,cis-9,12,15-octadeca-trienoic acid, 18:03;9,12,15

Linoleic and Linolenic acid are known as essential fatty acids.

Triacylglycerols (Triglycerides)

These are the esters of glycerol with fatty acids. The fats and oils that are widely distributed in both plants and animals are triglycerides. These are insoluble in water and are commonly known as natural fats.

Triglycerols are the most abundant group of lipids that works as fuel reserve of animals. The fat reserve of a normal human is sufficient to meet the body caloric requirement for 2-3 months [1].

Structure of Acylglycerols

Based on the number of fatty acids present, acylglycerols are of three types, *e.g.* monoacylglycerols (having one fatty acid), diacylglycerols (two fatty acids) and triacylglycerols (three fatty acids).

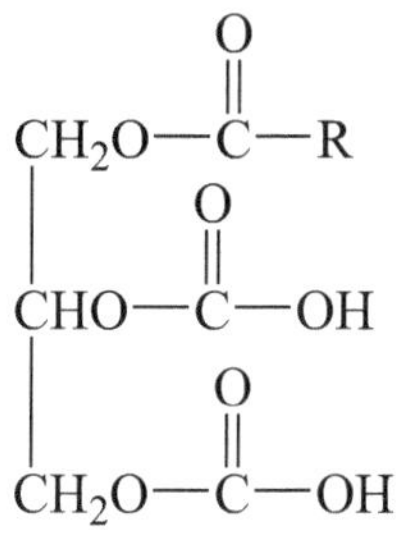

$$\begin{array}{lll}
CH_2O-\overset{\overset{O}{\|}}{C}-R & CH_2O-\overset{\overset{O}{\|}}{C}-R_1 & CH_2O-\overset{\overset{O}{\|}}{C}-R_1 \\
\vert & \vert & \vert \\
CHO-\overset{\overset{O}{\|}}{C}-OH & CHO-\overset{\overset{O}{\|}}{C}-R_2 & CHO-\overset{\overset{O}{\|}}{C}-R_2 \\
\vert & \vert & \vert \\
CH_2O-\overset{\overset{O}{\|}}{C}-OH & CH_2O-\overset{\overset{O}{\|}}{C}-OH & CH_2O-\overset{\overset{O}{\|}}{C}-R_3 \\
\text{Monoacylglycerol} & \text{Diacylglycerol} & \text{Triacylglycerol}
\end{array}$$

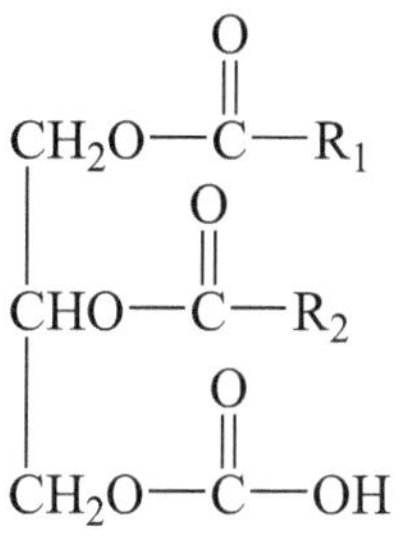

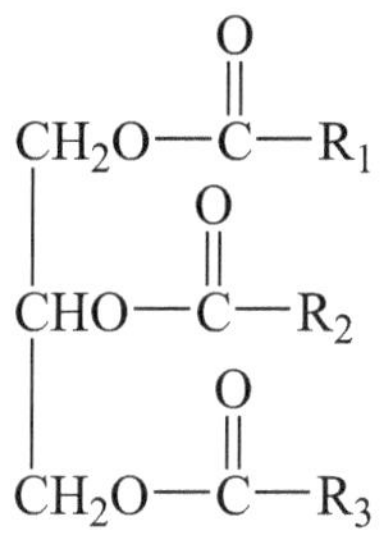

Among the above three, the triacylglycerols are the most important. Triacylglycerols are of two types-

- Simple triacylglycerols or simple triglycerides: They contain the same type of fatty acid residue at all the three-carbon, *e.g.* tristearin
- Mixed triacylglycerols or mixed triglycerides: They contain two or three types of fatty acid residue.

Properties of Triacylglycerols

- Hydrolysis: Triglycerides on hydrolysis provides fatty acids and glycerol, *i.e.*

Due to partial hydrolysis of oils and fats, a trace amount of fatty acids are always present in oils and fats. The estimation of free fatty acids present in fats/oils can be determined by its acid value.

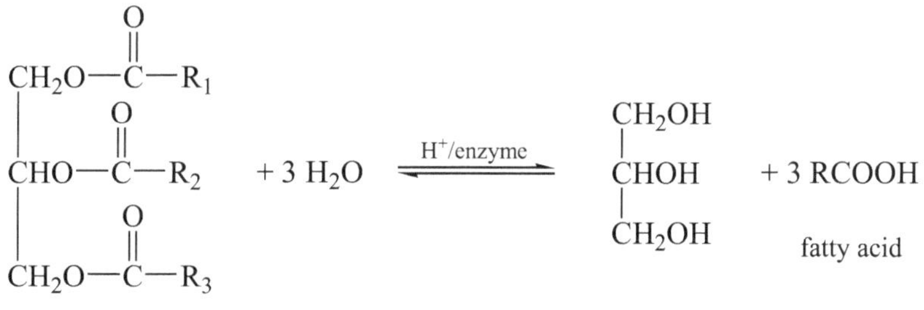

$$\begin{array}{l}
CH_2O-\overset{\overset{O}{\|}}{C}-R_1 \\
\vert \\
CHO-\overset{\overset{O}{\|}}{C}-R_2 \\
\vert \\
CH_2O-\overset{\overset{O}{\|}}{C}-R_3 \\
\text{Triacylglycerol}
\end{array}
+ 3\,H_2O \xrightleftharpoons{H^+/\text{enzyme}}
\begin{array}{l}
CH_2OH \\
\vert \\
CHOH \\
\vert \\
CH_2OH \\
\text{Glycerol}
\end{array}
+ \underset{\text{fatty acid}}{3\,RCOOH}$$

Acid value = Amount of KOH in milligrams used to neutralize the free acid (fatty acid) present in 1.0 g of oil/fat.

Oils on decomposition give free fatty acids. In normal circumstances, refined oil should be free from any free fatty acids. Therefore, oils with an increased acid number are unsafe for human consumption.

• **Saponification**: The hydrolysis of triglycerides with alkali to produce glycerols and soaps is known as saponification.

$$\begin{array}{l} CH_2O{-}\overset{O}{\overset{\|}{C}}{-}R \\ |\\ CHO{-}\overset{O}{\overset{\|}{C}}{-}R \\ | \\ CH_2O{-}\overset{O}{\overset{\|}{C}}{-}R \end{array} + 3\,NaOH \text{ or } KOH \longrightarrow \begin{array}{l} CH_2OH \\ | \\ CHOH \\ | \\ CH_2OH \end{array} + 3\,RCOONa$$

Triacylglycerol — Glycerol — Sodium or potasium salt of fatty acids (soap)

Saponification value or saponification number: The saponification value of oils or fats is defined as the milligrams of KOH required to hydrolyze (saponify) one gram of fat or oil [1]. It is a measure of the average molecular size of the fatty acid present in oils/fats. *e.g.* butter: 230-240, coconut oil: 250-260.

• **Hydrogenation reaction**: Hydrogenation of unsaturated fatty acids present in triglycerides gives saturated fatty acids *i.e.*

$$\underset{H\quad H}{-C{=}C-} + \underset{\text{Hydrogen gas}}{H\text{-}H} \xrightarrow{\text{Pt/Ni Catalyst}} \begin{array}{c} H\quad H \\ |\quad | \\ -C{-}C- \\ |\quad | \\ H\quad H \end{array}$$

Unsaurated Vegetable oil liquid at room temp. — Saturated Vegetable Ghee (Vanaspati Ghee) solid at room temp.

$$\begin{array}{l} CH_2O{-}\overset{O}{\overset{\|}{C}}{-}C_{17}H_{33} \\ | \\ CHO{-}\overset{O}{\overset{\|}{C}}{-}C_{17}H_{33} \\ | \\ CH_2O{-}\overset{O}{\overset{\|}{C}}{-}C_{17}H_{33} \end{array} + 3\,H_2 \longrightarrow \begin{array}{l} CH_2O{-}\overset{O}{\overset{\|}{C}}{-}C_{17}H_{35} \\ | \\ CHO{-}\overset{O}{\overset{\|}{C}}{-}C_{17}H_{35} \\ | \\ CH_2O{-}\overset{O}{\overset{\|}{C}}{-}C_{17}H_{35} \end{array}$$

Triolein (Unsaturated) Oils (lliquids) Double bond at C-9 position — Tristearin (Saturated) Fat (solid)

The number of double bonds (unsaturated) can be determined by using the **Iodine value**.

$—CH{=}CH— + I_2 \longrightarrow —CHI—CHI—$

Unsaurated Di-iodo compound

$—CH{=}CH—CH_2—CH{=}CH— + 2\,I_2 \longrightarrow —CHI—CHI—CH_2—CHI—CHI—$

Unsaurated Tetra-iodo compound

Iodine value of oils and fats is defined as the grams of iodine absorbed by 100 g of fats or oils. Iodine value is useful to know the relative unsaturation of fat and is directly proportional to the content of unsaturated fatty acids. The iodine number of some common oils/fat can be seen as follows- Coconut oil = 7-10, Butter = 25-30, sunflower oil = 125-135.

Phospholipids

In addition to alcohol and fatty acids, these contain phosphoric acid and a nitrogenous base. Phospholipids are of two types: Glycerophospholipids (Phosphoglycerides) and Sphingophospholipids (Sphingomyelins).

1. **Glycerophospholipids**: These are the major lipids that occur in biological membranes. These are the glycerol 3-phosphate esterified at C_1 and C_2 with fatty acids. Generally, C_1 contains saturated fatty acid while C_2 contains an unsaturated fatty acid.

• **Phosphatidic acid**: This is the simplest phospholipid. It is an intermediate in the synthesis of triacylglycerols and phospholipids, *e.g.*

1 $CH_2O—C(=O)—$ (16 C) (Saturated fatty acid) e.g. Palmitic acid

2 $CHO—C(=O)—$ (9; 18 C) (Unsaturated fatty acid) e.g. Oleic acid

3 $CH_2O—P(=O)(O^{\ominus})—O—X$

X = H, Phosphatidic acid

• **Lecithins (Phosphatidylcholine)**: These are the most abundant group of phospholipids in the cell membranes. It is a phosphatidic acid with choline as the base, *e.g.*

$$X = \text{Choline}, \quad -\overset{H_2}{C}-\overset{H_2}{C}-\overset{\ominus}{N}(CH_3)_3$$

From two fatty acids

The plus charge on the N is balanced by a negative ion- usually Chloride

From Glycerol

From Phosphate

From choline

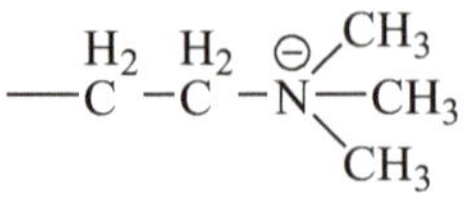

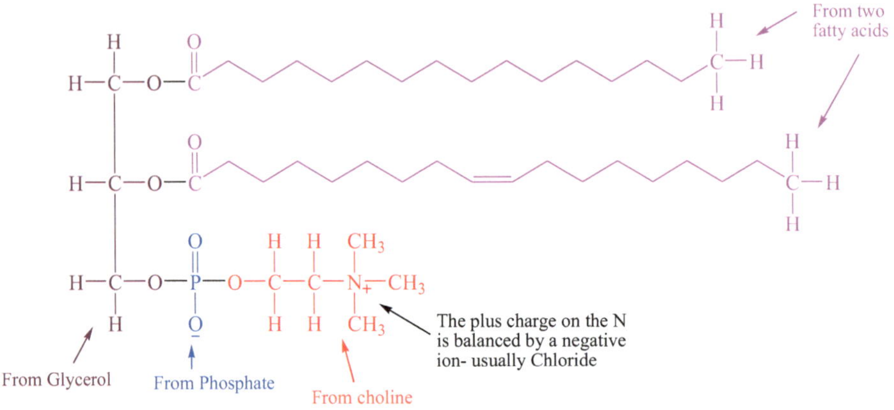

• **Cephalins (Posphatidylethanolamine)**: The nitrogenous base present in Cephalins is ethanolamine, *e.g.*

$$X = \text{Ethanolamine}, \quad -\overset{H_2}{C}-\overset{H_2}{C}-\overset{\ominus}{N}H_3$$

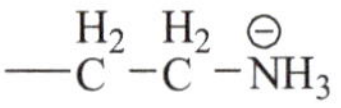

• **Phosphatidylserine**: Serine is present in this group of phospholipids, *e.g.*

$$X = \text{Serine}, \quad -\overset{H_2}{C}-\underset{COO^{\ominus}}{\overset{H}{C}}-\overset{\oplus}{N}H_3$$

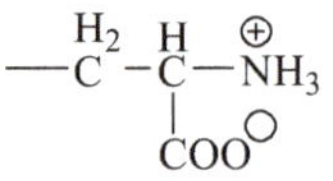

• **Phosphatidylinositol**: The stereoisomer myo-inositol is attached to phosphatidic acid to give Phosphatidylinositol (PI). This is an important component of cell membranes. The action of certain hormones (*e.g.* oxytocin, vasopressin) is mediated through PI.

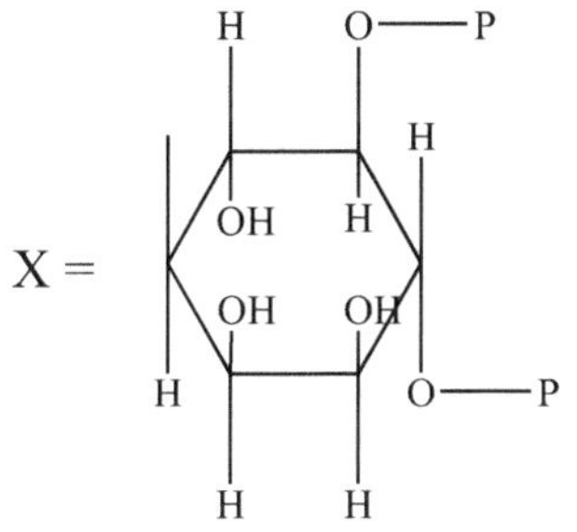

X =

H O—P
H
OH H
OH OH
H O—P
H H

myo-Inositol4,5-biphosphate

• **Cardiolipin**: It is so named because it was first isolated from the heart muscle. Structurally, cardiolipin consists of two molecules of phosphatidic acid held by additional glycerol through phosphate groups. It is an important component of the inner mitochondrial membrane and essential for mitochondrial function.

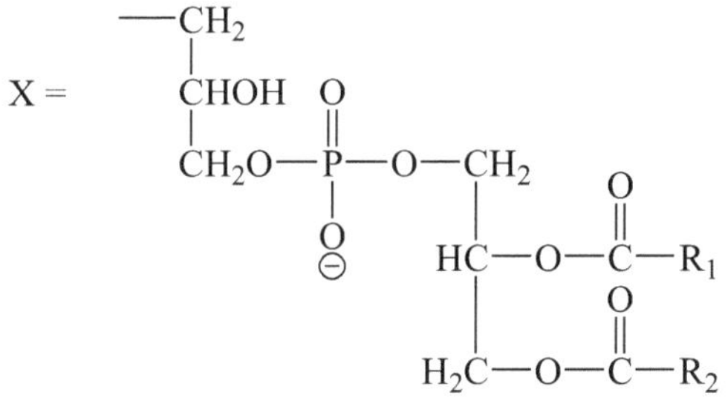

X =

—CH_2
CHOH O
CH_2O—P—O—CH_2 O
O⊖ HC—O—C—R_1
O
H_2C—O—C—R_2

Cardiolipin (diphosphatidylglycerol)

2. **Sphingophospholipids (Sphingomyelins)**: These phospholipids do not contain glycerol, instead sphingosine, amino alcohol, is present in sphingomyelins. Sphingosine is attached by an amide linkage to a fatty acid to produce ceramide. The alcohol group of sphingosine is bound to phosphocholine in sphingomyelin structure *i.e.*

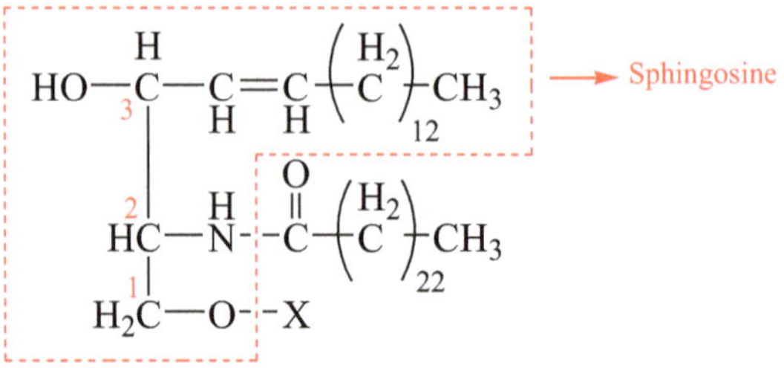

Sphingosine

X= H = Ceramide

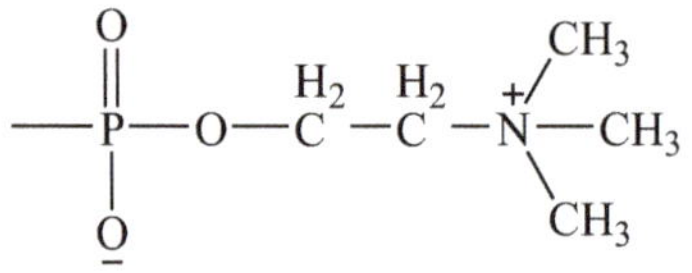

X= = Sphingomyelin

Phosphodoline

Sphingomyelins are important constituents of myelin and are found in good quantity in the brain and nervous tissue.

Functions of Phospholipids:

- In associations with proteins, phospholipids form the structural component of membranes and regulate membrane permeability.
- Phospholipids participate in the absorption of fat from the intestine.
- Phospholipids are essential for the synthesis of different lipoproteins.

Steroids (Non saponifiable lipid):

Steroids are the compounds containing a cyclic steroid nucleus namely cyclopentanoperhydrophenanthrene (CPPP). It consists of a phenanthrene nucleus (A, B, C) attached with a cyclopentane ring (D) [7].

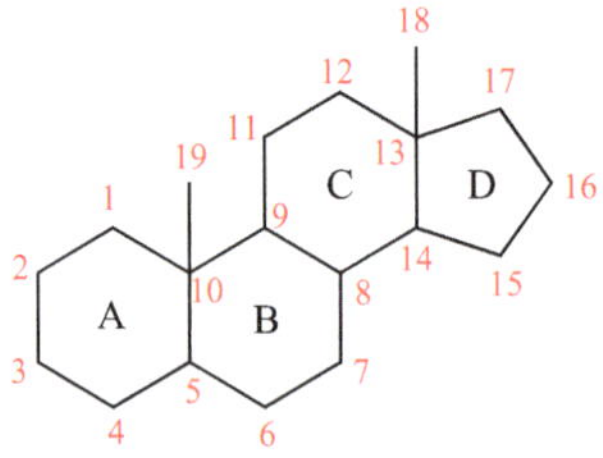

Steroid nucleus

There are several steroids in the biological systems. These include cholesterol, bile acids, Vitamin D, sex hormones, alkaloids, *etc.*

Cholesterol

It is a structural component of the cell membrane. As a structural component of plasma membranes, cholesterol is an important determinant of the permeability properties of the membrane. Our high cholesterol diet may be unhealthy because excess blood cholesterol leads to the deposition of cholesterol in the arteries that causes atherosclerosis [7]. The chemical structure of cholesterol is given below-

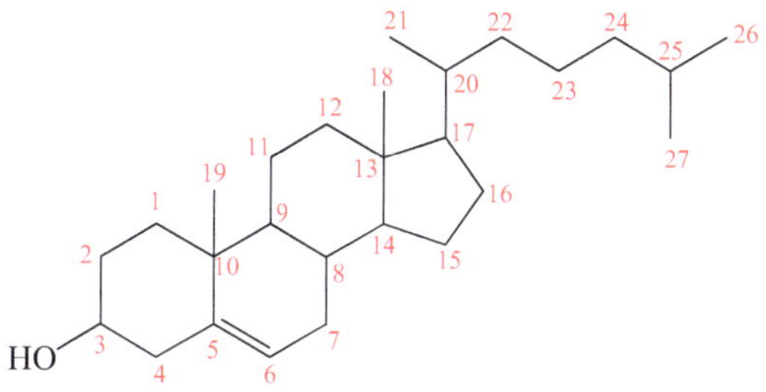

HO 1 2 3 4 5 6 7 8 9 10 11 12 13 14 15 16 17 18 19 20 21 22 23 24 25 26 27

Cholesterol

The structures of few important steroids can be seen as follows-

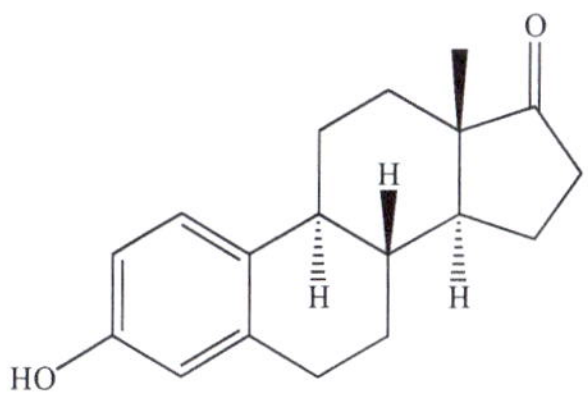

HO O H H H

Estrone (Estrogens)

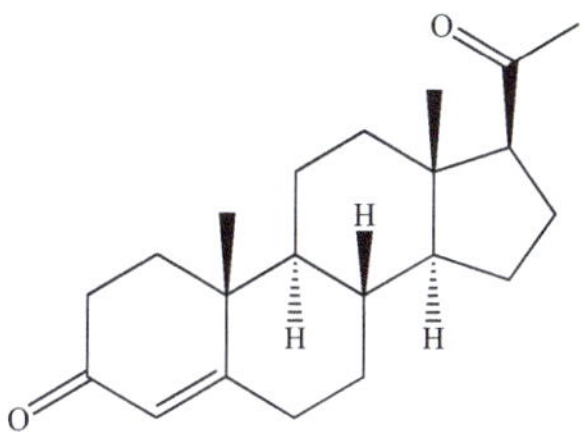

O O H H H

Progesterone

OH O H H H

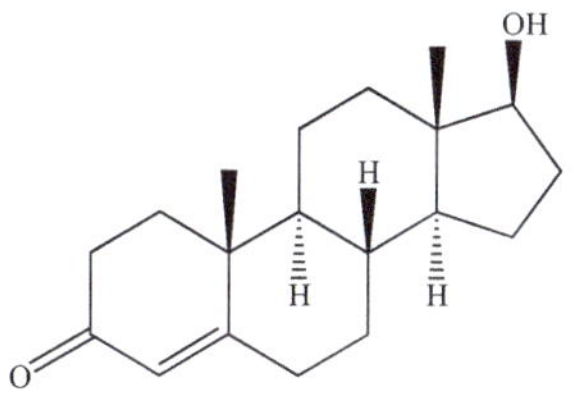

Testosterone

Brassinosteroids

These are the new group of plant hormones having significant growth-promoting activity. Brassinosteroids were first isolated from the pollen of Brassica napus.

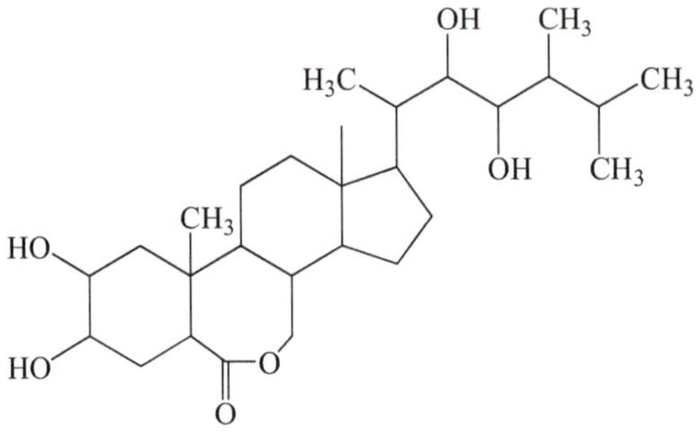

Brassinolide
(the first Brassinosteroids isolated)

Terpenes

It is a nonsaponifiable lipid found in plants. These are generally hydrocarbons formed by isoprenes [7].

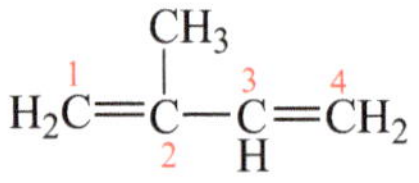

2-methylbuta-1,3-diene
or Isoprene (monomer of
the natural rubber)

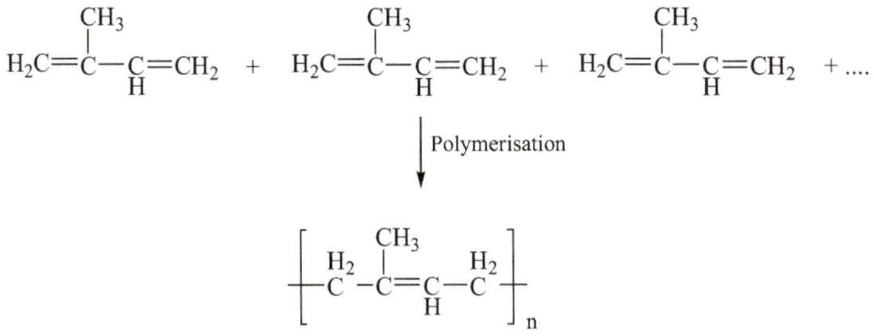

Polymerisation

Polyisoprene (Natural rubber)

Carotenoids

Carotenoids are tetraterpenes (4-isoprene units). These are widely distributed in both plant and animal kingdoms but are exclusive of plant origin. These are isoprene derivatives with a high degree of unsaturation and due to the presence of many conjugated double bonds, they are red or yellow coloured, *e.g.* Pigment of tomato (lycopene) = red, Pigment of carrot (α- and β-carotene) = red [7].

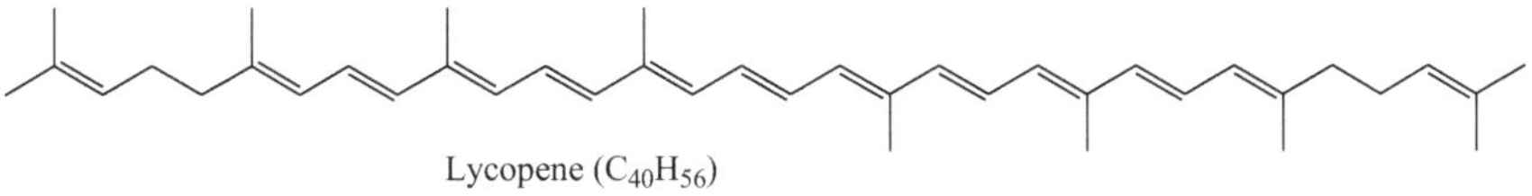

Lycopene ($C_{40}H_{56}$)

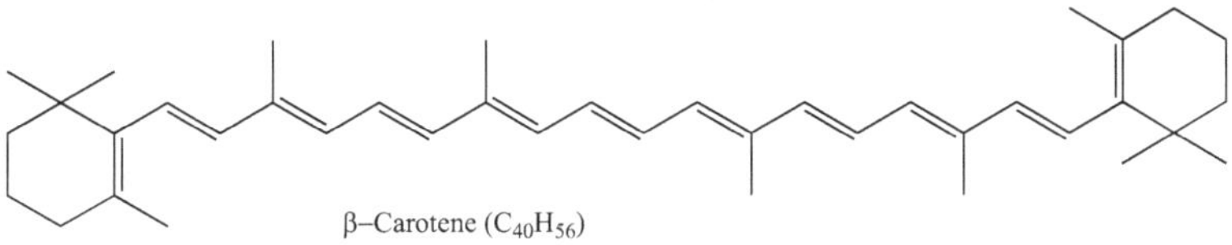

β–Carotene ($C_{40}H_{56}$)

Chlorophyll

It is the most abundant pigment in the chloroplast. Chlorophyll molecules are located in thylakoid membranes. It has an Mg^{-2} ion in the centre. It harvests energy (Photons) by absorbing a certain wavelength. Plants are green because the green wavelength is reflected back and not absorbed by chlorophyll. There are several types of chlorophyll and the important one is chlorophyll a and chlorophyll b (shown below) [7].

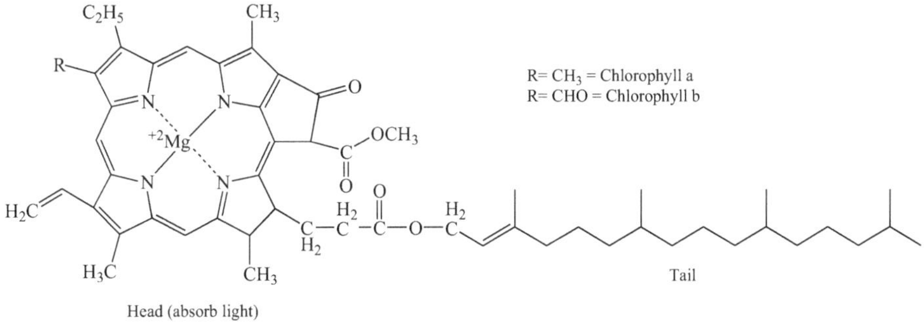

R= CH_3 = Chlorophyll a
R= CHO = Chlorophyll b

Head (absorb light)

Tail

BIO-MEMBRANES

A bio-membrane is a separating membrane and behaves as a selectively permeable barrier within living things. The bilayer of the membrane is formed due to the aggregation of the lipids in an aqueous solution. It is caused due to the hydrophobic effect *i.e.* the hydrophobic head comes into contact with each other away from water. This increases the hydrogen bonding between the hydrophilic heads and water.

Biomembranes consist of lipids, proteins and sugars (shown in Fig. **49**). The phospholipid bilayer is embedded with integral and peripheral proteins that are used in the communication and transportation of ions or chemicals [1].

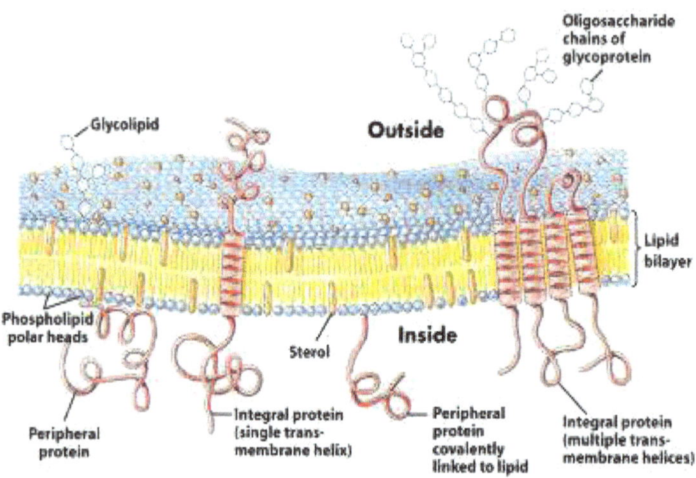

Fig. (49). Cell membrane (image courtesy: www.studyblue.com) [29].

Composition

The bilayer membrane consists of two layers- the outer leaflet and the inner leaflet. The bilayers are asymmetric to each other as the components are distributed unevenly between them. This asymmetry is important for cell functioning, like cell signaling. Due to asymmetry, the two layers have different functions.

The biomembranes are made up of lipids with hydrophilic (water-loving) heads and hydrophobic (water-hating) tails. The length and saturation of the hydrophobic tail are important to characterize the cell. The hydrophobic tail is made up of hydrocarbons *e.g.*, the bilayer of red blood cells or erythrocytes is composed of phospholipids and cholesterols. These also consist of phosphatidylserine, which plays an important role in blood clotting.

The bilayers contain proteins that have various functions and catalyze different chemical reactions. The integral proteins are strongly associated with the lipid bilayer which will detach only when the membrane breaks with some chemical treatment. But the peripheral proteins are weakly held to the surface of the bilayer and can be detached easily. Peripheral proteins are present on only one layer thus creating asymmetry in the membrane [1]. Examples of plasma membrane proteins with their functions are given below in Table **2.**

Table 2. Plasma membranes proteins with their functions

Functional Class	Protein	Specific Function
Enzymes	Adenylyl cyclase	In response to extracellular signals, it catalyzes the production of intracellular signaling molecule cyclic AMP
Receptors	Platelet-derived growth factor receptor	Binds extracellular PDGF and generates intracellular signals that cause the cell to grow and divide
Transporters	Na^+ pump	Pumps Na^+ out of cell and K^+ inside
Anchors	Integrins	The link between intracellular actin filaments and extracellular matrix protein

Different types of biomembranes have a different compositions of lipids and proteins. This composition of biomembranes defines their biological and physical properties.Some examples of different cell membranes and their compositions are shown below in Table **3.**

Table 3. Cell membranes and their compositions.

Membrane	Composition (% by weight)		
-	Lipid	Protein	Carbohydrate
Mitochondrial inner membrane	24	76	0
Human erythrocyte	43	49	8
Myelin	79	18	3
Amoeba	42	54	4
Mouse liver	52	44	4
Halobacterium purple membrane	25	75	0
Chloroplast spinach lamellae	30	70	0

Biomembranes contain glycolipids (sugar-containing lipid molecules). The sugar molecules of glycolipids can form hydrogen bonds as they are exposed to the surface of the membranes. They perform a large number of functions that include cell-cell adhesion and cell recognition. The sugar molecules are also bound to proteins and are known as glycoproteins which play an important role in the immune system.

Functions

• Selective Permeability:

It is the most important feature of a biomembrane. It is essential for the separation of a cell from its surroundings. Small hydrophobic molecules can easily cross the bilayer by diffusion. But some molecules cannot easily diffuse across the membrane. They are carried inside the cell by a membrane transport protein or through endocytosis [30].

• Fluidity:

The hydrophobic core of the bilayer remains in motion due to rotations around the bonds of the lipid tails. However, the hydrophilic head groups are in less motion as their rotation is constrained due to hydrogen bonds formed with water. As a result of this, there is an increase in viscosity towards hydrophilic head groups. The lipid bilayer loses fluidity and becomes a gel-like solid below a transition temperature. The transition temperature depends on the hydrocarbon chain length and saturation of the fatty acids. This helps the cold-blooded animals to maintain a constant fluidity by altering the fatty acid composition in accordance with varying temperatures.

Fluidity enables the membrane proteins to diffuse easily in the plane of the lipid bilayer and to interact with one another. It also allows the lipids and proteins to diffuse from their sites to other regions of the cell. It permits the membranes to fuse with each other and mix their molecules and ensure that during cell division the membrane molecules are evenly distributed between daughter cells. Without fluidity, it will be hard for a cell to live, grow, and reproduce [30].

KEYWORDS

Eukaryotic Cells

Cells consists of a nucleus and organelles and are enclosed by a plasma membrane.

Thylakoid Membrane

It is present in chloroplast; here the light reaction of photosynthesis takes place.

Integral Protein

A type of protein permanently attached to the biological membrane.

Peripheral Protein

A type of protein that is attached temporarily to the biological membrane with which it is associated.

Endocytosis

The transport of matter to a living cell through invagination of its membrane to form a vacuole.

SHORT-ANSWER TYPE QUESTION

1. Which carbon is anomeric? Distinguish between α- and β- type of anomers.
2. What do you mean by reducing and non-reducing sugars? Give examples.
3. Differentiate between enantiomers and diastereomers.
4. What is mutarotation?
5. Define proteins.
6. What is invert sugar?
7. What do you mean by Zwitter ion?
8. What is isoelectric pH?
9. Explain denaturation of protein.
10. What are simple, conjugate and derived proteins?
11. What do you mean by essential amino acid?
12. What is acid value? What is its significance?
13. How biomembrane is formed?

LONG-ANSWER TYPE QUESTIONS

1. Explain the classification of amino acids.
2. Explain the secondary structure of the protein.
3. Discuss the chemical properties shown by proteins and amino acids.
4. Distinguish between DNA and RNA.
5. Explain the structure of DNA.
6. Explain the different types of RNA.
7. What are steroids?
8. What are phospholipids? Explain the different types of phospholipids.
9. What are nucleic acids? Explain the different forms of structures of nucleic acids.

MULTIPLE CHOICE QUESTIONS

1. The metal cation (M+n) present in chlorophyll a is

a. Mn^{+2}

b. Fe^{+2}

c. Mg^{+2}

d. Co^{+2}

2. The amount of KOH in milligrams used to neutralize the free acid present in 1.0 g of oil, called

a. Iodine value

b. Saponification value

c. Acid value

d. pH value

3. The carbohydrate which can't be digested by the human digestive system is

a. Starch

b. Fructose

c. Glucose

d. Cellulose

4. Which of the following are non-essential amino acids?

a. Arginine

b. Alanine

c. Tryptophan

d. Threonine

5. The nature of glycosidic bond present in the structure of starch is

a. β-(1 → 4)-Glycosidic bond

b. α-(1 → 4)-Glycosidic bond and α-(1 → 6)-Glycosidic bond

c. α-(1 → 4)-Glycosidic bond

d. None of these

6. Which one of the following is not present in DNA?

a. Adenine

b. Uracil

c. Cytosine

d. Guanine

7. According to Chargaff's rule, the base pairs present in DNA is

a. A=T and G≡C

b. A≡T and G=C

c. A=U and G≡C

d. None

8. In sphingophospholipids the alcohol part instead of glycerol is

a. Sphingomyelins

b. Sphingosine

c. Phosphocholine

d. Choline

9. Adenosine consists of

a. Adenine and α-D-ribose

b. Adenine and 2-deoxy-ribose

c. Adenine and β-D-ribose

d. None of these

10. Which of the following is not present in nucleoside?

a. Nucleobase

b. Phosphate

c. Pentose sugar

d. None of these

11. The pH value at which the uncharged protein molecule does not migrate in an electric field is called

a. Optimum pH

b. Isoelectric pH

c. Minimum pH

d. Basic pH

12. Among the following, the protein which is a structural material is

a. Oxytocin

b. Albumin

c. Keratin

d. Haemoglobin

13. Biuret test is used to confirm the presence of

a. Amino acid

b. Aromatic amino acid

c. Peptide bond

d. Sulphur containing amino acid

14. In sucrose, the fructose unit consists of

a. A five-membered ring

b. An open chain structure

c. A six-membered ring

d. None of these

15. Which of the following is not a monosaccharide?

a. Ribose

b. Galactose

c. Glycogen

d. Fructose

16. Which of the following is a non-saponifiable lipid?

a. Testosterone

b. Triolein

c. Tristearin

d. Phosphatidic acid

17. Triglycerides are

a. Compound lipids

b. Phospholipids

c. Simple lipids

d. Derived lipids

18. The sugar molecule present in the structure of RNA is

a. α-D-ribose

b. β-D-deoxyribose

c. β-D-ribose

d. Glucose

19. Among the following, which is not optically active amino acid?

a. Glycine

b. Alanine

c. Lysine

d. Serine

20. Which of the following is not an aromatic amino acid?

a. Phenylalanine

b. Tryptophan

c. Tyrosine

d. Alanine

21. Aminoacid with a side chain containing sulphur atom is

a. Serine

b. Histidine

c. Cysteine

d. Lysine

22. The phenomenon of disorganization of native protein structure is called

a. Renaturation

b. Denaturation

c. Naturation

d. Rearrangement

23. Formation of Zwitter ion is characteristics of

a. Carbohydrate

b. Nucleic acid

c. Aminoacid

d. Lipid

24. The change in specific optical rotation of D-glucose in an aqueous solution with time to an equilibrium value is called

a. Optical rotation

b. Anomers

c. Mutarotation

d. None of these

25. Which of the following compound contains a fructose unit in its structure?

a. Raffinose

b. Maltose

c. Lactose

d. Starch

26. Among the following sugars, one having a reducing nature is

a. Sucrose

b. Fructose

c. Starch

d. Cellulose

27. Which of the following base is not present in RNA?

a. Adenine

b. Thymine

c. Guanine

d. Cytosine

28. A ketohexose contains the number of carbon atoms and functional group, respectively is

a. 5 carbon and aldehyde group

b. 6 carbon and aldehyde group

c. 6 carbon and ketone group

d. 5 carbon and ketone group

29. Which of the following is not a disaccharide?

a. Sucrose

b. Fructose

c. Maltose

d. Cellulose

30. The fluidity of biomembrane increases with an increase in

a. Phospholipid content in the membrane

b. Saturated fatty acid in the membrane

c. Glycolipid content in the membrane

d. Unsaturated fatty acids in the membrane

31. Glycolipids in biomembrane are usually located at

a. Inner leaflet of plasma membrane

b. Evenly distributed in both

c. The outer leaflet of plasma membrane

d. Cannot be predicted, it varies according to cell types

32. Which of the following statement about phospholipid molecules in the plasma membrane is wrong?

a. The non-polar tails face inwards

b. The polar heads are hydrophobic

c. The polar heads face outwards

d. The non-polar tails are hydrophobic

CONSENT FOR PUBLICATION

Not Applicable.

CONFLICT OF INTEREST

The authors confirm that this chapter contents have no conflict of interest.

ACKNOWLEDGEMENTS

Authors are thankful to Department of Chemistry, North Eastern Regional Institute of Science and Technology, Nirjuli, Itanagar for providing facilities to accomplish the work.

REFERENCES

[1] Nelson DL, Cox MM. Lehninger Principles of Biochemistry. 4th ed., New York: W. H. Freeman and company 2006.

[2] https://www.pexels.com/photo/basket-of-bread-1118332/

[3] https://www.pexels.com/photo/sliced-fruits-on-tray-1132047/

[4] https://www.pexels.com/photo/close-up-photo-of-rice-and-tacos-2087748/

[5] https://www.britannica.com/science/carbohydrate

[6] Finar IL. Organic Chemistry. 6th ed. Pearson Education 2009; p. 1.

[7] Finar IL. Organic Chemistry: Stereochemistry and the Chemistry of Natural Products. 5th ed. Pearson Education 2011; p. 2.

[8] https://www.pexels.com/photo/berry-close-up-cooking-delicious-141815/

[9] https://www.pexels.com/photo/clear-glass-jar-with-yellow-liquid-5634210

[10] https://www.pexels.com/photo/close-up-of-milk-against-blue-background-248412/

[11] https://www.pexels.com/photo/flour-in-a-jar-5765/

[12] Wade LG. Organic Chemistry. 8th ed.., Pearson Education 2013.

[13] Olujide OO, Maryam OO, James OO, Abidemi OO. Structural basis of antisickling effects of selected fda approved drugs: a drug repurposing study. Curr Computeraided Drug Des 2018; 14(2): 106-16. [http://dx.doi.org/10.2174/1573409914666180129163711]

[14] https://www.researchgate.net/figure/Illustrates-a-right-handed-alpha-helix-in-detail-lef--side-and-symbolic_fig5_328488525

[15] https://ib.bioninja.com.au/higher-level/topic-7-nucleic-acids/73-translation/protein-structure.html

[16] https://owlcation.com/stem/What-Are-Proteins-Part-3-of-3

[17] https://socratic.org/questions/5608054d581e2a2b1a6725c6

[18] https://ib.bioninja.com.au/standard-level/topic-2-molecular-biology/24-proteins/fibrous-vs-g-obular-protein.html

[19] Krieger M, Scott MP, Matsudaira PT, *et al.* 1: Structure of Nucleic Acids Molecular cell biology. New York: W.H. Freeman and CO 2004.

[20] Sedova A, Banavali NK. Geometric patterns for neighboring bases near the stacked state in nucleic acid strands. Biochemistry 2017; 56(10): 1426-43. [http://dx.doi.org/10.1021/acs.biochem.6b01101] [PMID: 28187685]

[21] https://www.wikiwand.com/en/Nucleic_acid_structure

[22] https://www.jing.fm/iclip/u2q8q8t4a9q8i1y3_dna-clipart-svg-dna-structure/

[23] https://www.chemguide.co.uk/organicprops/aminoacids/dna1.html

[24] Lee JC, Gutell RR. Diversity of base-pair conformations and their occurrence in rRNA structure and RNA structural motifs. J Mol Biol 2004; 344(5): 1225-49. [http://dx.doi.org/10.1016/j.jmb.2004.09.072] [PMID: 15561141]

[25] http://www.biologydiscussion.com/acids/nucleic-acid/trna-meaning-structure-and-initiator-transfer-nucleic-acids/35879

[26] Coleman RA, Lee DP. Enzymes of triacylglycerol synthesis and their regulation. Prog Lipid Res 2004; 43(2): 134-76.
[http://dx.doi.org/10.1016/S0163-7827(03)00051-1] [PMID: 14654091]

[27] https://chemistry.stackexchange.com/questions/60735/what-makes-trans-fats-more-harmfu--than-saturated-ones

[28] Rigaudy, J.; Klesney, S.P. Nomenclature of organic chemistry. "https://en.wikipedia.org/wiki/Pergamon"Pergamon 1979; "https://en.wikipedia.org/ wiki/ISBN_(identifier)" "https://en.wikipedia.org/wiki/Special:BookSources/978-0-08-022369-8"978-0-08-022369-8. OCLC 5008199

[29] https://www.studyblue.com/notes/note/n/biomembrane-structure/deck/11365594

[30] Watson H. Biological membranes. Essays in Biochemistry 2015. 59(0); 43–69.
[http://dx.doi.org/10.1042/bse0590043]

CHAPTER 6

Structure and Biological Function of Vitamins

Nagendra Nath Yadav[1], **Archana Pareek**[1,*] and **Sonam Tashi Khom**[1]

[1] *Department of Chemistry, North Eastern Regional Institute of Science And Technology (NERIST), Nirjuli, Itanagar-791109, Arunachal Pradesh, India*

Abstract: This chapter deals with the introduction, classification and biological functions of vitamins. The deficiencies caused by water-insoluble vitamins such as vitamin-A, vitamin-D, vitamin-E and vitamin-K and water-soluble vitamins such as vitamin-C and the vitamin-B complex have been discussed in detail. The structure and properties of various types of vitamins are also part of this subject.

Keywords: Ascorbic acid, Biotin, Niacin, Pyridoxine, Riboflavin, Tocopherols, Vitamin.

INTRODUCTION

Along with some biologically important compounds like proteins, carbohydrates and lipids, some organic chemical compounds are known as vitamins which are also required in small quantities for normal health and growth of animal life (including man). These organic compounds or vitamins cannot be synthesized by the body and hence must be obtained through diet or some other synthetic sources (shown in Fig. **1**). Due to this reason, vitamins are called essential nutrients. However, the word vitamin excludes dietary minerals, essential fatty acids and essential amino acids [1].

Vitamins are normally required in very small quantities by the body to complete its functions. These perform many biochemical functions as antioxidants, for *e.g.* vitamin E & C; as enzyme cofactors or their precursors, *e.g.* B-complex vitamins; hormone-like, *e.g.* vitamin D; regulators of mineral metabolism, regulators of cell and tissue growth and differentiation, *e.g.* some forms of vitamin A also facilitate or control vital chemical reactions in the body cells. The enzyme cofactors (as a catalyst) help enzymes in their metabolism. Here vitamins either may be tightly bound to enzymes as part of a prosthetic group like biotin, which is a part of

* **Corresponding author Meera Yadav:** Department of Chemistry, North Eastern Regional Institute of Science And Technology (NERIST), Nirjuli, Itanagar-791109, Arunachal Pradesh, India; E-mail: nny@nerist.ac.in

Meera Yadav & Prof. Hardeo Singh Yadav (Eds.)

enzymes that are involved in making fatty acids, or maybe less tightly bound to enzyme catalysts such as coenzymes that carry chemical groups or electrons between molecules, *e.g.* folic acid carries methyl, formyl and methylene groups in the cell [3]. These roles of vitamins in enzyme-substrate reactions are the best-known, although other functions are equally important. Lack of any vitamin in our body may develop a specific deficiency.

Fig. (1). Some vitamins-rich foods [2].

Till now, many vitamins have been isolated. Initially, they are designated by some selected letters of the alphabet (as shown in Fig. **2**), but as soon as their structure has been established, the vitamins are renamed. Each vitamin name (word vitamin followed by an alphabet) refers to a number of vitamers that show the same biological activity due to similarity in their structure [4]. For example, retinal, retinol and four known carotenoids qualify as "vitamin A", each with slightly different properties. Here vitamin A is the "generic descriptor" of the vitamin. Vitamers can be converted to the active form of the vitamin in the body and sometimes these are even inter-convertible to one another.

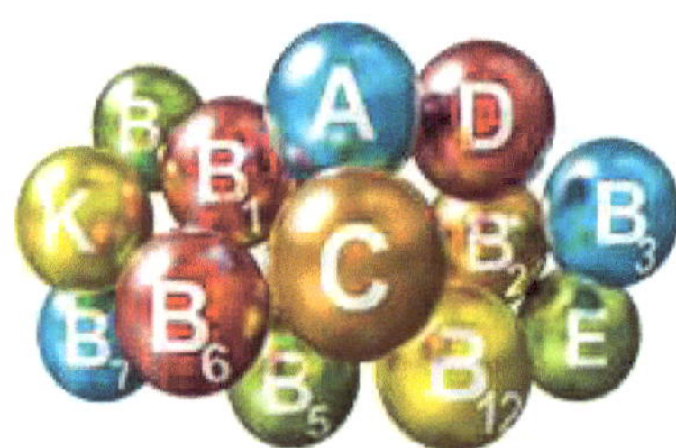

Fig. (2). Vitamins [5].

HISTORY

Long before the vitamins were identified, it was recognized that certain foods are necessary to maintain health. In ancient Egypt, night blindness (a disease caused by the deficiency of vitamin A) was cured by feeding animal (goat, pig, *etc.*) liver to a person [6]. In 1747, James Lind, the Scottish surgeon discovered that scurvy (a deadly disease in which collagen is not properly formed, causing poor wound healing, bleeding gums, severe pain and death) can be prevented by using citrus foods [6]. In 1753, Lind published his *Treatise on the Scurvy,* where the use of lemons and limes was recommended to avoid scurvy; however, it was not widely accepted [6].

During the late 18th and early 19th centuries, a number of vitamins were isolated and identified by scientists. Rickets in rats was cured by using lipid from fish oil and was called "antirachitic A". Thus, the first vitamin ever isolated that cured rickets was initially named "vitamin A"; however, now it is known as "vitamin D" [7]. In 1881, at the University of Tartu (now Estonia), Nikolai Lunin, a Russian surgeon, studied the effects of scurvy. He made an artificial mixture of all the constituents of milk (proteins, fats, carbohydrates and salts known at that time) and fed mice with this mixture while other mice were fed with milk. The mice fed with the mixture died, and the mice fed with milk developed normally. Therefore, he concluded that "milk must contain small quantities of unknown substances which are essential for life along with the known ingredients". However, his advisor Gustav von Bunge did not accept his conclusions. In 1905, a similar result was published in a Dutch medical journal by Cornelius Pekelharing but not widely reported [8]. In 1890, Christiaan Eijkman found that a nerve disease (polyneuritis) broke out among his laboratory chickens. The disease was similar to the polyneuritis associated with the nutritional disorder beriberi. In 1897, he discovered that when the chickens were fed with unpolished rice instead of polished rice, it helped to prevent beriberi in chickens [9]. In 1906-07, Frederick Gowland Hopkins, a British biochemist, observed that certain amino acids cannot be synthesized by the animals and concluded that in addition to proteins, carbohydrates, fats, *etc*., some foods contain 'accessory factors' that are necessary to support growth and proper functioning of the body [6]. In 1912, he published the same. In the same year, Casimir Funk, a polish scientist, observed that polyneuritis in pigeons could be cured by increasing their diet with a concentrate made from rice bran, present in the outer husk of the rice, which was removed during polishing. Thus he proposed that polyneuritis occurred due to lack of a vital factor found in rice bran in the bird's diet. Similarly, he believed that some human diseases like beriberi, scurvy and pellagra were also caused by deficiencies of some vital factors and could be cured if the same is supplied to them in their diet. Funk called these vital factors as "vital amines", later shortened to "vitamines", as each of the factors contained an amine [10]. Later it was found that all the vitamins do not contain nitrogen, hence all are not amines; therefore, the final 'e' was dropped from 'vitamine' to de-emphasize the amine reference [11].

CLASSIFICATION

According to the National Institute of Health, for a normal healthy body, 13 vitamins are required which include vitamins A, C, D, E, K and B complex. Vitamins have been classified on the basis of their solubility. Traditionally vitamins were classified into two groups, the water-soluble vitamins and the fat-soluble vitamins.

Water-soluble vitamins are soluble in water and, therefore, are easily excreted out from the body. They could not be stored in the body and hence need to be replenished every day through diet. Due to their solubility in water, one cannot be overdosed with these vitamins as the excess amount of these will simply get excreted [12]. Vitamin B-complexes and vitamin C fall under this category.

The vitamins that are soluble in fats and lipids are known as fat-soluble vitamins. They are absorbed into the body through the intestinal tract, specifically through the small intestine. In the body, fat-soluble vitamins are stored in the liver and adipose tissues. Vitamins A, D, E and K fall under this category. These vitamins (except K) are stored in the body for long periods of time; hence, overdosing on them could lead to hypervitaminosis [13].

VITAMIN A

Vitamin A is the name of a group of nutritional organic compounds which include retinol, retinal, retinoic acid, retinyl esters and several provitamin A carotenoids [14] (shown in Fig. **3**). Vitamin A is a powerful antioxidant. It is involved in growth and development, maintenance of the immune system, building strong bones, reproduction, cellular communication, neurological function, healthy skin and good vision [15]. It helps in reducing inflammation by fighting free radical damage. It is essential for the retina of the eye in the form of retinal that combines with the protein opsin forming rhodopsin, a light-absorbing protein in retinal receptors also necessary for color vision [16] (shown in Fig. **4**).

Fig. (3). Vitamin A rich foods [2, 5].

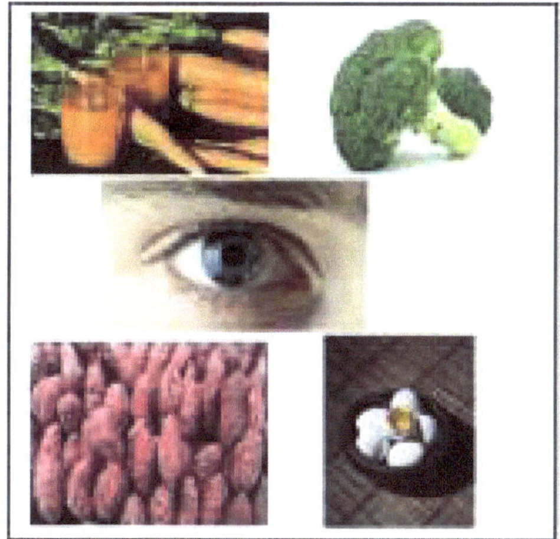

Fig. (4). Vitamin A is good for the retina [2].

Vitamin A is found in two forms in the human diet: preformed vitamin A (as an ester, mainly retinyl palmitate and retinol) and provitamin A (carotenoids). Preformed vitamin A is found in animal-derived foods like fish, meat and dairy products. It is the storage form of the vitamin and can be converted to and from the retinal when needed [17] (as shown in Fig. **5**).

Retinyl ester

RCOOH ⇅ H_2O

Retinol

NADH ⇅ NAD^+

Retinal

Fig. (5). Conversion of retinyl ester to retinal.

Provitamin A carotenoids may be alpha-carotene, beta-carotene, gamma-carotene and xanthophyll beta-cryptoxanthin, all of which contain beta-ionone rings (shown in Fig. **6**). These are obtained from colorful fruits and vegetables and are converted to retinol after the food is ingested by the body. All forms of vitamin A are stored in the liver in the form of retinyl esters.

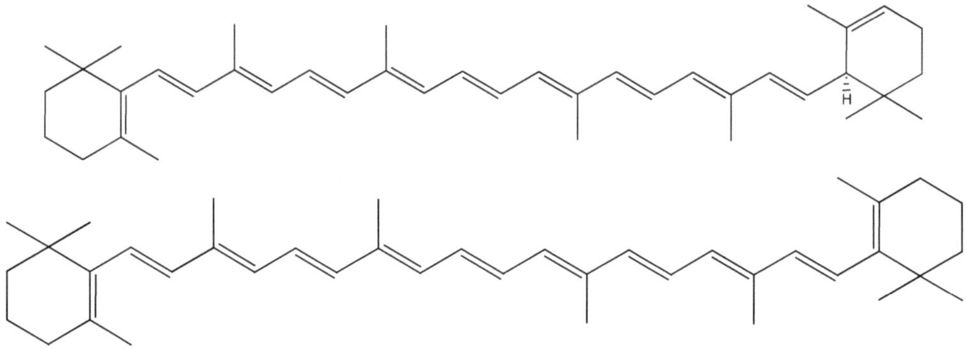

H

beta-carotene, with beta-ionone ring at both ends

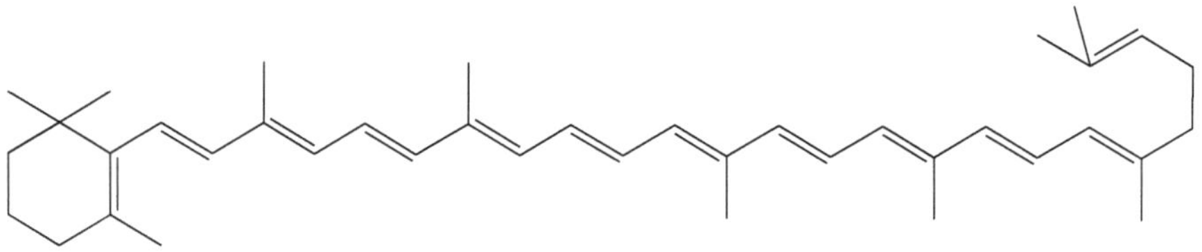

gamma-carotene

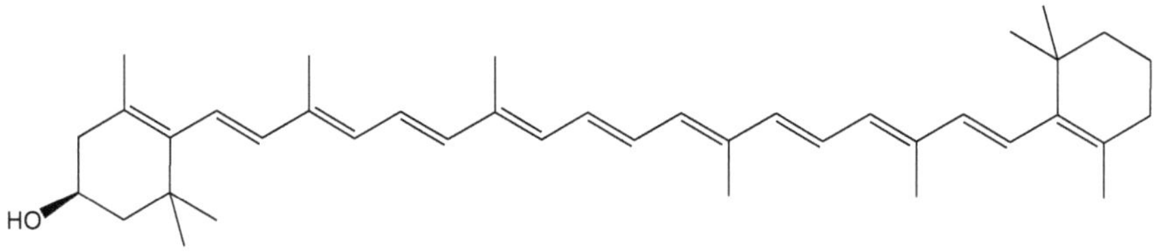

HO

beta-cryptoxanthin

Fig. (6). Structures of α-, β-, γ-carotene and β-cryptoxanthin, respectively.

There are two structural features in all the forms of vitamin A, *i.e.beta-ionone ring* to which an *isoprenoid chain (known as retinyl group)* is attached and both these features are essential for their activity [14]. All the other carotenes do not have vitamin activity as they lack these two structural features.

Sources of Vitamin A

Some best sources of vitamin A are cod liver oil (30000µg), beef liver (6500µg), carrots (835µg), sweet potato (961µg), kale (681µg), spinach (469µg), apricots (96µg), broccoli leaf (800µg), milk (28µg), butter (684µg), ghee (3069µg), egg (140µg), cantaloupe melon (169µg), mango (38µg), capsicum (2081µg), dandelion greens (508µg), pumpkin (426µg), collard greens (333µg), papaya (55µg), tomatoes (42µg), peas (38µg), broccoli florets (31µg) (the values given in the brackets are retinol activity equivalences per 100 grams of the food) [18].

Deficiencies of Vitamin A

Vitamin A deficiency is of two types: primary and secondary deficiency. Primary vitamin A deficiency occurs among individuals who do not consume a diet rich in provitamin A carotenoids, *i.e.* fruits and vegetables, or meat, fish and dairy products. Secondary vitamin A deficiency occurs when a person is subjected to long-term malabsorption of lipids, long-term exposure to oxidants like cigarette smoke, long-term alcoholism and impaired bile production and release [10].

The most common health problem due to deficiency of vitamin A is poor eye health (shown in Fig. 7). Retinal has a unique function as a visual chromophore; therefore, the earliest symptoms of vitamin A deficiency include impaired vision, particularly in reduced light, *i.e.* night blindness. If the deficiency continues to persist, then it will give rise to a series of changes, most of which will occur in the eyes. It can lead to thickening of cornea and even to keratomalacia (a condition affecting both eyes), including extreme dryness in the eyes. Eyesight will be followed by wrinkling, cloudiness, softening of corneas leading to infected corneas and a degenerative tissue change causing blindness [20].

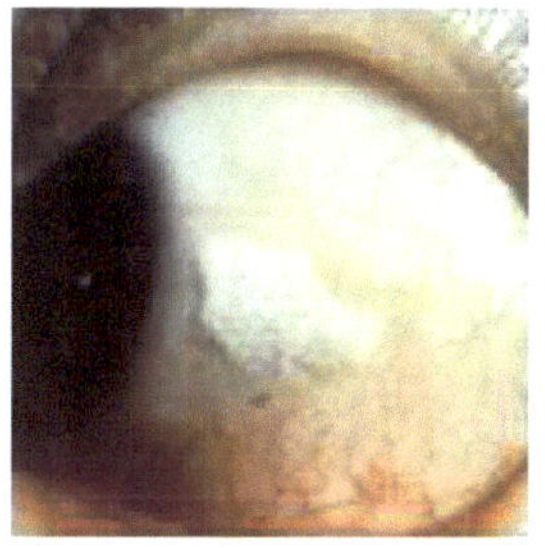
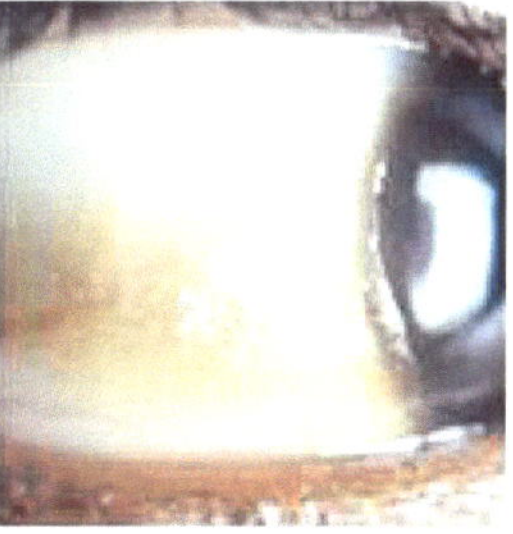

Fig. (7). Bitot's spots, caused by vitamin A deficiency [19].

Other than impaired vision, deficiency of vitamin A also causes premature skin damage, *i.e.* drying, scaling, follicular thickening of the skin and respiratory infections (shown in Fig. **8**).

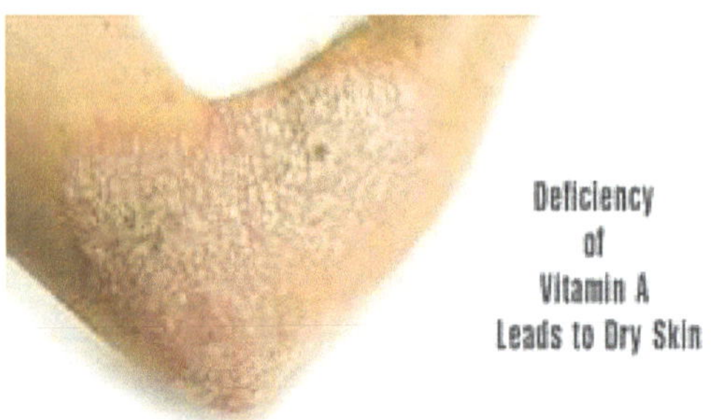

Fig. (8). Vitamin A deficiency [21].

Health Benefits of Vitamin A

- Proper vision: Vitamin A gets attached to opsin protein and forms a rhodopsin molecule that activates when light falls on the retina. This sends a signal to the brain, resulting in the enhancement of vision. Beta-carotene form of vitamin A plays an important role in preventing macular degeneration (the main cause of age-related blindness) [22].
- Better immune function: Vitamin A is also known as an important immune-boosting vitamin. Vitamin A regulates the genes involved in immune responses. It is not only essential for fighting serious diseases like cancer but also essential in curing the illnesses like common cold or flu [22].
- Fights inflammation: Sometimes, the immune system overreacts to food proteins, resulting in food allergies and eventually inflammation. Vitamin A intake can prevent the cells from this overreaction. Vitamin A also has antioxidants properties that neutralize the free radicals in the body, preventing tissue and cellular damage [22].
- Helps to prevent cancer: Retinoic acid helps in cell development and differentiation as well as in cancer treatment because it has the ability to control malignant cells in the body. Several forms of cancer like lung, prostate, breast, ovarian, bladder, oral and skin cancers have been found to be suppressed by retinoic acid [22].
- Vitamin A reduces the occurrence of acne. Acne leaves spots and blackheads on the area it occurs. These spots are physically harmless but may affect a person mentally, *i.e.* leading to depression, anxiety and low self-esteem [22].

Side Effects due to Excess of Vitamin A

Being fat-soluble, the body stores the excess amount of vitamin A mainly in the liver. The major side effects do not arise due to the excess amount of provitamin A carotenoids, but its chronic intake may lead to carotenodermia in which the skin becomes yellow-orange in color. It is a harmless condition and can be reversed by discontinuing its ingestion. However, ingestion of excess preformed vitamin A can cause significant toxicity (primarily liver toxicity) [23].

Excessive consumption of vitamin A can also lead to nausea, vomiting, irritability, anorexia, drowsiness, hair loss, headaches, pain in joints and bones, blurry vision, weakness and altered mental status [22]. All these symptoms will resolve if the excessive intake of vitamin A is reduced.

VITAMIN D

Vitamin D is a fat-soluble vitamin that includes vitamin D_1 (a mixture of ergocalciferol and lumisterol, 1:1), D_2 (ergocalciferol), D_3 (cholecalciferol), D_4 (22-dihydroergocalciferol) and D_5 (sitocalciferol), out of which D_2 and D_3 are the most important for humans (Fig. **9**). Vitamin D is also sometimes known as 'sunshine vitamin' as it can be synthesized in our body when ultraviolet rays from the sun strike the skin [24]. It is estimated that 5-10 minutes of sun exposure on bare skin 2-3 times a day is sufficient for most people to attain vitamin D. It is also naturally available but in very few foods.

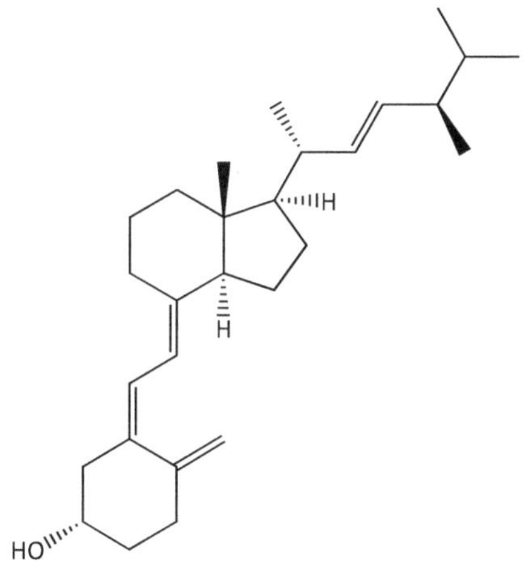

Vitamin D_2, Ergocalciferol

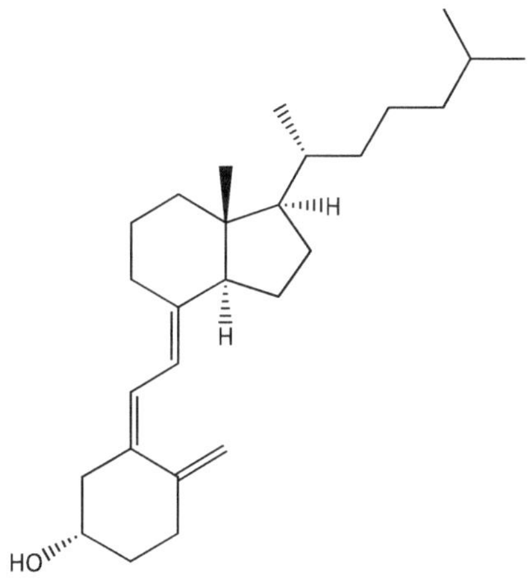

Vitamin D_3, Cholecalciferol

Fig. (9). Structure of Vit. D_2 and Vit. D_3.

Vitamin D obtained from sun exposure, diet or supplements is biologically inactive. For its activation, it must undergo hydroxylation twice in the body. The first hydroxylation occurs in the liver that converts vitamin D to calcidiol (25-hydroxyvitamin D) and the second hydroxylation occurs in the kidney converting it to calcitriol (1,25-dihydroxy vitamin D), also written as 25(OH)D and 1,25$(OH)_2$D respectively [25]. The mechanism is shown below (Fig. **10**).

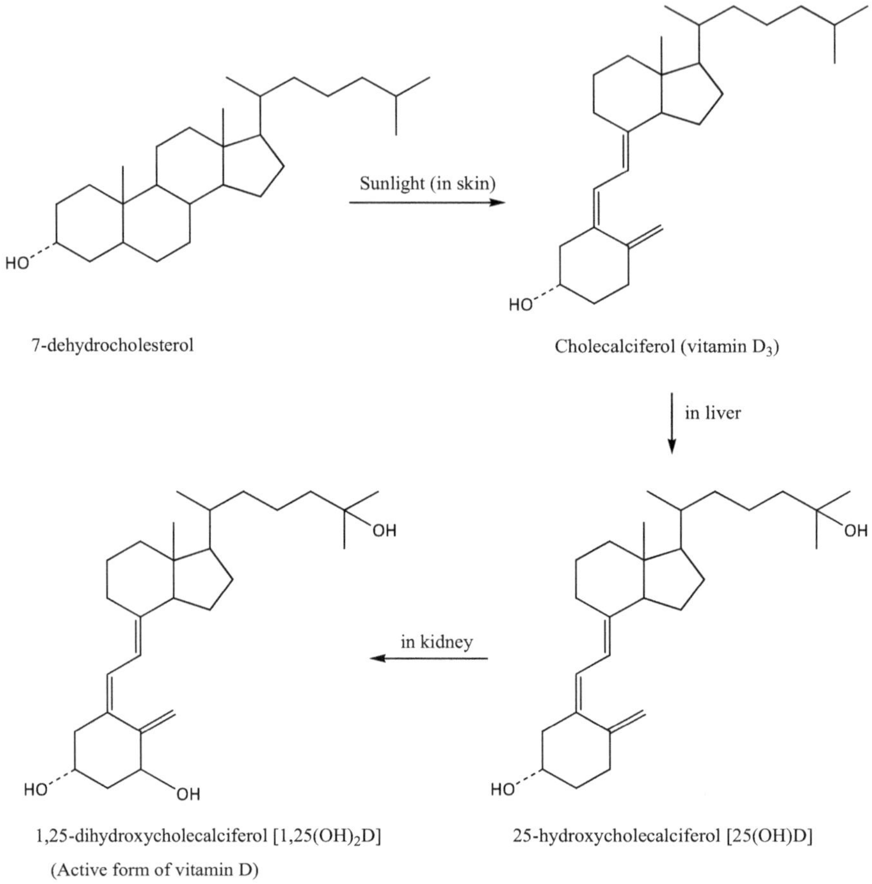

Fig. (10). Mechanism of action of vitamin D.

As we know vitamins are nutrients that cannot be synthesized by the body hence must be supplied through diet but vitamin D can be synthesized in the presence of sunlight, therefore, it is considered a pro-hormone and not a vitamin [24]. The most important function of vitamin D is to regulate the absorption of calcium and phosphorus and to facilitate normal immune function. It also helps in the modulation of cell growth and reduction of inflammation.

Sources of Vitamin D

Sunlight is the most efficient source of vitamin D, other than this there are very few foods that contain vitamin D. These are fish liver oil, fatty fish like salmon, tuna, mackerel, beef liver, cheese, egg yolk, orange juice, yogurt, milk, mushrooms and plant milk (soy milk, almond milk, rice milk, coconut milk) [26].

Deficiencies of Vitamin D

Vitamin D deficiency can occur due to various reasons like lower intake than the recommended level, limited exposure to sunlight, inability to convert vitamin D to its active form by the body, or inadequate absorption of vitamin D from the intestine. The deficiency symptoms include tiredness, aches, severe bones and muscle pain, weakness, stress fractures, a sense of not feeling well [27].

Vitamin D deficiency causes rickets in children, causing softening of bones and skeletal deformities (shown in Fig. (**11**). Children with this deficiency generally have bow legs (shown in Fig. **12**). The same condition in adults is known as osteomalacia that leads to a bending of the spine, bowing of legs and increased risk of fractures [27].

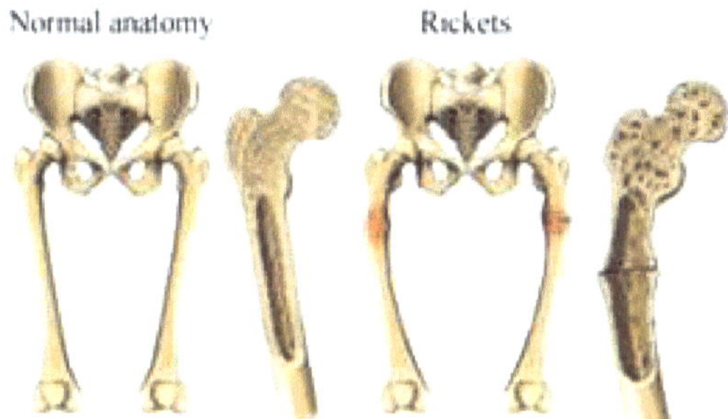

Fig. (11). Vitamin D deficiency [28].

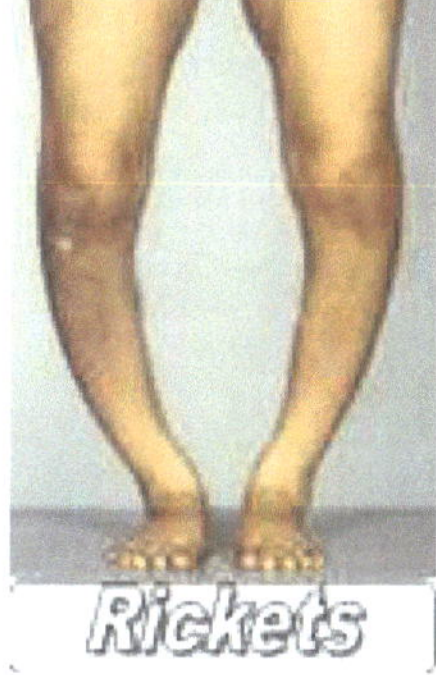

Fig. (12). Legs with rickets [29].

Health Benefits of Vitamin D

- Vitamin D plays a role in reducing multiple sclerosis, preventing flu, reducing depression, decreasing the chance of developing heart diseases, prevention of colon, prostate and breast cancer [30].
- It helps to improve bone health, boost weight loss, and decrease the risk of respiratory tract infections [30].
- It supports the health of the brain, nervous, immune system and also regulates the insulin levels in the body [30].

Side Effects due to Excess of Vitamin D

Toxicity from vitamin D is rare. It does not result from excessive exposure to the sun as the heat on the skin photodegrade previtamin D_3 and vitamin D_3, also thermal activation of previtamin D_3 causes to form various non-vitamin D forms, that suppress the formation of vitamin D_3 itself. Toxicity however occurs from a high intake of dietary supplements that consist of vitamin D [27].

Overdose of vitamin D can cause hypercalcemia resulting in excess deposits of calcium in the kidney, liver and heart causing subsequent organ damage. The common symptoms of vitamin D overdose are headache, nausea, vomiting, dry mouth, a metallic taste, a loss of appetite, weight loss, mental retardation in young children, abnormal bone growth, diarrhea and severe depression. The toxicity can be treated by discontinuing vitamin D supplementation and calcium intake [27].

VITAMIN E

Vitamin E is a group of fat-soluble vitamins that includes both tocopherols and tocotrienols. Naturally, it exists in eight different forms α-, β-, γ- and δ-tocopherol and α-, β-, γ- and δ- tocotrienol. Out of this, γ-tocopherol is the most common form found in the American diet while α-tocopherol is the most biologically active and the second most common form of vitamin E *i.e.* α-tocopherol (structure shown in Fig. **13**) is the only form that meets human requirements [31]. It is a fat-soluble antioxidant that interrupts the propagation of reactive oxygen species (Vit. E capsules are shown in Fig. **14**). It was first discovered in 1922 then isolated in 1935, and for the first time, it was synthesized in 1938.

HO

O

alpha-tocopherol form of vitamin E

Fig. (13). Structure of α-tocopherol.

Fig. (14). Vitamin E capsules.

Sources of Vitamin E

Vitamin E is mainly found in plant foods (Fig. **15**). Some sources of vitamin E with the amount per 100 grams of the food is given below:

Wheat germ oil (150mg), hazelnut oil (47mg), rapeseed oil (44mg), sunflower oil (41.1mg), safflower oil (34.1mg), almond oil (39.2mg), grapeseed oil (28.8mg), sunflower seed kernels (26.1mg), almonds (25.6mg), almond butter (24.2mg), wheat germ (19mg), canola oil (17.5mg), palm oil (15.9mg), peanut oil (15.7mg), hazelnuts (15.3mg), corn oil (14.8mg), olive oil (14.3mg), soybean oil (12.1mg), pine nuts (9.3mg), peanut butter (9mg), peanuts (8.3mg), popcorn (5mg) and many others [26].

Fig. (15). Some food rich in vitamin E [32].

Deficiencies of Vitamin E

Vitamin E deficiency is found to be very rare in human beings. It is believed that this deficiency does not occur from a diet low in vitamin E. People with fat-malabsorption disorders are somehow likely to become vitamin E deficient. A premature infant can be at risk of vitamin E deficiency; however, it can be treated by a pediatrician [33].

The deficiency symptoms are peripheral neuropathy, ataxia, skeletal myopathy, retinopathy, impairment of the immune responses and red blood cell destruction [33].

Health Benefits of Vitamin E

- Being an antioxidant, vitamin E can disable the production of damaging free radicals in tissues. It reacts with the free radical to form a tocopheryl free radical which is then reduced to its reduced state by a hydrogen donor like vitamin C [34].
- It inhibits oxidation of biomaterials and medical devices like ultra-high molecular weight polythene that is used in hip and knee implants and synthetic membranes in hemodialysis [34].
- It inhibits smooth muscle growth by deactivating the protein kinase C.
- It inhibits platelet coagulation and plays a role in eye and neurological functions.
- It prevents the oxidation of polyunsaturated fatty acids.
- It plays a crucial role in balancing the endocrine and nervous systems.
- Vitamin E can prevent or delay coronary heart disease.
- It has an anti-inflammatory activity which may help in protection against Alzheimer's disease [34].

Side Effects due to Excess of Vitamin E

High doses of vitamin E could lead to heart failure in people with diabetes. Being anticoagulant, if taken in excess, it can worsen bleeding disorders and can increase bleeding during and after surgery. It may increase the chances of head, neck and prostate cancer reoccurring. Sometimes it may cause diarrhea, blurred vision, nausea rash, bruising and stomach cramps [33].

VITAMIN K

Vitamin K is a group of fat-soluble vitamins that plays a crucial role in blood clotting and bone metabolism (binding with calcium in bones and other tissues). Vitamin K is required by the body to produce prothrombin, a protein and clotting factor, without which calcium ions cannot bind to the bones [35]. The family of vitamin K comprises of 2-methyl-1,4-naphthoquinone-3-derivatives. It consists of two vitamers: vitamin K_1 and vitamin K_2. Vitamin K_2 also consists of a number of subtypes ($K_{2(30)}$, $K_{2(35)}$, $K_{2(45)}$, *etc.*) with the difference in the length of carbon side chains (shown in Fig. **16**). Vitamin K_1 or phylloquinone is found in plants particularly in green leafy vegetables and is the main type of dietary vitamin K whereas Vitamin K_2 or menaquinone is found in some animal-based and fermented food (fermented by bacteria) and is the important form of vitamin K for people. Even the ingested vitamin K_1 is converted to vitamin K_2 by bacteria present in the large intestine which is then absorbed in the small intestine and stored in the fatty tissues and liver [36].

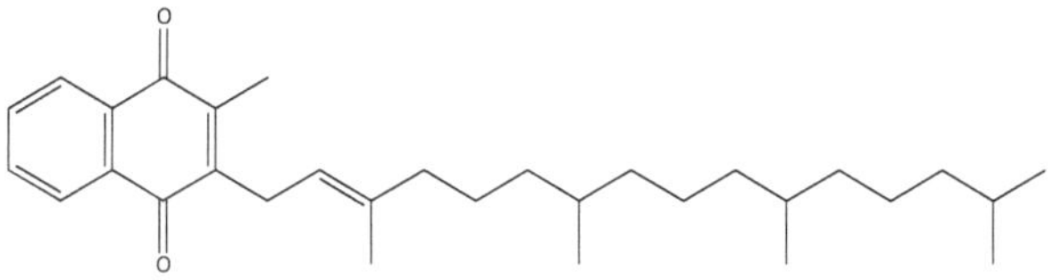

Vitamin K_1

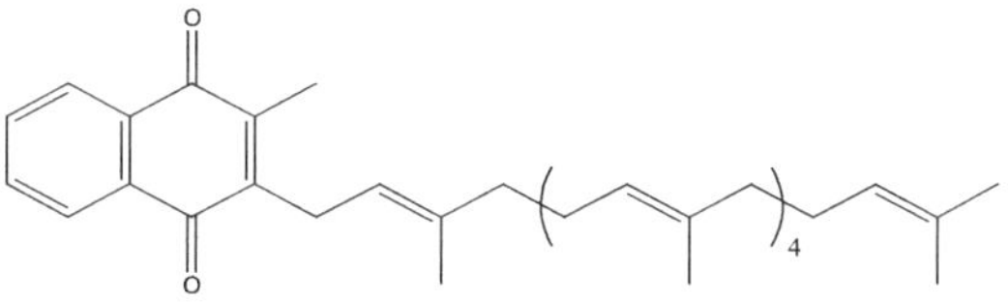

Vitamin $K_{2(30)}$

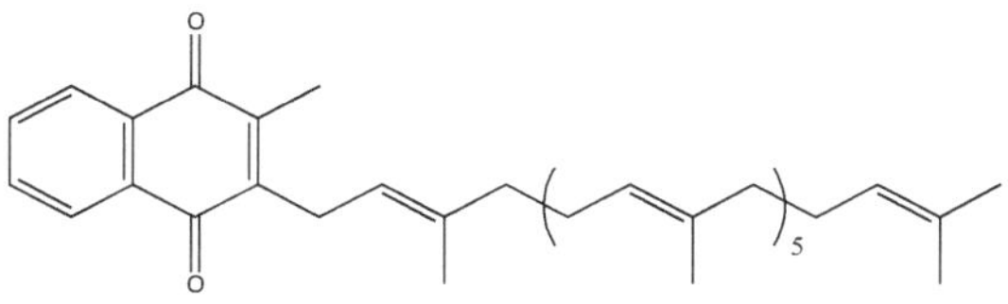

Vitamin $K_{2(35)}$

Fig. (16). Structures of different forms of vitamin K.

Sources of Vitamin K

Vitamin K_1 is chiefly available in green leafy vegetables like kale, spinach, collards, swiss chard, mustard greens, turnip greens, broccoli, brussels sprouts, cabbage, asparagus, *etc.* (shown in Fig. **17**). The absorption of vitamin K_1 is greater if these are consumed along with fats like butter or oil. Vitamin K_2 is available in some animal-based food such as Nattō, goose liver pâté, Australian emu oil, hard and soft cheese, egg yolk, goose leg, cheddar cheese, chicken, *etc.* [37].

Fig. (17). Sources of vitamin K [38].

Deficiencies of Vitamin K

Vitamin K deficiency is very rare as an average diet does not lack this vitamin; also vitamin K_2 is synthesized in the gut itself by the bacteria. However, deficiency can occur to those who suffer from liver damage or disease, cystic fibrosis or inflammatory bowel diseases, or in people with bulimia or those taking anticoagulants. The symptoms for vitamin K deficiency are bleeding nose (shown in Fig. **18**) and bleeding of gums, anemia, bruising and heavy menstrual bleeding [40].

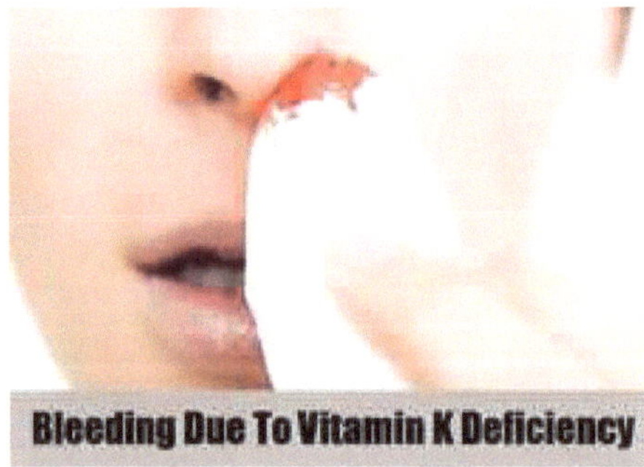

Fig. (18). Deficiency of vitamin K [39].

Health Benefits of Vitamin K

- Vitamin K may decrease the risk of fractures as it helps in maintaining strong bones and improves bone density [41].
- It prevents calcification of arteries, thus helps to prevent heart attack [41].
- It helps to reduce menstrual cramps by regulating the function of hormones.
- It may slow tumor growth.
- Its plays an important role in healing bruise and cuts fast.
- It also helps in maintaining healthy gums and teeth [41].

Side Effects due to Excess of Vitamin K

Toxicity is very rare with high doses of vitamin K_1 and K_2 and therefore no tolerable upper limit is determined. However, a diabetic person needs to monitor blood sugar levels closely, and in a kidney patient who is receiving dialysis, excess of vitamin K is harmful [42]. Vitamin K_1 and K_2 are safe and natural but a synthetic form of vitamin K *i.e.* vitamin K_3 is available and is highly toxic at high levels [35].

VITAMIN C

Vitamin C, another name is L-ascorbic acid, (shown in Fig. **19**) is a water-soluble vitamin that is naturally found in some food and also available as a dietary supplement. It is one of the safest and most important essential nutrients and is involved in the repairing of tissues. It acts as an antioxidant and helps in immune system function. It helps in the enzymatic production of certain neurotransmitters [42].

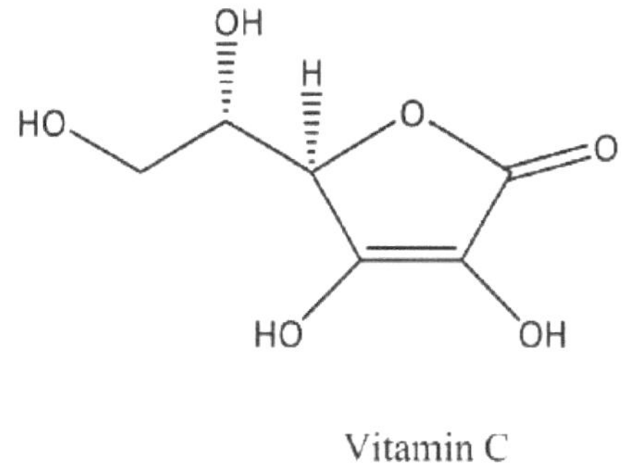

Vitamin C

Fig. (19). Structure of L-ascorbic acid.

Sources of Vitamin C

Plants foods are a very good source of vitamin C (shown in Fig. **20**). Some of them are: kadu plum (1000-5300mg); acerola (1677mg); seabuckthom (695mg);

Indian gooseberry (445mg); rose hip (426mg); guava (228mg); blackcurrant (200mg); capsicum (128mg); kale (120mg); kiwifruit and broccoli (90mg); papaya, strawberry, orange, lemon, pineapple, cauliflower (40-60mg); grape, raspberry, spinach, mango, lime, potato (20-30mg); *etc*. [values given in the brackets indicates the amount of vitamin C in mg per 100 gram of the food] [26].

Fig. (20). Some foods rich in vitamin C [43].

Animal foods do not consist of much vitamin C and even if there is some then the heat involved in the cooking process destroys it.

Deficiencies of Vitamin C

Vitamin C deficiency leads to scurvy because in the absence of vitamin C, collagen becomes unstable to perform its function. The symptoms of scurvy appear within one month of low vitamin C intake, *i.e.* below 10 mg per day. Initial symptoms include fatigue, inflammation of gums (as shown in Fig. **21**), brown spots on the skin mostly on thighs and legs, bleeding from mucous membranes. If the deficiency continues, then it leads to impaired collagen synthesis and weak connective tissues causing joint pain, poor wound healing, loss of teeth, hyperkeratosis and corkscrew hair [44].

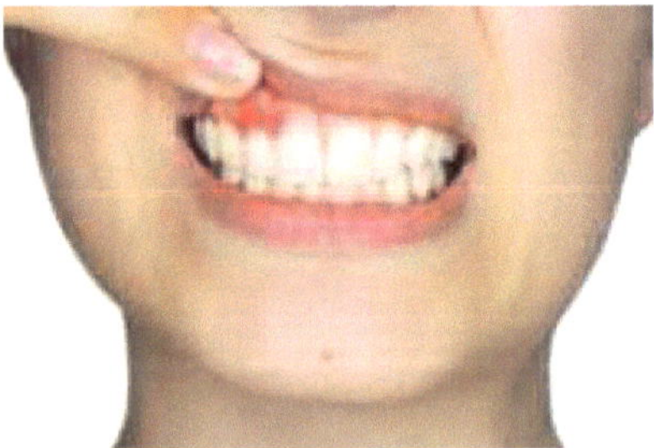

Fig. (21). Symptoms of deficiency of vitamin C [44].

Health Benefits of Vitamin C

- As deficiency of vitamin C causes scurvy, it can be treated by food containing vitamin C or dietary supplements [45].
- Vitamin C may not be the cure for the common cold, but it may shorten its duration [45].
- It has been found that vitamin C can limit the formation of carcinogens and regulate the immune responses and also being an antioxidant, it can reduce the oxidative damage that can lead to cancer [45].
- It reduces the risk of cardiovascular disease, eye disease and skin wrinkling [45].

Side Effects due to Excess of Vitamin C

Vitamin C is water-soluble therefore the body cannot store it, if there is any extra intake by the body then it gets excreted in the urine. Hence it is necessary to consume vitamin C-rich foods daily. However, if daily intake is more than two to three grams then it may lead to adverse effects. If taken with an empty stomach it may cause indigestion. Symptoms include nausea, abdominal cramps and diarrhea. It may increase the risk of kidney stones in individuals with renal disease [46].

Vitamin C can bind with non-heme iron, found in plant food (which is not efficiently absorbed by our body), and makes it much easier to be absorbed by the body. This does not affect a healthy individual but to individuals with hereditary hemochromatosis, it may cause iron overload resulting in tissue damage, causing serious damage to the heart, liver, pancreas, thyroid and central nervous system [46].

VITAMIN B-COMPLEX

The B-vitamin group is a group of water-soluble vitamins that are essential for various metabolic processes. Though they have some general similarities, each of them has unique and important functions in the body. Being water-soluble they can't be stored in the body and hence should be supplied regularly through diet. They are easily destroyed by extended cooking, food processing and alcohol. There are eight types of vitamin B; each of them is a cofactor for enzymatic reactions or a precursor for it. Lack of B-vitamin reduces the energy of the body as they help the body to use the energy gained from energy-yielding nutrients.

The different types of B-vitamin are B_1, B_2, B_3, B_5, B_6, B_7, B_9 and B_{12}.

Vitamin B_1 or Thiamine

Fig. (22). Structure of thiamine.

VITAMIN B_1

It is also known as thiamine or thiamin (shown in Fig. **22**). It is the first B-vitamin discovered, hence the name 'vitamin B_1'. It is a coenzyme that helps in food metabolism to produce energy in the form of ATP and also maintains the heart and nerve functions. It regulates important functions of the cardiovascular, digestive and endocrine systems. It is found in many foods and also can be taken as a dietary supplement. It is used to treat disorders like beriberi, korsakoff's syndrome, korsakoff's psychosis, maple syrup urine disease and Leigh's disease [47].

Sources of Vitamin B_1

Thiamine is available in many foods like beef, liver, dried milk, oats, oranges, eggs, pork, peas, nuts, dried beans, soybeans, whole grain cereals, lentils, legumes, bread, rice, yeast and pasta (shown in Fig. **23**) [26].

Fig. (23). Some sources of vitamin B1 [48].

Deficiencies of Vitamin B_1

Vitamin B_1 deficiency is rare, only in developed countries; it generally occurs in those countries whose staple food is white rice. Alcoholic people or those who are undergoing dialysis or taking loop diuretics are also at risk of thiamine deficiency.

Thiamine deficiency can cause two major diseases: Beriberi and Wernicke-Korsakoff; the former leading to abnormal nerve function, heart failure and swelling of legs while the later leads to memory loss, confusion and difficulty in balance (as shown in Fig. **24**). These diseases can be treated with thiamine injections or supplements [47].

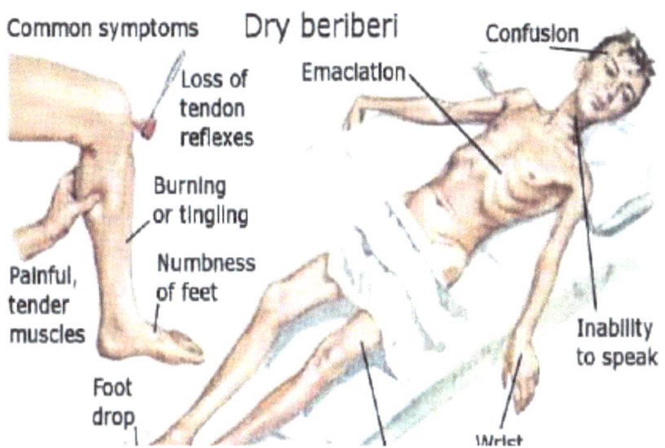

Fig. (24). Some symptoms of beriberi [49].

Some symptoms of thiamine deficiency are rapid weight loss, poor appetite, ongoing digestive problems, nerve damage, neuritis, fatigue, confusion, irritability, muscle weakness, depression, cardiovascular effects like enlarged heart and memory loss [47].

Health Benefits of Vitamin B_1

- It helps to maintain a healthy metabolism as it is used in the conversion of carbohydrates into glucose, producing energy in the form of ATP [47].
- It helps to prevent nerve damage, maintains a proper cardiovascular system and boosts immunity.
- It helps to prevent memory loss, enhance learning abilities and defends against depression [47].

Side Effects Due to Excess Vitamin B_1

Being water-soluble in nature overdosing of thiamine is extremely rare. However, sometimes it may cause allergies like difficulty in breathing, swelling of the face, lips, tongue, or throat. Symptoms like nausea, sweating, feeling warm, rash, itching and restlessness may occur. Severe side effects include chest pain, coughing of blood, blue-colored lips, black or bloody stools, *etc*. [47].

Vitamin B_2 or Riboflavin

Fig. (25). Structure of riboflavin.

VITAMIN B_2

Vitamin B_2, also known as riboflavin (shown in Fig. **25**), is a very important vitamin required by the body for cellular respiration, maintaining healthy blood cells and facilitating healthy metabolism. It prevents free radical damage thus acts as an antioxidant, helps to prevent migraines and protects skin and eye health. It acts as a coenzyme, helping the enzymes to perform normal physiological functions [50].

Sources of Vitamin B_2

Food sources that contain riboflavin are milk, cheese, eggs, leafy vegetables, mushrooms, certain nuts like almonds, beans, legumes, fish, meat, organ meat, *etc*. [26]. (Shown in Fig. **26**).

Fig. (26). Some food sources of riboflavin [51]. (image courtesy: pinterest.com).

Deficiencies of Vitamin B_2

Riboflavin deficiency is very rare as it is present in wheat flour, bread, pasta, cornmeal, or rice, which is the staple food of many countries.

But, if a person is on a poor diet then vitamin B_2 deficiency will occur as it is a water-soluble vitamin and hence cannot be stored in the body. A person with riboflavin deficiency lacks other vitamins too.

The symptoms of this deficiency are cracked lips, cracks at the corner of the mouth, dry skin, mouth ulcers, inflammation of the tongue, the lining of the mouth, red lips, sore throat, red and itchy eyes, anemia, and scrotal dermatitis [50].

Health Benefits of Vitamin B_2

- Helps to treat migraine headaches, and reduces the number of episode of migraines [50].
- It helps to improve eye health, prevents eye disorders like cataracts, keratoconus and glaucoma [50].
- It helps to treat anemia and may prevent cancer.
- Boosts energy, act as antioxidants and reduces depression.
- It maintains collagen levels thus making skin and hair healthy [50].

Side Effects due to Excess of Vitamin B_2

Overdose of vitamin B_2 is unlikely to occur, however, in some cases, it may cause diarrhea and increased urination (color of the urine changes to yellow-orange) [50].

VITAMIN B_3

Vitamin B_3, also known as niacin, comes in three different forms: Nicotinic acid, Niacinamide and Inositol Hexaniacinate (shown in Fig. **27**). The body breaks those forms and produces NAD and NADP, which are important for a variety of chemical reactions and plays a role in cell metabolism. It supports the functions of more than 200 enzymes in the body and also plays an important role in maintaining skin, digestive and mental health [52].

i) Nicotinic acid ii) Niacinamide iii) Inositol Hexaniacinate

Fig. (27). Structures of nicotinic acid, niacinamide and inositol hexaniacinate.

Fig. (28). Group of foods rich in Vitamin B_3 [53].

Sources of Vitamin B_3

Food sources of niacin are beef liver, chicken breast, tuna, salmon, lamb, sunflower seeds, peas, mushrooms, baked potato, sesame seeds, apricots, oats, soya milk, cheese, pumpkin seeds, yeast, eggs, nuts and green vegetables [26]. (Shown in Fig. **28**).

Deficiencies of Vitamin B_3

Niacin deficiency may occur if a person does not get enough vitamin B_3 through its diet. It generally occurs in the regions where the major part of the diet is maize because the body cannot absorb the vitamin unless being treated with alkali. It also occurs if a person suffers from diarrhea, hartnup disease, liver disease, or alcoholism [52].

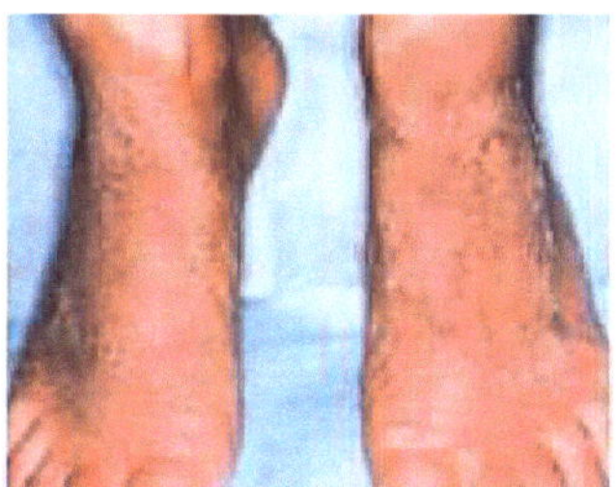

Fig. (29). Symptoms of vitamin B3 deficiency [54]. (image courtesy: healthline.com).

Mild deficiency of niacin may cause the following symptoms to occur: irritated or red skin, fatigue, headaches, dizziness, poor circulation, anxiety or depression, digestive problems and unable to concentrate. However severe deficiency can cause a condition known as pellagra (shown in Fig. **29**), which affects the skin, nervous system, digestive system, and mucous membranes [52].

Health Benefits of Vitamin B_3

- It produces certain hormones in the adrenal glands and also helps to remove harmful chemicals from the liver [52].
- It helps to reduce diarrhea-related to cholera and also treats the headache caused by migraine, dizziness and circulation problems.
- It helps to lower the LDL (bad) cholesterol and triglycerides in the body while raising the HDL (good) cholesterol by more than 30 percent [52].
- It lowers cardiovascular disease risk and maintains proper brain function.
- Treats pellagra, arthritis, diabetes and maintains skin health [52].

Side Effects Due to Excess of Vitamin B_3

The most common side effects of niacin are flushing (a feeling of warmth, redness, itching or tingling under the skin), usually harmless and subside within 1-2 hours. Other side effects include low blood pressure, irregular heartbeat, worsening of liver disease, gallbladder problems, stomach upset, intestinal gas, headache, nausea, vomiting, diarrhea and rash [52].

Vitamin B_5 or Pantothenic acid

Fig. (30). Structure of vitamin B5.

VITAMIN B_5

Also known as pantothenic acid (Shown in Fig. **30**), like other B-vitamins, it also plays a role in breaking down of fats, carbohydrates and proteins, producing energy for the cells. It helps in the production of red blood cells, steroids, neurotransmitters and stress-related hormones, thus maintaining the health of the nervous system, circulation and cholesterol levels. It also helps to maintain a healthy digestive tract. Vitamin B_5 is used to synthesize coenzyme-A that acts as an acyl group carrier, thus transporting carbon atoms within the cells [52].

Sources of Vitamin B_5

The name ‘Pantothenic acid’ originates from the Greek word 'pantos' which means 'everywhere, which means this vitamin is found in every food even in small quantities. The major sources are beef liver, avocado, sunflower seeds, duck, eggs, salmon, yogurt, broccoli and dried mushrooms (shown in Fig. **31**). This vitamin is sensitive to vinegar, baking soda and other forms of alkali and is lost from the food during the process of cooking, canning, roasting and milling [26].

Fig. (31). Food rich in vitamin B5 [55]. (image courtesy: healthline.com).

Deficiencies of Vitamin B_5

Since Pantothenic acid is available in almost every food, therefore, its deficiency is very uncommon and rare, a healthy and varied diet is sufficient. If somehow there is a deficiency then its symptoms are similar to other B-vitamin deficiencies, *i.e.* fatigue, insomnia, depression, vomiting, stomach pain, burning feet, respiratory infections, muscle cramps, *etc.* [52].

Health Benefits of Vitamin B_5

- Pantothenic acid is used by the body to synthesize and regulate cholesterol, maintaining its levels in the arteries and preventing dangerous plaque build-up that can lead to heart attack or stroke.
- Pantothenic acid helps vitamin B_2 to prevent and treat anemia.
- It converts carbohydrates to glucose, thus produces energy.
- It helps to create acetylcholine that allows the nervous system to communicate within the organs of the body, thus maintaining healthy nerve function.
- It helps in wound healing, makes skin healthy, improves mental performance and aids in immune function [52].

Side Effects Due to Excess of Vitamin B_5

Since this vitamin is water-soluble hence the excess of it simply gets excreted from the body through urine. Therefore, overdose is unlikely. Even no tolerable upper-level intake is established. However, in case if massive doses (10 gram per day) are ingested then mild diarrhea may occur [52].

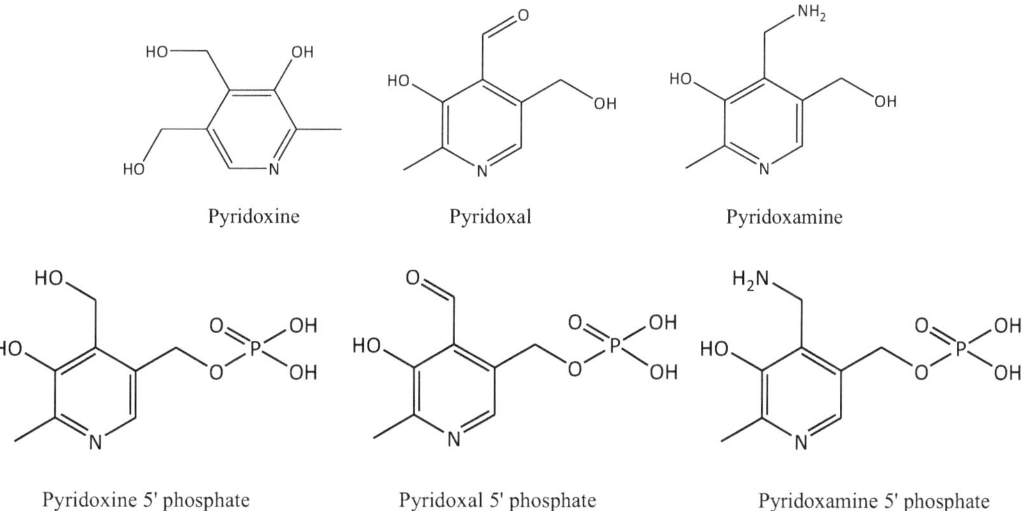

Fig. (32). Vitamers B6.

VITAMIN B_6

Vitamin B_6, also known as pyridoxine, is naturally available in many foods. It refers to a group of six vitamers (as shown in Fig. **32**) with similar activity which can be inter-converted to each other in the biological system. These six vitamers are pyridoxine, pyridoxal, pyridoxamine and their respective 5'-phosphate esters. Among these pyridoxal 5, phosphate (PLP) and pyridoxamine 5' phosphate (PMP) are the active forms of vitamin B_6. They act as a coenzyme in several enzyme reactions like amino acid, glucose and lipid metabolism [52].

Fig. (33). Foods rich in vitamin B6 [56]. (image courtesy: livescience.com).

Sources of Vitamin B_6

Plants food mostly contains pyridoxine while animal food mostly consists of pyridoxal or pyridoxamine. Cooking, storage and processing make the food lose vitamin B_6, which is least in plant food as pyridoxine is much more stable than pyridoxal or pyridoxamine.

It is widely distributed in foods. Some of the sources are fortified breakfast cereals, pork, turkey, beef, bananas, potatoes, pistachio nuts, tuna, chicken breast, avocado, sunflower seeds, sesame seeds, chickpeas, *etc.* [26] (shown in Fig. **33**).

Deficiencies of Vitamin B_6

Vitamin B6 deficiency is rare, the elderly people and alcoholics are mostly at risk of this deficiency. People with end-stage renal disease, chronic renal insufficiency and other kidney diseases can also suffer from vitamin B_6 deficiency. Since it is important for nerve function, therefore, its deficiency can cause neuropsychiatric disorder that includes seizures, migraines and depression. Other symptoms of this deficiency are confusion, muscle pains, fatigue and irritability [52].

Health Benefits of Vitamin B_6

- It regulates the level of homocysteine in the blood, thus reducing the chance of heart attack, as the high level of homocysteine is linked to inflammation, development of heart disease and blood vessel disease [52].

- Manages blood pressure and cholesterol levels.
- Vitamin B6 helps to control mood by playing an important role in synthesizing 'happy hormones' *i.e.* serotonin and norepinephrine.
- Helps to treat anemia and improves eye health.

Side Effects Due to Excess Vitamin B_6

Side effects never occur from food sources, but if one is taking vitamin B_6 supplements then overdosing side effects may occur. Overdose over long periods causes irreversible neurological problems. Excess of vitamin B_6 also causes oversensitivity to sunlight leading to skin rashes and numbness, nausea, vomiting, abdominal pain and loss of appetite [52].

VITAMIN B_7

Vitamin B_7, commonly referred to as biotin (shown in Fig. **34**), is also known as vitamin H or coenzyme R, is a coenzyme for carboxylase enzymes that is used in the synthesis of fatty acids, isoleucine, valine, and in gluconeogenesis. The letter 'H' (in vitamin H) stands for "Haar" and "Haut", German words for hair and skin. Its structure consists of an ureido ring with a tetrahydrothiophene ring. It supports adrenal function and helps to maintain a healthy nervous system [52].

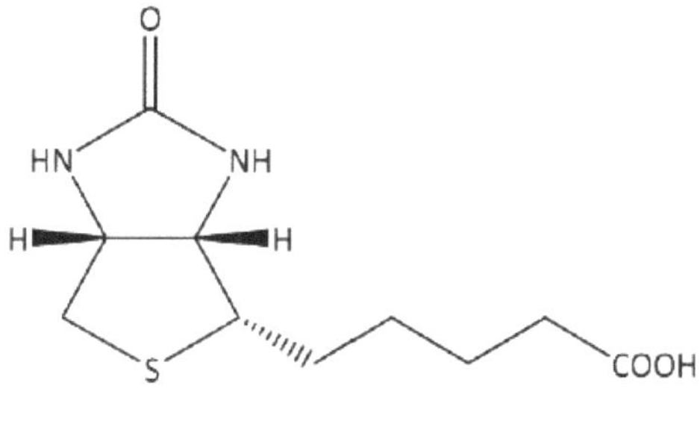

Biotin or Vitamin B_7 or Vitamin H

Fig. (34). Structure of biotin.

Sources of Vitamin B_7

As biotin is a water-soluble vitamin, hence the excess gets excreted through urine. Therefore, it is essential to meet the requirement through a daily diet for the optimal functioning of the body. Food rich in biotin is yeast, organ meat, egg yolk, soybeans, nuts and cereals. Some other sources are whole wheat bread, avocado, salmon, cauliflower, cheese, peanuts, bananas, mushrooms and chocolates [26]. (Shown in Fig. **35**).

Fig. (35). Sources of vitamin B7 [57]. (image courtesy: pixabay.com)

Deficiencies of Vitamin B_7

Biotin deficiency is rare and may occur due to a poor diet. Consumption of raw egg whites continuously for the long term may also result in biotin deficiency, as it contains avidin that inhibits the absorption of biotin in the body. Some other causes of biotin deficiency are long-term use of anti-seizure medications and antibiotics, intestinal malabsorption and intravenous feeding. Deficiency symptoms (shown in Fig. **36**) include hair loss, dry eyes, a scaly red rash around the eyes, nose, mouth and genitals, brittle and thin fingernails, depression, hallucination, numbness in hands and feet, lethargy, cracked lips, fatigue and loss of appetite [52].

Fig. (36). Symptoms of vitamin B7 deficiency [58]. (image courtesy: pixabay.com).

Health Benefits of Vitamin B_7

• Like other B-vitamins biotin helps the body in protein, fats and carbohydrates metabolism to produce energy, and helps to transfer carbon dioxide [52].

- It helps to maintain hair, skin and mucous membranes.
- It can help to control blood sugar levels in people with type 2 diabetes [52].
- It also maintains the appropriate function of the nervous system.
- It contributes to normal psychological function [52].

Side Effects Due to Excess Vitamin B_7

Biotin is a nontoxic vitamin and being water soluble the excess gets excreted out from the body. Hence there are no serious side effects of biotin. However overdosing may cause some side effects like causing acne and allergic reactions, frequent urine, diarrhea and nausea. Pregnant women are at a higher risk of biotin overdosing because it may increase the risk of miscarriage [52].

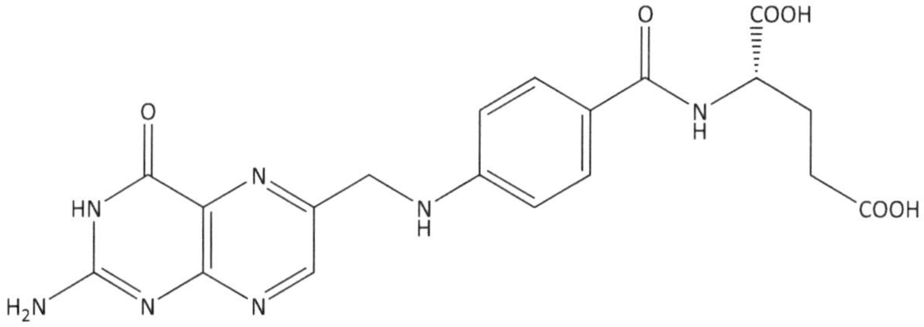

Folic acid

Fig. (37). Structure of vitamin B9.

VITAMIN B_9

Vitamin B_9 is also known as folate or folic acid (shown in Fig. **37**). The name folate originates from the Latin word 'folium' which means leaf. Folate is naturally present mostly in green leafy plants. Folic acid is the synthetic form of folate. Like other B-vitamins, it also plays an important role in cellular metabolism and energy production. The synthetic form, folic acid, has no physiological activity until it is converted into folates. Therefore, folic acid is converted into tetrahydrofolate (THF) in the liver by the enzyme dihydrofolate reductase [52].

Fig. (38). Foods rich in vitamin B9 [59]. (figure courtesy: pixabay.com).

Sources of Vitamin B_9

Folate is naturally available in a wide variety of foods mainly in green leafy vegetables (shown in Fig. **38**). Good sources of folate are spinach, green vegetables, beans, avocado, beetroot, liver, yeast, asparagus, kale and Brussels sprouts. Other sources include fruits, fruit juices, nuts, soybeans, chickpeas, dairy products, poultry and meat, eggs, seafood, grains and fortified products like enriched flour used to produce pasta, bread, breakfast cereals, *etc.* [26].

Deficiencies of Vitamin B_9

A healthy diet never leads to vitamin B9 deficiency. However, it may occur if dietary intake or folate absorption is inadequate or if the body excretes more folate. Pregnant women, alcoholics, infants, children and adolescents are at risk of this deficiency. The symptoms are tiredness, irritability, loss of appetite, weight loss, weakness, headaches, sore tongue and megaloblastic anemia. Pregnant women with folate deficiency may give birth to low birth weight premature infants and infants with neural tube defects [52].

Health Benefits of Vitamin B_9

- It plays an important role in producing nucleic acids [52].
- In pregnant women, folic acid supports the growth of the placenta and fetus. It also helps to prevent different types of birth defects.
- Folic acid has efficient free radical scavenging activity, hence can act as an antioxidant [52].
- It supports red blood cell replication and division and helps to treat anemia [52].

- It reduces the risk of stroke and cardiovascular disease.
- It is necessary to promote fertility in both men and women.
- It promotes a healthy immune system [52].

Side Effects Due to Excess Vitamin B_9

Side effects due to folic acid are generally rare. A few side effects can occur only if it is taken in a high amount. These are nausea, gas, poor appetite, feeling depressed and sleep disturbance. However, extremely high levels of folic acid may cause a risk of developing colon or rectum cancer [52].

VITAMIN B_{12}

Vitamin B_{12} (shown in Fig. **39**) exists in different forms and consists of the metal ion 'cobalt' therefore it is also known as 'cobalamin'. Among the different forms of cobalamins, methylcobalamin and 5-deoxyadenosylcobalamin form are active in human metabolism [52].

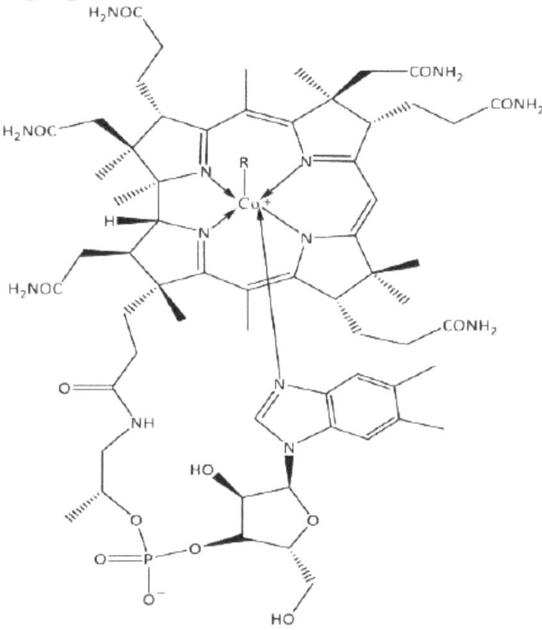

R= 5'-deoxyadenosyl, Me, OH, CN

Vitamin B_{12} or Cobalamin

It plays an important role in the normal functioning of the nervous system through myelinogenesis. It also acts as a cofactor to methionine synthase and L-methylmalonyl-CoA mutase [52].

Naturally, vitamin B_{12} can only be synthesized by bacteria and archaea; no plants, fungi, or animals can synthesize it as they lack the enzyme needed for its synthesis. It has the largest and complicated structure, with the cobalt atom at the centre of a corrin ring [52].

It is mainly produced as hydroxocobalamin by the bacteria and is converted into different forms of the vitamin in the body after consumption. Industrially, it can be produced by bacterial fermentation synthesis and through total synthesis [52].

Sources of Vitamin B_{12}

Vitamin B_{12} is naturally available in animal foods and also can be obtained from fortified foods (shown in Fig. **40**). In nature, it can only be produced by some prokaryotes and by some gut bacteria in humans and other animals. In humans, the vitamin is produced in the colon hence they can't absorb the vitamin synthesized in their body, but animals like cows and sheep can absorb the vitamin produced in their guts. Vitamin B_{12} is stored in the liver and muscles of the animals and is also present in their eggs and milk.

Fig. (40). Foods rich in vitamin B_{12} [60]. (image courtesy: pexels.com).

Food sources for vitamin B_{12} are crab meat, beef, lamb, liver, turkey, mackerel, eggs, milk, fish and fortified food products like breakfast cereals, soy products, energy bars and nutritional yeast [26].

Deficiency of Vitamin B_{12}

This deficiency occurs due to poor diet. It can also occur if one has undergone weight loss surgery or is involved in the prolonged intake of acid-reducing medication, and if one is an alcoholic or has some intestinal disorders. Aged people suffer from this deficiency as it becomes harder for them to absorb this vitamin. Vitamin B_{12} is mostly available in animal foods and is rare in plant foods; hence people who are vegetarians suffer from this deficiency unless they take vitamin B_{12} supplements [52].

Even a slight deficiency (slightly lower than the normal required amount) of vitamin B_{12} can cause a range of symptoms like depression, fatigue, lethargy, headaches, breathlessness, poor memory, soreness of the mouth or tongue, constipation and loss of appetite (shown in Fig. **41**). Severe deficiency can cause irreversible damage to the brain and nervous system [52].

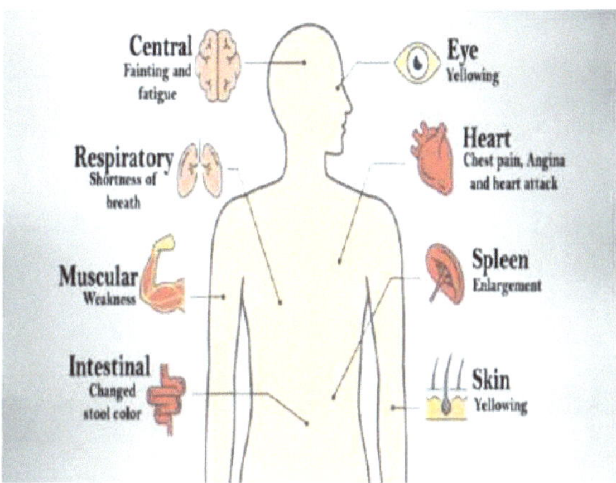

Fig. (41). Symptoms of vitamin B_{12} deficiency [61]. (image courtesy: emitpost.com).

This deficiency can be treated if one is given vitamin B_{12} injections or oral supplements.

Health Benefits of Vitamin B_{12}

- Like other B-vitamins, it helps to maintain energy levels by converting carbohydrates into glucose in the body.
- It helps to regulate the proper functioning of the nervous system, prevents mood disorder and improves memory [52].
- Vitamin B_{12} helps to lower down the homocysteine levels in the blood, thus preventing heart diseases like a heart attack or stroke.
- Vitamin B_{12} has a crucial role in cell reproduction; therefore, it is essential for maintaining skin and hair health [52].
- It can also improve sleep-wake rhythm disorder.
- It may cure Alzheimer's disease.

Side Effects Due to Excess Vitamin B_{12}

Vitamin B_{12} is not toxic even at high levels. No upper level has established its toxicity so far. However 'cyanocobalamin' (an injectable form of a supplement) can cause some side effects as it consists of traces of cyanide. The common side effects are headache, itching, swelling, nervousness, low potassium in blood, diarrhea, nausea and numbness [52].

KEYWORDS

Prosthetic Group

A non-protein group that combines with a protein and assists them in their work.

Collagen

A structural protein that is found in skin and also in other connective tissues.

Polyneuritis

Peripheral nerve disorder

Multiple Sclerosis

A disease that causes the insulating cover of the spinal cord and brain to damage.

Ataxia

Lack of control and coordination of muscle movements.

Skeletal Myopathy

A disorder of skeletal muscles.

Retinopathy

It describes the damage of the retina in the eyes causing impaired vision.

Bulimia

It is an eating disorder followed by purging, *i.e.* eating a lot of food at a time and then trying to get rid of that food consumed, say through vomiting.

Korsakoff's Syndrome

A memory disorder.

Leigh's Disease

A neurological disorder characterizes by loss of movement and mental abilities finally leading to death within 2-3 years.

Maple Serum Urine Disease

Is a metabolic disorder mostly affecting infants.

Keratoconus

A disease-causing thinning of cornea in the eye.

Glaucoma

A disease-causing the fluid pressure of the eye to rise.

Hartnup Disease

It is a metabolic disorder that affects the absorption of nonpolar amino acids. It is also known as pellagra.

Megaloblastic Anemia

Anemia resulting from inhibition of DNA synthesis when red blood cell is being produced.

VERY SHORT ANSWER TYPE QUESTION

1. Which vitamins are useful for the brain?
2. Name the vitamin that produces collagen.
3. Which vitamin can be produced in our gut?
4. What do you mean by anti-vitamin?
5. What do you mean by vitamin?
6. Write the other name of provitamin D_2.
7. Name the fat-soluble vitamins.
8. Name the water-soluble vitamins.
9. What are the food sources that contain vitamin K?
10. Name the vitamins that protect the human body from scurvy and rickets disease.
11. Which is the rich food source of vitamin A?
12. What is the chemical name of vitamin B_7?
13. Name the disease caused by an overdose of vitamin B_3.
14. Name the vitamin that can be produced in our skin.
15. Name the vitamin that is present mainly in animal source foods.

SHORT-ANSWER TYPE QUESTIONS

1. What is the role of vitamins?
2. What do you mean by macronutrient and micronutrient?
3. Differentiate between water-soluble and fat-soluble vitamins. How fat-soluble vitamins can cause harm if ingested in excess?
4. What are the problems caused by vitamin A deficiency? What role vitamin A plays in the physiology of vision?
5. What do you mean by megaloblastic anemia? How it is caused?
6. Vitamin C-rich foods should not be cooked. Why?
7. How vitamin D is formed in our body?
8. What is the importance of folic acid? What are the diseases caused by its deficiency?
9. What is the role of pantothenic acid in the body?
10. Write the symptoms of vitamin B_{12} deficiency.

LONG-ANSWER TYPE QUESTIONS

1. Which vitamins fall under the group of B-complex? What are the problems caused by their deficiencies?
2. Explain the health benefits of vitamin A.
3. Explain the classification of vitamins along with their functions.
4. Explain the structure of vitamin B_{12}.

MULTIPLE CHOICE QUESTIONS

1. Which of these deficiency cause scurvy?

(a) Vitamin A

(b) Vitamin D

(c) Vitamin C

(d) Vitamin K

2. The fat-soluble vitamin among the following is:

(a) Vitamin E

(b) Vitamin A and D

(c) Vitamin K

(d) All of the above

3. Which among the following is a water-soluble vitamin?

(a) Vitamin D

(b) Vitamin C

(c) Vitamin E

(d) Vitamin A

4. Which of the following is also known as vitamin A?

(a) Riboflavin

(b) Retinol

(c) Thiamine

(d) Pyridoxin

5. Which of the following is also known as vitamin B1?

(a) Thiamine

(b) Niacin

(c) Retinol

(d) Pyridoxin

6. Biotin is also known as

(a) Vitamin B_8

(b) Vitamin B_7

(c) Vitamin C

(d) None above the above

7. Thiamine is present in

(a) Liver, eggs

(b) Brown rice

(c) Potatoes

(d) All the above

8. Niacin is present in

(a) Fish

(b) Meat, Eggs

(c) Vegetables

(d) All of the above

9. Vitamin that maintains proper eye health is

(a) Vitamin C

(b) Vitamin D

(c) Vitamin A

(d) Vitamin E

10. Which vitamin deficiency causes beriberi?

(a) Vitamin B_1

(b) Vitamin C

(c) Vitamin K

(d) Vitamin B_7

11. Which vitamin can prevent pellagra disease?

(a) Vitamin B_7

(b) Vitamin B_5

(c) Vitamin B_3

(d) Vitamin A

12. Which vitamin is known as sunshine's vitamin?

(a) Vitamin A

(b) Vitamin D

(c) Vitamin E

(d) None of the above

13. Vitamins are grouped as

(a) Elements and compounds

(b) Essential and non-essential

(c) Fat-soluble and water-soluble

(d) Organic and inorganic

14. Vitamin A deficiency can cause

(a) Anemia

(b) Scurvy

(c) Night blindness

(d) Jaundice

15. Which of the following is the main active form of vitamin D in the body?

(a) Retinol

(b) Carotenoids

(c) Choline

(d) $1,25(OH)_2D$

16. Which of the following causes calcification of soft tissues?

(a) Vitamin K deficiency

(b) Vitamin D deficiency

(c) Vitamin D excess

(d) Vitamin A excess

17. Which are the best sources of carotenoids?

(a) Pumpkin, carrots, squash, sweet potatoes, apricots

(b) Corn, peas, beans

(c) Lean meat, poultry, fish, legumes

(d) Whole grains, nuts, seeds, egg yolk, plant oils

18. Which vitamin contains cobalt ion?

(a) Vitamin B_1

(b) Vitamin B_7

(c) Vitamin B_9

(d) Vitamin B_{12}

19. The incorrect match among the following is

(a) Vitamin B4- Pellagra

(b) Vitamin B_{12}- Pernicious anaemia

(c) Vitamin B6- Beriberi

(d) Vitamin C- Scurvy

20. Which of the following is a steroid vitamin?

(a) Vitamin A

(b) Vitamin D

(c) Vitamin E

(d) Vitamin K

21. Vitamin E helps to prevent

(a) Formation of vitamin D in the skin

(b) Keratinisation of epidermal cells

(c) Both a and b

(d) None of these

22. Vitamins were first discovered by

(a) Jenner

(b) Funk

(c) Calvin

(d) Mellanby

23. Vitamin C deficiency symptoms are

(a) Fatigue

(b) Muscle weakness

(c) Both a and b

(d) None of these

24. Vitamin B plays a role in

(a) Maintaining good eyesight

(b) Maintaining cell health

(c) Maintaining smooth skin

(d) None of these

25. Which of the following vitamin helps in blood clotting?

(a) Vitamin K

(b) Vitamin B_{12}

(c) Vitamin D3

(d) Vitamin E

26. Vitamin C is not present in

(a) Milk

(b) Lemon

(c) Orange

(d) Guava

27. Which of the following is correct?

(a) Vitamin A- Biotin

(b) Vitamin D- Riboflavin

(c) Vitamin B- Calciferol

(d) Vitamin E- Tocopherol

28. Which of the following is incorrect?

(a) Vitamin K- Beriberi

(b) Vitamin C- Scurvy

(c) Vitamin D- Rickets

(d) Vitamin A- Xerophthalmia

29. What is the common function of Thiamin, Riboflavin and Niacin?

(a) Help to strengthen blood vessels walls

(b) Used in the synthesis of blood clotting proteins

(c) Work as a part of a coenzyme used in energy metabolism

(d) Used to stabilize cell membranes

30. Vitamin B_{12} is also known as

(a) Riboflavin

(b) Cobalamin

(c) Niacin

(d) None of these

CONSENT FOR PUBLICATION

Not Applicable.

CONFLICT OF INTEREST

The authors confirm that this chapter contents have no conflict of interest.

ACKNOWLEDGEMENTS

Authors are thankful to Department of Chemistry, North Eastern Regional Institute of Science and Technology,Nirjuli, Itanagar for providing facilities to accomplish the work.

REFERENCES

[1] Anthea M, Hopkins J, Mc Laughlin CW, *et al.* Human biology and health. Englewood Cliffs, New Jersey, USA: Prenice Hall. 1993; ISBN 978-0-13-981176-0. OCLC 32308337.

[2] www.pexels.com

[3] Bender DA. Nutritional biochemistry of the vitamins. Cambridge, U.K.: Cambridge University Press 2003.

[4] Cheldelin VH. Nomenclature of the vitamins. Nutr Rev 1951; 9(10): 289-92. [http://dx.doi.org/10.1111/j.1753-4887.1951.tb02499.x] [PMID: 14882675]

[5] www.pngall.com/vitamin-png

[6] Lanska DJ. Handbook of clinical neurology] history of neurology. Historical aspects of the major neurological vitamin deficiency disorders 2009; 95(3): 435-44. [http://dx.doi.org/10.1016/s0072-9752(08)02129-5]

[7] Rajakumar K. Vitamin D, cod-liver oil, sunlight, and rickets: a historical perspective. Pediatrics 2003; 112(2): e132-5. [http://dx.doi.org/10.1542/peds.112.2.e132]

[8] Gratzer W. Terrors of the table: the curious history of nutrition. The quarry run to earth. Oxford: Oxford University Press 2006.

[9] Wendt D. Packed full of questions: Who benefits from dietary supplements? Distillations Magazine 2015; 1(3): 41-5.

[10] Semba RD. The discovery of the vitamins. International Journal for Vitamin and Nutrition Research 2012; 82(5): 310-5. [http://dx.doi.org/10.1024/0300-9831/a000124]

[11] Rosenfeld L. Vitamine--vitamin. The early years of discovery. Clin Chem 1997; 43(4): 680-5. [http://dx.doi.org/10.1093/clinchem/43.4.680] [PMID: 9105273]

[12] Fukuwatari T, Shibata K. Urinary water-soluble vitamins and their metabolite contents as nutritional markers for evaluating vitamin intakes in young Japanese women. J Nutr Sci Vitaminol (Tokyo) 2008; 54(3): 223-9. [http://dx.doi.org/10.3177/jnsv.54.223] [PMID: 18635909]

[13] Maqbool A, Stallings VA. Update on fat-soluble vitamins in cystic fibrosis. Curr Opin Pulm Med 2008; 14(6): 574-81. [http://dx.doi.org/10.1097/MCP.0b013e3283136787] [PMID: 18812835]

[14] Fennema O. Fennama's Food Chemistry. CRC Press/Taylor and Francis 2008; pp. 454-5.

[15] Tanumihardjo SA. Vitamin A: biomarkers of nutrition for development. Am J Clin Nutr 2011; 94(2): 658S-65S. [http://dx.doi.org/10.3945/ajcn.110.005777] [PMID: 21715511]

[16] Wolf G. The discovery of the visual function of vitamin A. J Nutr 2001; 131(6): 1647-50. [http://dx.doi.org/10.1093/jn/131.6.1647] [PMID: 11385047]

[17] https://www.healthline.com/nutrition/vitamin-a-benefits

[18] Booth SL, Johns T, Kuhnlein HV. Natural food sources of vitamin a and provitamin a. Food and Nutrition Bulletin 1992; 14(1): 1-15.
[http://dx.doi.org/10.1177/156482659201400115]

[19] Jagat R, Jitender J. Bitot's Spots. New England Journal of Medicine 2018; 379(9): 869-9. Community eye health journal. 2010; 23(72).. 2010;
[http://dx.doi.org/10.1056/NEJMicm1715354]

[20] Roncone DP. Xerophthalmia secondary to alcohol-induced malnutrition. Optometry 2006; 77(3): 124-33.
[http://dx.doi.org/10.1016/j.optm.2006.01.005] [PMID: 16513513]

[21] https://www.medindia.net/patients/patientinfo/medical-management-of-vitamin-deficiencies.htm

[22] https://www.healthline.com/nutrition/vitamin-a-benefits#section7

[23] Takita Y, Ichimiya M, Hamamoto Y, Muto M. A case of carotenemia associated with ingestion of nutrient supplements. J Dermatol 2006; 33(2): 132-4.
[http://dx.doi.org/10.1111/j.1346-8138.2006.00028.x] [PMID: 16556283]

[24] Norman AW. From vitamin D to hormone D: fundamentals of the vitamin D endocrine system essential for good health. Am J Clin Nutr 2008; 88(2): 491S-9S.
[http://dx.doi.org/10.1093/ajcn/88.2.491S] [PMID: 18689389]

[25] Holick MF, Schnoes HK, DeLuca HF, Suda T, Cousins RJ. Isolation and identification of 1,25-dihydroxycholecalciferol. A metabolite of vitamin D active in intestine. Biochemistry 1971; 10(14): 2799-804.
[http://dx.doi.org/10.1021/bi00790a023] [PMID: 4326883]

[26] U.S. Department of Agriculture, Agricultural Research Service. Food Data Central 2019.

[27] Holick MF. Vitamin D deficiency. N Engl J Med 2007; 357(3): 266-81.
[http://dx.doi.org/10.1056/NEJMra070553] [PMID: 17634462]

[28] https://www.kauveryhospital.com/blog/wp-content/uploads/2016/08/rickets.jpg

[29] https://edu.glogster.com/glog/rickets/20okri0h6ns

[30] https://www.medicalnewstoday.com/articles/161618.php

[31] Reboul E, Richelle M, Perrot E, Desmoulins-Malezet C, Pirisi V, Borel P. Bioaccessibility of carotenoids and vitamin E from their main dietary sources. J Agric Food Chem 2006; 54(23): 8749-55.
[http://dx.doi.org/10.1021/jf061818s] [PMID: 17090117]

[32] https://www.medicalnewstoday.com/articles/319689.php#what-are-tocotrienols

[33] Sokol RJ. Vitamin E deficiency and neurologic disease. Annual Review of Nutrition. 1988; 8: pp. (1)351-73.
[http://dx.doi.org/10.1146/annurev.nu.08.070188.002031]

[34] https://www.medicalnewstoday.com/articles/318168.php

[35] Vitamin K. Corvallis. OR: Micronutrient Information Center, Linus Pauling Institute, Oregon State University 2014.

[36] Ferland G, Vitamin K. Present Knowledge in Nutrition. Washington, DC: Wiley-Blackwell 2012; pp. 230-47.
[http://dx.doi.org/10.1002/9781119946045.ch15]

[37] Nutrition facts, calories in food, labels, nutritional information and analysis. Nutritiondata.com. 2008

[38] https://bujint.com/vitamin-k/

[39] https://in.pinterest.com/pin/348466089901488847/

[40] Shearer MJ, Vitamin K. Vitamin K. Lancet 1995; 345(8944): 229-34.

[http://dx.doi.org/10.1016/S0140-6736(95)90227-9] [PMID: 7823718]

[41] Suttie JW, Vitamin K. Encyclopedia of Dietary Supplements. London, New York: Informa Healthcare 2010; pp. 851-60.
[http://dx.doi.org/10.1201/b14669-95]

[42] Rasmussen SE, Andersen NL, Dragsted LO, Larsen JC. A safe strategy for addition of vitamins and minerals to foods. Eur J Nutr 2006; 45(3): 123-35.
[http://dx.doi.org/10.1007/s00394-005-0580-9] [PMID: 16200467]

[43] https://www.pexels.com/photo/photo-of-slices-of-kiwi-lime-and-orange-fruits-1414134/

[44] https://www.onlymyhealth.com/health-slideshow/vitamin-c-deficiency-signs-and-sympt-ms-you-must-know-1554953867.html

[45] Li Y, Schellhorn HE. New developments and novel therapeutic perspectives for vitamin C. J Nutr 2007; 137(10): 2171-84.
[http://dx.doi.org/10.1093/jn/137.10.2171] [PMID: 17884994]

[46] Spoelstra-de Man, Angélique ME, Elbers Paul WG, Oudemans-Van S, Heleen M. Vitamin C. Current Opinion in Critical Care 2018; 1
[http://dx.doi.org/10.1097/MCC.0000000000000510]

[47] Fattal-Valevski A. Thiamin (vitamin B1). J Evid Based Complementary Altern Med 2011; 16(1): 12-20.
[http://dx.doi.org/10.1177/1533210110392941]

[48] https://ayureveryday.com/vitamin-b1-thiamine-benefits-deficiency-sources/

[49] http://www.chm.bris.ac.uk/motm/vitaminB1/vitaminb1js.htm

[50] Nittiya S, Ijad K, Axel P, Radostina G, Hans B. The health benefits of a forgotten natural vitamin. International Journal of Molecular Sciences 2020; 21(3): 950-71.
[http://dx.doi.org/10.3390/ijms21030950]

[51] https://in.pinterest.com/pin/279856564332553495/?lp=true

[52] Institute of Medicine. Food and Nutrition Board.Dietary Reference Intakes for Tjiamine, Riboflavin, Niacin, Vitamin B6, Folate, Vitamin B_{12}, Pantothenic Acid, Biotin and Choline.. Washington, D.C.: National Academy Press 1998.

[53] https://www.livescience.com/51825-niacin-benefits.html

[54] https://www.healthline.com/health/pellagra

[55] https://www.healthline.com/health/vitamin-watch-what-does-b5-do

[56] https://www.livescience.com/51920-vitamin-b6.html

[57] https://pixabay.com/photos/vegetables-meal-vitamins-healthy-419638/

[58] https://pixabay.com/photos/problem-hair-comb-brush-lost-5146271/

[59] https://pixabay.com/photos/vegetables-garden-harvest-organic-790022/

[60] https://in.pinterest.com/pin/855683997927586415/?lp=true

[61] https://www.pexels.com/photo/food-plate-healthy-restaurant-7936686/

CHAPTER 7

Enzymes (Biocatalyst)

Nagendra Nath Yadav[1,*], **Archana Pareek**[1] and **Kamlesh Singh Yadav**[2]

[1] *Department of Chemistry, Nerist, Nirjuli, Itanagar-791109 (AP), India*

[2] *DDU Gorakhpur University, Gorakhpur-273009, Uttar Pradesh, India*

Abstract: In this chapter, brief accounts for the classification, nomenclature, and physiochemical nature of enzymes have been discussed. This chapter also deals with the action mechanism of an enzyme, enzyme kinetics, and factor affecting enzyme activity. A brief discussion about coenzymes, cofactors, and their role in biological systems has been addressed in this chapter.

Keywords: And enzyme kinetics, Biological catalyst, Coenzymes, Cofactor, Denaturation, Reaction specificity.

INTRODUCTION

The term enzyme was coined by Kuhne in 1878 for the biologically derived catalysts. These are protein molecules which work as a catalyst. They increase the rate of chemical reactions in the body. Enzymes work on the substrate and convert them into products [1]. Some important features of enzymes are as follows-

1. All enzymes are proteinous in nature, but all proteins are not enzymes.

2. They act as a catalyst to accelerate a chemical reaction.

3. They are very specific in their action.

4. Their names end in –ase, *e.g.*, sucrase, lactase, maltase, *etc.*

5. Enzymes work by weakening chemical bonds, which lower the activation energy.

6. Enzymes are key to life, *i.e.*, no enzyme, no life.

7. Enzymes contain active sites.

* **Corresponding author Nagendra Nath Yadav:** Department of Chemistry, NERIST, Nirjuli, Itanagar-791109 (AP), India. E-mail:nny@nerist.ac.in

Meera Yadav & Prof. Hardeo Singh Yadav (Eds.)

ENZYME STRUCTURE

Enzymes are made up of long chains of amino acids that arefolded into a very precise three-dimensional shape and contained an active site. An active site may be defined as a region on the surface of an enzyme to which substrate will bind and catalyse a biochemical reaction. There are various enzymes, and their names predict their activity. Enzymes not only break large chemicals into smaller ones but also build large chemicals from smaller ones along with many other chemical tasks. For example, an enzyme found in saliva (shown in Fig. **1**), amylase, breaks down starch molecules into glucose and maltose. Lipase breaks down fats into smaller molecules [1].

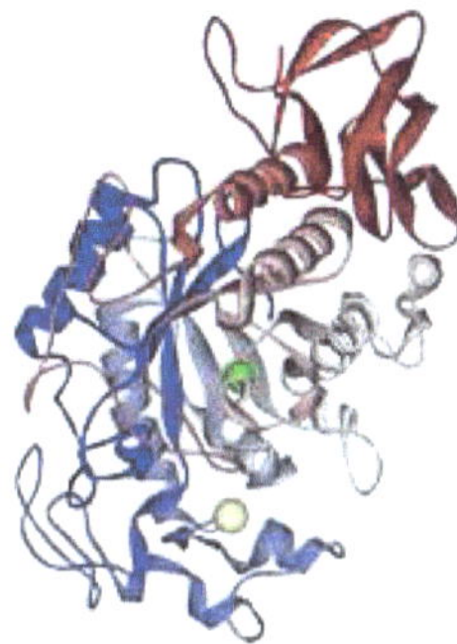

Fig. (1). Salivary amylase, chloride ion green: calcium beige [2].

Some chemicals are called activators, which help to enhance enzyme activity. Some chemicals can slow down the activity of an enzyme or even stop any further activity of the enzyme, and such chemicals are called inhibitors. Most drugs are such types of chemicals that either enhance or slow down the activity of some enzymes in the human body [1].

NOMENCLATURE AND CLASSIFICATION

Each enzyme is allocated two names. The first one is a trivial name or also called recommended name, which is very suitable for everyday use. The second one is a systematic name as projected by the International Union of Biochemistry (IUB).

Recommended Name

Most commonly used enzyme names end with the suffix '-ase' attached to the substrate, type of chemical reaction, *etc*. [3].

A. Based on the type of substrate as given in Table **1** [3]:

Table 1. Enzyme with their substrate.

Substrate	Enzyme (Substrate + Suffix 'ase')
Carbohydrates	Carbohydrases
Proteins	Proteinases
Lipids	Lipases
Nucleic acids	Nucleases
Maltose	Maltase
Sucrose	Sucrase
Urea	Urease
Aspartic acid	Aspartase
Fumaric acid	Fumarase

B. Based on the type of reaction which they catalyzed are given in Table **2** [3]:

Table 2. Type of chemical reaction catalyzed by an enzyme.

Reaction	Enzyme
Hydrolysis	Hydrolases
Isomerization	Isomerases
Oxidation	Oxidases
Reduction	Reductase
Dehydrogenation	Dehydrogenases
Decarboxylation	Decarboxylases

C. Based on the type of reaction catalyzed along with the type of substrate are given in Table **3** [3]:

Table 3. Type of substrate and type of reaction catalyzed by an enzyme.

Substrate	Reaction	Enzyme
Alcohol	Dehydrogenation	Alcohol dehydrogenase (removal of hydrogen from alcohol)
Succinic acid	Dehydrogenation	Succinic dehydrogenase
Glucose	Oxidation	Glucose oxidase
Glycogen	Synthesis	Glycogen synthetase
Phosphoglycerate	Isomerization	Phosphoglyceromutase

D. Based on product form:

L-Malate → Fumarate → Fumarase (enzyme)

E. Names that do not relate to the substrate or type of reaction are given in Table 4.

Table 4. Type of reaction catalysed by an enzyme.

Enzymes	Activity
Amylase	Hydrolysis of carbohydrates
Catalase	Decomposition of H_2O_2
Chymotrypsin	Hydrolysis of protein
Lysozyme	Hydrolysis of carbohydrate
Pepsin	Hydrolysis of protein

Systematic Name

The international union of biochemistry (IUB) appointed an enzyme commission in 1961. According to IUB system of enzyme classification, enzymes are divided into six major classes. Each class of enzyme represents a specific type of reaction [4].

- Oxidoreductases
- Transferases
- Hydrolases
- Lyases
- Isomerases
- Ligases

According to enzyme commission, all enzymes are allocated with an 'EC' number. EC number consists of four digits, *e.g.*, a.b.c.d, where 'a' indicates the class, 'b' indicates the subclass, 'c' indicates the sub-subclass, and 'd' digit indicates the sub-sub-sub class. The letters 'b' and 'c' of the digit illustrate the type of reaction, while the 'd' digit is used to distinguish between different enzymes of the same role based on the authentic substrate in the given reaction [4]. *e.g.*, For alcohol: NAD^+ oxidoreductase EC number is 1.1.1.1.

EC 1: Oxidoreductase (oxidation-reduction reaction): Enzymes involved in oxidation-reduction reactions *i.e.*

$$A_{red} + B_{oxi} \rightarrow A_{oxi} + B_{red}$$

EC 2: Transferase: This group contains enzymes that catalyze the group transfer reaction (except hydrogen or oxygen) *i.e.*

$$A\text{-}X + B \rightarrow BX + A$$

EC 3: Hydrolases : This group contains enzymes that catalyze the hydrolysis reactions. These include lipases, esterases, nitrilases, *etc.*

$$A\text{-}X + H_2O \rightarrow AOH + HX$$

EC 4: Lyases : This group contains enzymes that catalyze the addition or elimination reactions. These include decarboxylases and aldolases in the elimination reaction and synthases in the addition reaction.

$$A\text{-}B + C \rightarrow A{=}B + C^1$$

EC 5: Isomerases : This contains enzymes that catalyze isomerization reactions, including racemizations and cis-trans isomerization, *i.e.*, epimerases, mutases, *etc.*

EC 6: Ligases : This group contains the enzymes that catalyze the synthetic reactions (Greek: ligate means to bind) where two molecules are joined together by utilizing ATP, *i.e.*, synthetase, carboxylase, *etc.*

$$A + B + ATP \rightarrow A\text{-}B + ADP + P$$

FACTORS AFFECTING ENZYME ACTIVITY

Concentration of Enzyme

The velocity of the enzymatic reaction is directly proportional to the concentration of the enzyme, *i.e.*, Rate of reaction = [E]

The velocity of a bio-chemical reaction increases on increasing the concentration of the respective enzyme, accordingly [5] (as shown in Fig. **2**).

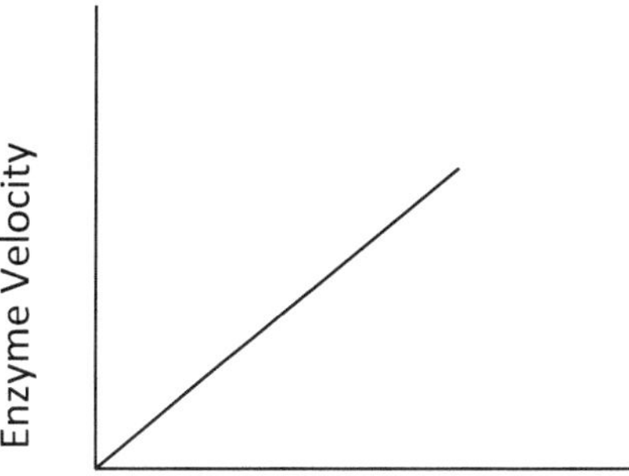

Fig. (2). Effect of enzyme concentration on the velocity of a reaction.

Substrate Concentration

As we increase the substrate concentration, the velocity of enzyme catalysed reaction gradually increases up to a certain limit (V_{max}). At V_{max} the reaction attains maximum velocity. After this point, there is no further increase in the rate of the reaction even if the concentration of substrate increases because at this point, the enzyme molecules and the substrate molecules are completely saturated with each other. A rectangular hyperbola is obtained (as shown in Fig. **3**) when velocity of the reaction is plotted against the concentration of substrate [5] *i.e.*

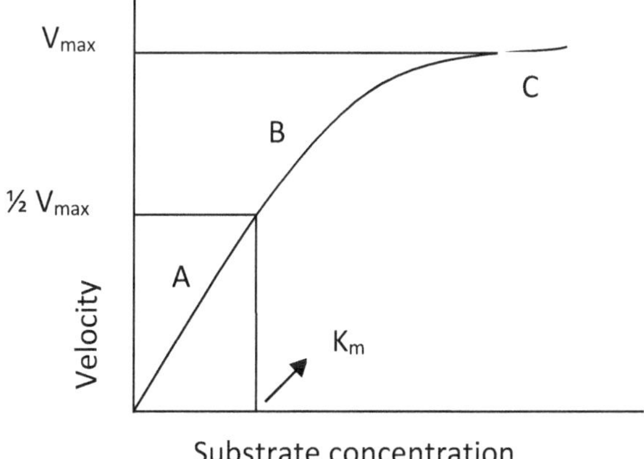

Fig. (3). Effects of concentration of substrate on velocity of enzyme reaction (A-linear, B-curve, C-almost unchanged).

K_m is the substrate concentration at which the enzymatic reaction proceeds at its half-maximal velocity, *i.e.*, ½ V_{max}. An enzyme with a low K_m value has a high affinity for the substrate and acts even at low substrate concentration in its half-maximal velocity. Enzymes with a high K_m value have a low affinity for the substrate and therefore need a high concentration of the substrate to act at its maximal velocity.

Effect of Temperature

Enzymes are inactive at a very low temperature. The velocity of an enzymatic reaction increases with an increase in temperature up to a maximum and then decreases (as shown in Fig. **4**). That maximum temperature is called the optimum temperature at which the enzyme has its maximal activity. This optimum temperature ranges from 37 - 40 °C in humans. A bell-shaped curve is usually observed.

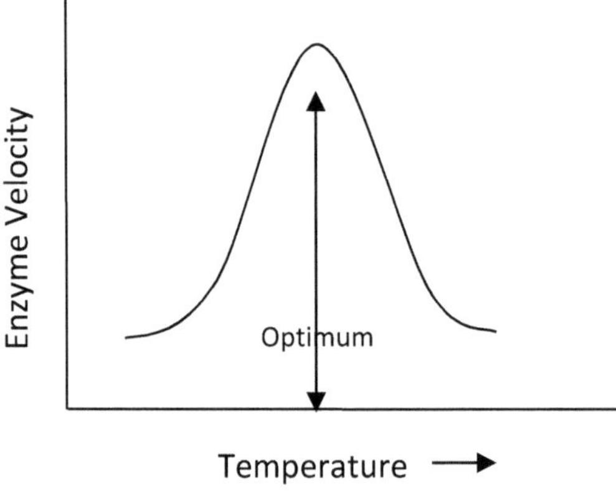

Fig. (4). Effect of temperature on enzyme velocity.

Increasing temperature increases the rate of a reaction because- 1) It increases the substrate energy therefore decreases the activation energy, and lower down the activation energy of the reaction. 2) Increase in temperature also raises the collisions between the molecules. However, if the temperature rises above the optimum temperature then denaturation of the enzymes occurs, as enzymes are proteineous in nature, leading to loss of their catalytic activity [5].

Temperature of Coefficient (Q_{10})

Temperature coefficient may be defined as the changes in the velocity of reaction when temperature is raised by 10°C. For most of enzyme Q_{10} = 2 between 0°C→4°C. It means increase in temperature by 10°C resulted in doubled reaction velocity. As temperature is increased, the activation energy of the substrate molecules increases and more molecular collision for the reaction to proceed faster.

For a vast of enzymes, the optimum temperature is in between 35°C→40°C. Generally, when enzymes are exposed to higher temperature *i.e.* above 50°C, denaturation of protein happens which causes the change in native structure of protein.

Effect of pH

Enzymes are very specific towards pH. Each enzyme shows an optimum pH at which the velocity of reaction is maximum (as shown in Fig. **5**). Below and above optimum pH, the activity of enzyme is much slower and at high pH, the enzyme becomes completely inactive. This is because pH other then optimum pH, changes the ionization state of substrate or/and enzyme and at extreme pH change, denaturation of enzyme occurs [5].

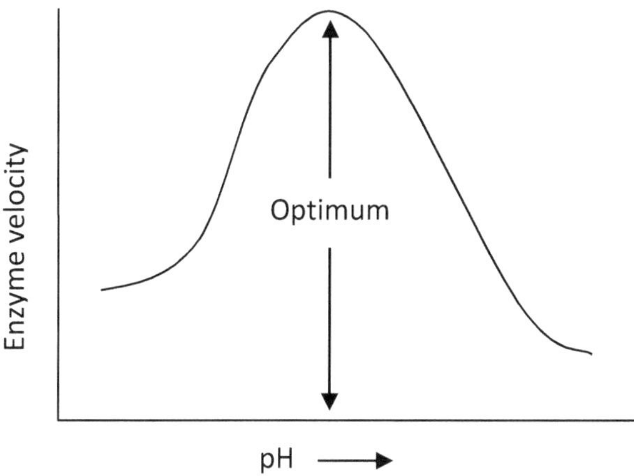

Fig. (5). Effect of pH on enzyme activity.

Effect of the Concentration of Product

The accumulation of product in an enzymatic reaction generally decreases the velocity of an enzyme. In case of few enzymes, the product combines with the active site of enzyme and inhibits the enzyme activity.

ENZYME KINETICS

It is the study of the enzyme catalysed chemical reactions. In enzyme kinetics, the rate of the reaction is measured and effects of changes in the conditions of the reaction are studied. The substrate (S) combine with enzyme (E) to give unstable enzyme-substrate complex (ES), which on decomposition produced the product (P) [5].

$$E + S \underset{k_2}{\overset{k_1}{\rightleftharpoons}} ES \xrightarrow{k_3} E + P$$

k_1, k_2 and k_3 are the velocity constant for the given reactions.

The mathematical equation used to defines the quantitative connection between the rate of an enzyme catalysed reaction, and the concentration of substrate [s] is given by the equation, called Michaelis-Menten equation.

$$V = \frac{V_{max}\,[S]}{k_m + [S]}$$

where, V= measured velocity/observed velocity at the given [S]

k_m = Michaelis-Menten constant $k_2 + k_3/k_1$

V_{max} = Maximum velocity

[S]= Substrate concentration

Let us assume that the measured velocity (V) is equal to $\frac{1}{2}V_{max}$ Then the above equation may be substituted as follows-

$$\frac{1}{2}V_{max} = \frac{V_{max}\,[S]}{k_m+[S]}$$

$$k_m + [S] = \frac{2V_{max}\,[S]}{V_{max}}$$

$$k_m + [S] = 2[S]$$

$$k_m = [S]$$

k_m or the Michaelis-Menten constant is defined as "the substrate concentration given in moles/L to produce half-maximum velocity in an enzyme-catalyzed reaction.

The value of k_m is a constant and it defines the characteristic feature of a specified enzyme. It is helpfull for measuring the strength of enzyme-substrate (ES) complex. A low value for k_m specifies a strong affinity among the enzyme and substrate [5].

Mechanism of Enzyme Action

Enzymes are biocatalyst and the nature of catalysis in biological system is complementary to that of normal reaction *i.e.* non-biological catalysis [3]. In all chemical reactions, reactants first converted to an activated state or transition state by absorbing energy and then converting to the product *i.e.*

$$\underset{\text{Reactant}}{A} \xrightarrow{\text{energy}} \underset{\text{T.S}}{A^*} \longrightarrow \underset{\text{Product}}{P}$$

The energy required by the reactant to undergo the reaction is known as activation energy. The catalyst (enzyme) reduces the activation energy, which results the reaction to proceed at a lower temperature [3]. The energy level diagram of enzyme catalysis is shown below Fig. (**6**).

The enzymes lower the activation energy of the reactants so that all the chemical reactions in biological system occur at ambient temperature (below 40°C).

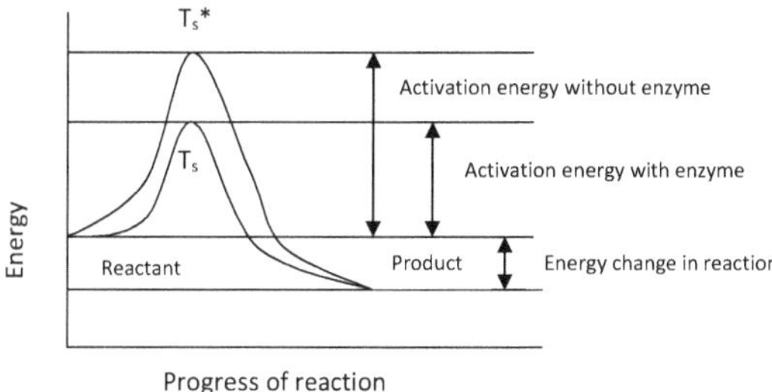

Fig. (6). Energy level diagram of enzyme catalysis.

Enzyme-substrate Complex Formation

The enzymatic reactions take place by combining the substrate with active site of the enzyme molecules by several weak bonds [3]. *i.e.*

$$\underset{\text{Enzyme}}{E} + \underset{\text{Substrate}}{S} \rightleftharpoons \underset{\text{E.S. complex}}{ES} \longrightarrow \underset{\text{Enzyme}}{E} + \underset{\text{Product}}{P}$$

Enzymes bind with substrate to give enzyme-substrate complex which ultimately results in product and enzyme. The mechanism of enzyme-substrate complex formation can be explain by the following theories-

Fisher's Lock and Key Model or Fisher's Template Theory

The German biochemist, Emil Fischer (1898) proposed this hypothesis and it was the first model proposed to describe the mechanism of an enzyme catalyzed reaction.

According to this theory, the structure of enzyme is pre-shaped and rigid in nature. The substrate, whose structure is resembles with active sites fits to the active site of enzyme in same manner as a key is completely fits into the suitable lock (as shown in Fig. **7**). The active site of enzyme is very specific towards the substrate [6].

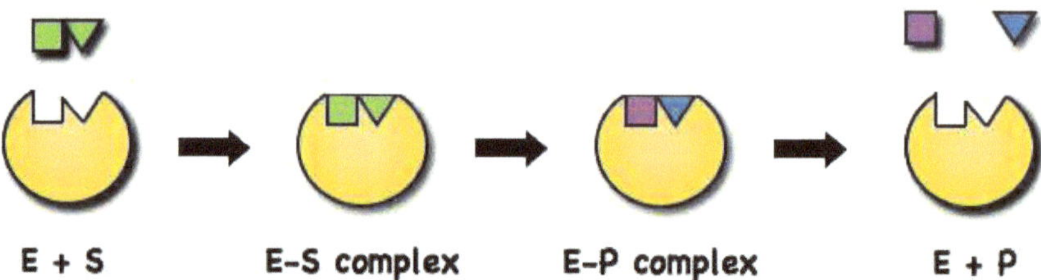

Fig. (7). Mechanism of enzyme substrate (ES) complex formation by lock and key model [7].

Drawback of Fisher's Model

This model does not explain the flexible nature of enzymes and fails to explain many facts of enzymatic reactions [8].

Koshland's Induced Fit Model

Koshland in 1958, proposed by a more acceptable and realistic model for enzyme-substrate complex formation.

This model explains that the active site of enzyme is not rigid and preshaped [9]. When there is an interaction between the substrate and enzyme, and then there is some configurational or geometrical changes in the enzyme, which results in the creation of a strong binding site for substrate. Further, due to induced fit, the specific amino acids of the enzyme are repositioned to provide the active site for substrate molecule and completed the catalysis (as shown in Fig. **8**) [10].

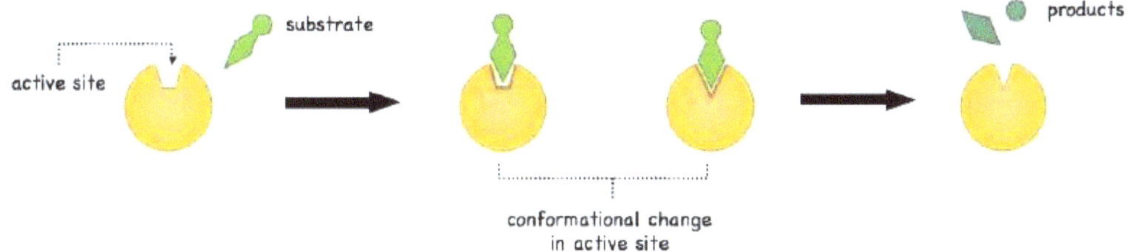

Fig. (8). Induced fit hypothesis [7].

ENZYME INHIBITION

Any substance which binds with enzyme and reduces the catalytic activity of that enzyme is called enzyme inhibitor and such a process is known as enzyme inhibition [3]. The inhibitor may be of organic in nature or inorganic in nature. The types of inhibition are as follows:

- Reversible inhibition
- Irreversible inhibition

Reversible Inhibition

This type of inhibitor binds with the active site of enzyme *via*non-covalent bond and the enzyme inhibition can be upturned if the inhibitor is detached [3]. The reversible inhibition is of two types-

- Competitive inhibition
- Non-competitive inhibition.

Competitive Inhibition

Competitive inhibitors are substrate that is closely resembles the structure of real substrate (s) and compete with substrate for the active site of the enzyme but does not undergo any catalysis.

$$E \underset{}{\overset{+S}{\rightleftharpoons}} ES \longrightarrow P + E$$

$$E \underset{}{\overset{+I}{\rightleftharpoons}} \underset{\text{(inactive)}}{EI} \longrightarrow \text{No reaction}$$

As long as the competitive inhibitor is occupying the active site, the enzyme is not accessible for the substrate to bind. Such type of inhibition can be conquering with high concentration of the substrate. In some enzymatic reaction, the inhibitor combines a site other than the active site of an enzyme and creates an allosteric effect that changes the shape of the natural active-site [3].

The relative concentration of the substrate and inhibitor and their respective affinity with the enzymes determine the degree of competitive inhibition [3]. Examples of competitive inhibition can be seen as follows-

An enzyme succinic acid dehydrogenase catalyses the conversion of succinic acid to fumaric acid *i.e.*

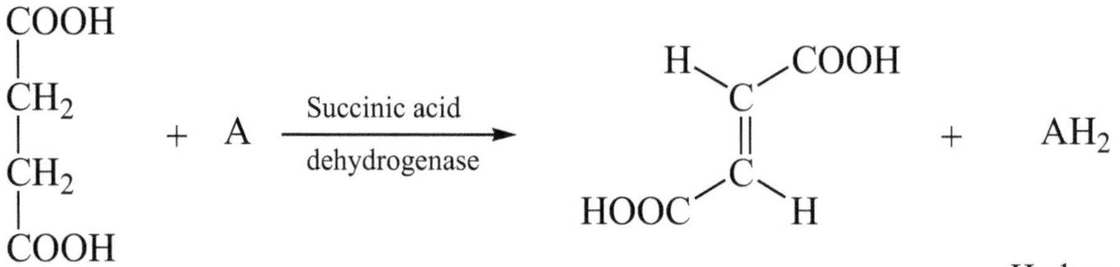

$$\underset{\text{Succinic acid}}{HOOC{-}CH_2{-}CH_2{-}COOH} + A \xrightarrow[\text{dehydrogenase}]{\text{Succinic acid}} \underset{\text{Fumaric acid}}{H(HOOC)C{=}C(COOH)H} + \underset{\text{Hydrogenated acceptor}}{AH_2}$$

There are several organic compounds which are structurally related to succinic acid, combine with the enzyme and inhibit the reaction *i.e.*

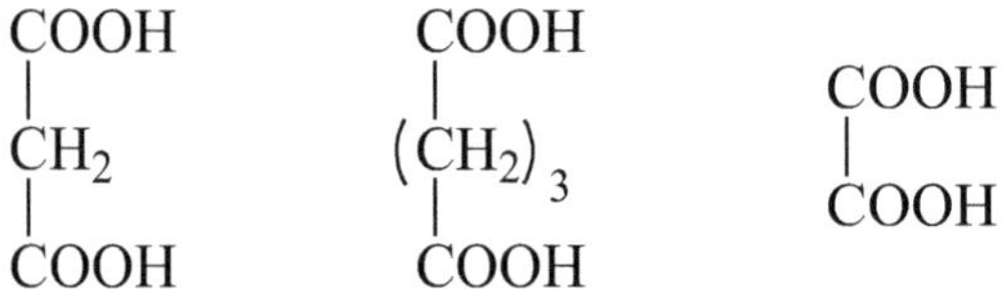

$$\underset{\text{Malonic acid}}{HOOC{-}CH_2{-}COOH} \qquad \underset{\text{Glutaric acid}}{HOOC{-}(CH_2)_3{-}COOH} \qquad \underset{\text{Oxalic acid}}{HOOC{-}COOH}$$

Malonic acid is the most efficient inhibitor

Non-competitive Inhibition

In this type of inhibition there is no competition between the substrate (s) ands inhibitor for the active site of enzyme. Therefore, this type of inhibition cannot be overcome by high concentration of the substrate. The inhibitor has little or no structural resemblance with the substrate and it binds with the enzyme at a place other than the active site (allosteric site). Both I and S may combine at different site of enzyme to form EI and ESI complexes, which break down to produce the reaction product [3] *i.e.*

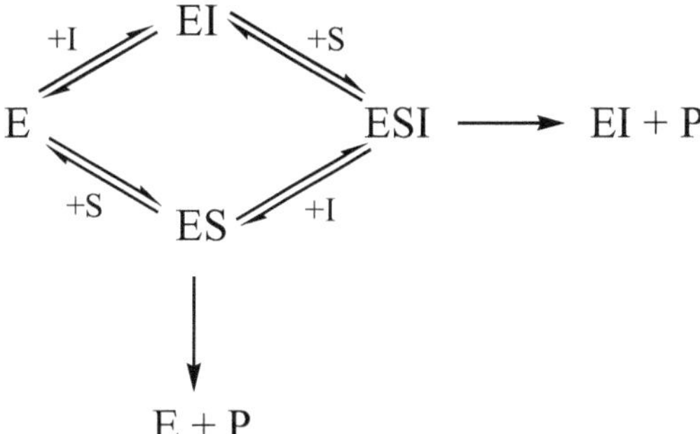

Irreversible Inhibition

In this type of inhibition, inhibitors bind covalently with the enzymes and inactive them, which is irreversible. These inhibitors are usually toxic substance that poisons enzymes. Example: Diisopropyl fluorophosphates (DFP) is a well known nerve gas, which was discovered by the German scientist during Second World War. Diisopropyl fluorophosphates irreversibly binds with enzymes having an amino acid Serine at the active site. Ex. Serine Proteases, acetyl choline esterase. (Sarin is another example). Cyanide inhibits cytochrome oxidase (binds to iron atom) of electron transport system [11].

In many organisms, inhibitors act as a part of the reaction mechanism. For *e.g.* if one substance is produced in large amounts by an enzyme in an organism then that substance at the beginning may act as an inhibitor for the enzyme that produces it. Sufficient amount of inhibitor may either slows down or stops the production of the substance [1].

PHYSICO-CHEMICAL NATURE OF ENZYME

All the enzymes are always proteins. In recent years, however a little RNA molecule have been exposed to functions as enzymes. Each enzyme has its own tertiary structure and precise conformation, which is very important for its catalytic action [3].

CHARACTERISTICS OF ENZYMES

1. The functional unit of any enzyme is called holoenzyme which is generally made up of apoenzyme *i.e.* the proteineous part and a coenzyme *i.e.* non-protein part [3, 5].

Holoenzyme	⟶	Apoenzyme + Coenzyme	
(active enzyme)		(protein part)	(non-protein part)

2. A prosthetic group is a non-protein structure tightly (covalently) binds with the apoenzyme. The coenzyme can be detached by dialysis from enzyme whereas the phosphate group cannot be detached from enzyme.

3. Enzymes are colloidal in nature. They are macromolecule (polypeptide) and possess high molecular weight ranging from 12000 to over 1 million.

4. Enzymes, which is composed of a single polypeptide, called monomeric enzymes *e.g.* ribonuclease, trypsin, *etc.* while those possess more than one polypeptide chain are known as oligomeric enzymes *e.g.* lactate dehydrogenase, *etc.* [3, 5].

5. Catalytic nature of enzyme or effectiveness: The catalytic power of an enzyme is calculated by the term "turnover number" (molecular activity) and defined as- "The number of substrate molecules transformed into product per unit of time, when the enzyme is fully saturated with substrate". *e.g.* A single molecule of catalase can convert 5,000,000 H_2O_2 molecules into H_2O and O_2 in a minute [3].

6. Enzyme specificity: Enzymes are highly specific in their action as compare with the catalysts. There are three major type of enzyme specificity known-

- Stereo specificity
- Reaction specificity
- Substrate specificity

Stereo specificity or optical specificity : The enzymes act only on one isomer and therefore exhibit stereo specificity [3].

$$\text{L-amino acids} \xrightarrow{\text{L-amino acid oxidase}} \text{Product}$$

$$\text{D-amino acids} \xrightarrow{\text{D-amino acid oxidase}} \text{Product}$$

$$\underset{\text{Cellulose}}{(C6H10O5)_n} + n\ H2O \xrightarrow{\text{Cellulase}} \underset{\text{Glucose}}{n\ C_6H_{12}O_6}$$

Cleavage of β–glycosidic bond

Reaction specificity : If same substrate can go through different type of reaction by utilizing different enzyme. This is called as reaction specificity [3]. *e.g.*

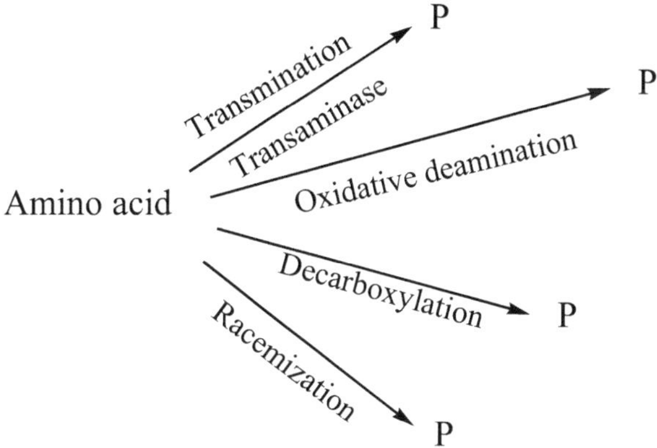

Substrate specificity : The substrate specificity differes from enzyme to enzyme. It may be absolute, relative or broad [3].

- **Absolute substrate specificity** : Certain enzyme be active on one substrate *e.g.*

$$\text{Glucose} \xrightarrow{\text{Glucopinase}} \text{Glucose-6-Phosphate}$$

$$\text{Urea} \xrightarrow{\text{Urease}} NH_3 + CO_2$$

- **Relative substrate specificity** : Certain enzymes can act on structurally associated substances. These types of enzymes are specific towards nature of group or bond present. *e.g.*

$$\text{Peptide (Arg-AA or Lys-AA)} \longrightarrow \text{Arginine + Peptide}$$

Trypsin hydrolyses peptide linkage concerning arginine or lysine, Chymotrypsin cleaves peptide bond attached to aromatic amino acids (Phenylamine, tyrosin and tryptophan) [3].

- **Broad specificity** : Some enzymes act on many closely related substrates *e.g.* hexopinase acts on glucose, fructose, mannose and glucosamine but not on galactose [3].

COENZYMES

The non-protein, organic, low molecular weight and dialyzable substance related with enzyme function are called coenzyme. They are bound to an enzyme loosely or tightly. They help in transporting chemical groups between different enzymes.

Examples: NADH, NADPH and adenosine triphosphate (ATP). The coenzymes are not synthesized in the body therefore acquired from diet through closely related compounds like vitamins [12].

Holoenzyme	⟶	Apoenzyme + Coenzyme	
Active enzyme or functional enzyme		Protein part	Non-protein part

Many enzymes require certain non-protein small additional factors, collectively known as cofactors. The term activator is used to inorganic cofactor (like Ca^{+2}, Mg^{+2}, Mn^{+2}, *etc.*) necessary to enhance the enzyme activity. Some important salient features of coenzymes can be seen as follows-

- Coenzymes are generally termed as the second substrate or co-substrate because they have affinity with the enzyme as comparable with the substrate [13].
- Coenzymes undergo alternations during the enzymatic reactions and they are regenerated during product formation.
- Coenzymes participate in various reactions involving transfer of atom or groups like hydrogen, aldehyde, keto, amino, acyl, *etc.*
- Most of the enzymes are the derivatives of water-soluble vitamin-B complex. The biochemical functions of B-complex vitamins are in the form of coenzymes [12].
- All coenzymes are not vitamin derivative, instead here are some other organic substances which acts as coenzymes *i.e.* ATP, CDP, UDP, *etc.*
- Some of the coenzymes are nucleotides *i.e.* they contain sugar, nitrogenous base and phosphate *e.g.* NAD^+, $NADP^+$, FAD, coenzyme A, *etc.* [12].
- A particular coenzyme may participate in catalytic reactions with different enzymes *e.g.* NAD^+ acts as a coenzyme for lactate dehydrogenase and alcohol dehydrogenase. In both cases NAD^+ is involved in hydrogen transfer [13].
- The specificity of the enzyme mostly depends on the apoenzyme and not on the enzyme [12].

BIOLOGICAL FUNCTIONS OF COENZYME

Enzymes has a wide variety of functions in living organisms. They help in cell regulation and signal transduction through kinases and phosphatases. Some viruses contain enzymes to infect cells. They are also involved in exotic functions, like lucifrase in fireflies generates light. Enzymes play many important functions in the digestive systems of animals, *e.g.* amylases and proteases break down large molecules like starch and protein into smaller ones that can be absorbed by intestine [14]. The name of some important coenzyme and their biological functions can be seen as follows:

Adenosine triphosphate (ATP)

Donates phosphate, adenosine and adenosine monophosphate (AMP) moieties.

Thymine Pyrophosphate (TPP)

Transfer the aldehyde or keto group.

Flavin mononucleotide (FMN)

Transfer of hydrogen and electron.

Nicotinamide adenine dinucleotide (NAD^+)

Hydrogen and electron transfer.

Biocytin

Transfer of CO_2.

CONTROL ACTIVITY OF ENZYMES

Enzyme activity is controlled in the following five ways in the cell [1].

Regulation

Other molecules can either activate or inhibit enzymes. *e.g.*, in metabolic pathways, the end product often acts as an inhibitor for any of the first enzyme. It thus regulates the amount of the end product. Such type of mechanism is called a negative feedback mechanism as the concentration of the end product itself regulates its production [15].

Post-translational Modification (PTM)

It refers to the modification of proenzymes to form mature protein product. *e.g.*, conversion of prohormones to hormones, chymotrypsinogen (a digestive protease) is produced in the pancreas in an inactive form and is transported to the stomach where it is then converted to its active form chymotrypsin. This prevents the enzyme from digesting into the pancreas and other tissues before entering the gut [15].

Quantity

In response to the changes in the cell's environment, the production of an enzyme can either be enhanced or diminished. This type of regulation is called enzyme induction. Regulation of enzyme levels can also be done by changing the rate of enzyme degradation [1].

Subcellular Distribution

Enzymes can be divided into discrete sections in different cellular categories according to their different metabolic pathways. For example fatty acids are synthesized in Golgi, endoplasmic reticulum, and cytosol, but they are used as sources of energy by different enzymes in mitochondria [16].

Organ Specialization

Different enzymes are available for metabolism in cells present in different organs and tissues.

KEYWORDS

Catalyst

A substance that speeds up a chemical reaction.

Dialysis

The separation of particles in a liquid on the basis of differences in their ability to pass through a membrane.

Denaturation

Process in which proteins or nucleic acids lose their quaternary structure.

Biological Catalyst

Catalysts present in biological systems like in plants and animals.

SHORT-ANSWER TYPE QUESTIONS

1. Define the term enzyme.

2. What are isoenzymes? Give examples.

3. What do you mean by turn-over number?

4. What is the unit of measuring enzyme activity?

5. What is the importance of enzymes for living beings?

6. Enzyme action is considered to be highly active. Why?

LONG-ANSWER TYPE QUESTIONS

1. What is the I.U.B system of nomenclature of enzymes? What is the E.C code number? Write its significance.

2. Define active site. How does it participate in enzyme catalysis?

3. What are cofactors? Differentiate between coenzyme, cofactor, and prosthetic group.

4. What are the factors that affect the rate of enzyme catalysed reaction?

5. Distinguish between competitive and non-competitive inhibition.

6. What are the theories that explain enzyme-substrate complex formation?

MULTIPLE CHOICE QUESTIONS

1. All enzymes are

a. Nucleic acids

b. Aminoacids

c. Polypeptides

d. Carbohydrate

2. For enzymes, inhibitors are the substances that can

a. Reduce the catalytic activity

b. Increase the catalytic activity

c. Degrade the enzyme

d. No effect on enzyme activity

3. In competitive inhibition, the real substrate compete with the inhibitor for the

a. Active site of enzyme

b. Allosteric site of enzyme

c. Other part of enzyme

d. None of these

4. The increase in enzyme velocity, when the temperature is increased by 10 °C, called

a. Optimum temperature

b. Temperature coefficient

c. Minimum temperature

d. None

5. Substance that can reduce the catalytic activity of enzymes, called

a. Catalyst

b. Inhibitor

c. Activator

d. None

6. The velocity of the enzymatic reaction increases as the concentration of the enzyme

a. Decreases

b. Increases

c. keep constant

d. Zero

7. A particular pH value at which the rate of an enzymatic reaction is maximum, called

a. Optimum pH

b. Minimum pH

c. Isoelectric pH

d. High pH

8. The term enzyme is coined by

a. Buchner

b. Kuhne

c. Pasteur

d. Urey Miller

9. The enzyme that hydrolysed fat is

a. Lipase

b. Amylase

c. Pepsin

d. Trypsin

10. Apoenzyme is

a. Protein part of a conjugated enzyme

b. Inorganic cofactor of a conjugated enzyme

c. Organic cofactor of a conjugated enzyme

d. Simple enzyme

11. Enzymes

a. Do not change the requirement of activation energy

b. Decreases requirement of activation energy

c. Increases requirement of activation energy

d. Do not require activation energy

12. Enzymes that form peptide bond is

a. Peptidase

b. Peptidyl transferase

c. Carbonic anhydrase

d. Carbohydrase

13. Coenzyme is

a. Always a protein

b. Always an inorganic compound

c. Often a vitamin

d. Often a metal

14. Apoenzyme and coenzyme combines to produce

a. Enzyme substrate complex

b. Holoenzyme

c. Enzyme product complex

d. Prosthetic group

15. Catalyst is different from enzymes in

a. Not used up in reaction

b. Functional at high temperature

c. Having high rate diffusion

d. Being proteinaceous

CONSENT FOR PUBLICATION

Not Applicable.

CONFLICT OF INTEREST

The author confirms that this chapter contents have no conflict of interest.

ACKNOWLEDGEMENTS

Authors are thankful to Department of Chemistry, North Eastern Regional Institute of Science and Technology,Nirjuli, Itanagar for providing facilities to accomplish the work.

REFERENCES

[1] Stryer L, Berg JM, Tymoczko JL. Biochemistry. 5th ed., San Francisco: W.H. Freeman 2002.

[2] https://www.researchgate.net/publication/316937135_Behaviour_of_salivary_amylase_in_various_reaction_environments_with_reference_to_Km_and_Vmax_an_overview

[3] Nelson DL, Cox MM. Enzymes. Lehninger Principles of Biochemistry. 4th ed. New York: W. H. Freeman and company 2006; pp. 190-237.

[4] Tipton, K.; Boyce, S., History of the enzyme nomenclature system. Bioinformatics. 2000; 16(1); 34–40.
[http://dx.doi.org/10.1093/bioinformatics/16.1.34]

[5] http://www.worthington-biochem.com/introBiochem/Enzymes.pdf

[6] Fischer E. Einfluss der Configuration auf die Wirkung der Enzyme. Ber Dtsch Chem Ges 1894; 27(3): 2985-93.
[http://dx.doi.org/10.1002/cber.18940270364]

[7] http://www.vce.bioninja.com.au/aos-1-molecules-of-life/biochemical-processes/enzymes.html

[8] Cooper GM. The Central Role of Enzymes as Biological Catalysts The Cell: a Molecular Approach. 2nd ed., Washington, DC: ASM Press 2000.

[9] Koshland DE. Application of a Theory of Enzyme Specificity to Protein Synthesis. Proc Natl Acad Sci USA 1958; 44(2): 98-104.
[http://dx.doi.org/10.1073/pnas.44.2.98] [PMID: 16590179]

[10] Boyer R. Enzymes I, Reactions, Kinetics, and Inhibition Concepts in Biochemistry. 2nd ed. New York, Chichester, Weinheim, Brisbane, Singapore, Toronto: John Wiley & Sons, Inc. 2002; pp. 137-8.

[11] Strelow JM. A Perspective on the Kinetics of Covalent and Irreversible Inhibition. SLAS Discov 2017; 22(1): 3-20.
[http://dx.doi.org/10.1177/1087057116671509] [PMID: 27703080]

[12] Wagner AL. Vitamins and Coenzymes. Krieger Pub Co. 1975.

[13] BRENDA The Comprehensive Enzyme Information System. Technische Universität Braunschweig. 2015

[14] Crane FL. Biochemical functions of coenzyme Q10. J Am Coll Nutr 2001; 20(6): 591-8.
[http://dx.doi.org/10.1080/07315724.2001.10719063] [PMID: 11771674]

[15] Suzuki H. Control of Enzyme Activity How Enzymes Work: From Structure to Function. Boca Raton, FL: CRC Press 2015; pp. 141-69.
[http://dx.doi.org/10.1201/b18087-9]

[16] Faergeman NJ, Knudsen J. Role of long-chain fatty acyl-CoA esters in the regulation of metabolism and in cell signalling. Biochem J 1997; 323(Pt 1): 1-12.
[http://dx.doi.org/10.1042/bj3230001] [PMID: 9173866]

CHAPTER 8

Hormones

Nagendra Nath Yadav[1,*] and **Archana Pareek**[1]

[1] *Department of Chemistry, Nerist, Nirjuli, Itanagar-791109 (AP), India*

Abstract: This chapter includes an introduction, classification, and type of hormones, as well as their biological functions. This chapter also gives a brief discussion about the structure, properties, and regulatory action of hormones. Various plant hormones and their functions have been discussed in this chapter.

Keywords: Cortisol, Osteoporosis, Oxytocin, Plant hormones, Progesterone, Sex hormones, Testosterone, Vasopressin.

INTRODUCTION

The term 'hormone' is derived from the Greek word 'hormaein', which means "to set in motion". Hormones are specific molecules that act as chemical messengers. These are secreted directly into the blood by the endocrine glands. These are carried by the circulatory system to organs and tissues of the body to exert their functions [1]. The major body functions include monitoring of basic needs, hunger, to a complex mechanism like reproduction [2]. The hormones secreted into the blood come in contact with a number of cells, but only certain target cells are influenced by these. For each specific hormone, there are specific receptors in the target cells, which may be present on the surface of the cell membrane or inside the cell. As the hormone binds to the receptor, some changes occur, influencing the cellular function. Since the hormones influence the target cells over a distance hence this type of signaling is known as endocrine signaling [3]. Hormones also influence the neighboring cells when secreted into the interstitial fluid surrounding cells, which then diffuse to nearby target cells. This type of signaling is known as paracrine signaling. Sometimes the hormones cause changes to the same cell that releases them and do not enter the other cells, known as autocrine signaling [3].

* **Corresponding author Nagendra Nath Yadav:** Department of Chemistry, NERIST, Nirjuli, Itanagar-791109 (AP), India. E-mail:nny@nerist.ac.in

Meera Yadav & Prof. Hardeo Singh Yadav (Eds.)

The glands, part of the endocrine system shown in Fig. (**1**), that produce hormones are ductless; hence they are released directly into the blood. The glands are:

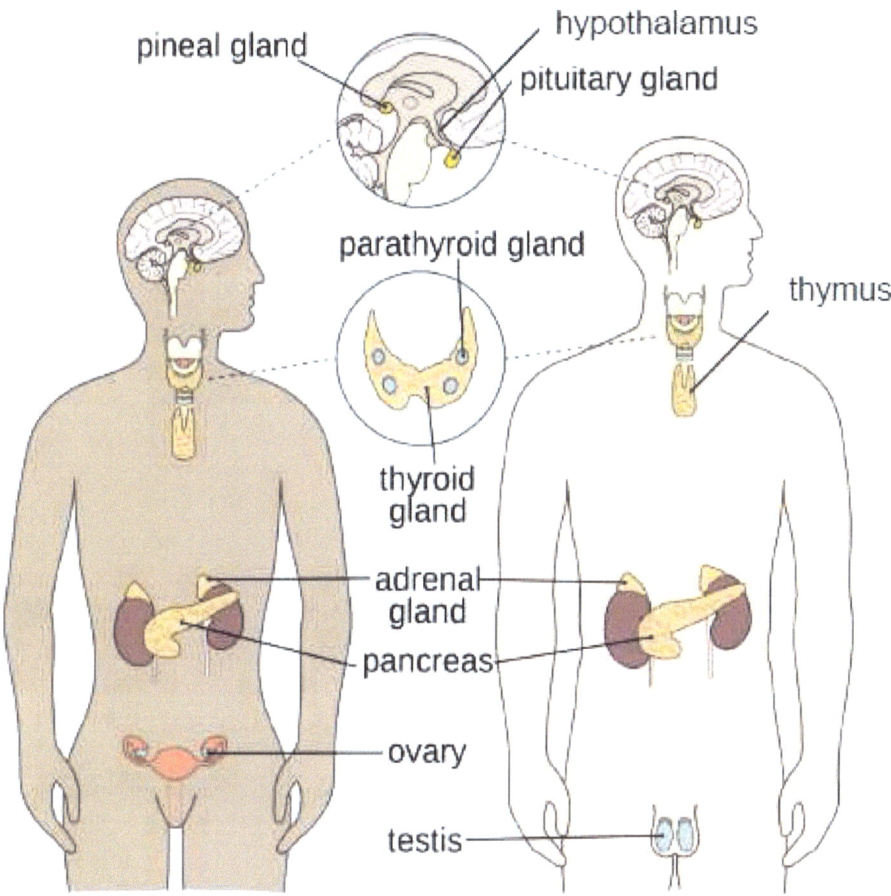

Fig. (1). A diagram of the endocrine system [4]. (image courtesy: scioly.org).

- **Hypothalamus**: It combines the endocrine and nervous systems. It initiates endocrine responses when it receives input from body and brain areas. The hormones produced by it play a role in hunger, moods, body temperature, and release of hormones from other glands and can also control thirst, sleep, and sex drive [5]. Hypothalamus includes the following hormones [6]-
  - Corticotropin-releasing hormone: It stimulates the anterior pituitary to release adrenocorticotropic hormone (ACTH).
  - Gonadotropin-releasing hormone: It stimulates the anterior pituitary to release luteinizing hormone and follicle-stimulating hormone.

- Thyrotropin-releasing hormone: It stimulates the anterior pituitary to release thyroid-stimulating hormone (TSH).
- Growth hormone-releasing hormone: It stimulates the anterior pituitary to release growth hormone (GH).
- Oxytocin: It is stored in the posterior pituitary. It induces labour during childbirth and controls bleeding after childbirth. It also induces the release of milk from mammary glands.
- Antidiuretic hormone (vasopressin): It is also stored in the posterior pituitary. It helps the kidneys to reabsorb water.

- **Parathyroid**: Releases hormones that balance the amount of calcium in the body [7].
- **Thymus**: Releases hormones that produce T-cells and help in the function of the adaptive immune system [8].
- **Pancreas**: It is located between the stomach and proximal portion of the small intestine. It helps to control blood sugar levels by producing hormones like insulin and glucagon. Insulin promotes the uptake of glucose by liver and muscle cells and converts it to glycogen, thus decreasing blood sugar levels. But glucagon promotes the breakdown of glycogen, stored in liver and muscle cells, and releases glucose, thus increasing blood sugar levels [9].
- **Thyroid**: It is located in the neck and is regulated by the hypothalamus-pituitary axis. Hormones produced by this gland are associated with regulating metabolism and growth, calorie-burning, and heart rate. Hormones produced by this gland are thyroxine (T_4) and triiodothyronine (T_3) [10].
- **Adrenal**: It consists of two glands located in each kidney. It consists of two parts adrenal cortex and adrenal medulla, each producing different sets of hormones.
  - Adrenal cortex: It releases mineralocorticoids, *e.g.* aldosterone, and glucocorticoids, *e.g.* cortisol. Aldosterone helps to increase reabsorption by kidneys and regulates water balance. Cortisol is released when blood sugar levels decrease due to long-term stress. It increases the blood glucose levels through glycogenesis by liver cells. It also aids in the metabolism of proteins, carbohydrates, and fats [11].
  - Adrenal medulla: It releases epinephrine, *e.g.* adrenaline, and norepinephrine, *e.g.* noradrenaline. These are released as a result of short-term stress and cause an increase in heart rate, breathing rate, contraction of cardiac muscle, blood glucose level and blood pressure [11].
- **Pituitary**: It is also known as the "master control gland" because it controls other glands and releases hormones that trigger growth. It is present at the base of the brain and is attached through the pituitary stalk to the hypothalamus. It consists of two distinct regions: the anterior pituitary gland and the posterior pituitary gland [12].

  - Anterior pituitary: It is regulated by hormones that are produced by the hypothalamus. The hypothalamus signals the anterior pituitary to release the following hormones: adrenocorticotropic hormone (ACTH), growth hormone, follicle-stimulating hormone (FSH), luteinizing hormone (LH), thyroid-stimulating hormone and prolactin [12].
  - Posterior pituitary: It is regulated by neurosecretory cells produced by the hypothalamus. It releases the following hormones: antidiuretic hormone and oxytocin [12].
- **Pineal**: It is also known as the thalamus. It releases hormones that affect sleep [13].
- **Ovaries**: It (present only in females) releases female sex hormones *i.e.* estrogen, progesterone, and testosterone. It regulates the maintenance of ovarian and menstrual cycles and also prepares the uterus for pregnancy [14].
- **Testes**: It (present only in males) releases male sex hormones *i.e.*, testosterone, and also regulates the production of sperm.

A wide range of regulatory systems creates specific biochemical signals in response to which hormones are secreted. For example, insulin synthesis is affected by blood sugar; parathyroid hormone synthesis is affected by serum calcium concentration; the amount of gastric juice and pancreatic juice (released by stomach and pancreas) is the input of the small intestine; hence the small intestine releases hormones to stimulate or inhibit the stomach and pancreas. Some hormones are water-soluble, while some are lipid-soluble. Water-soluble hormones are rapidly transported through the circulatory system, and lipid-soluble hormones combine with carrier plasma glycoproteins to form lipid-protein complexes. Some of the hormones, like insulin or growth hormones, become completely active as soon as these are released in the bloodstream, while some are prohormones that are activated in specific cells *via* a series of activation steps.

CLASSIFICATION OF HORMONES

Hormones are classified into five categories [15]:

1. According to chemical nature
2. On the basis of the mechanism of action
3. According to the nature of an action
4. According to effect
5. On the basis of stimulation of Endocrine Glands

According to the Chemical Nature

- Steroid hormones: Hormones like testosterone, estrogen, progesterone falls under this category. They are derived from cholesterol.

- Amine hormones: As the name indicates, these are made up of amines, mainly the derivative of the amino acid tyrosine. *e.g.* T_3, T_4, epinephrine, norepinephrine.
- Peptide hormones: These are simple linear chains made up of few amino acid residues. *e.g.* Oxytocin and vasopressin.
- Protein hormones: These are present in primary, secondary, and tertiary configurations and are made up of a large number of amino acid residues. *e.g.* insulin, glucagon, STH, *etc.*
- Glycoprotein hormone: These are conjugated proteins with carbohydrates like mannose, galactose, *etc. e.g.* LH, FSH, TSH, *etc.*
- Eicosanoid hormones: These are derivatives of small fatty acids with a variety of arachidonic acids. *e.g.* prostaglandins [15].

On the Basis of the Mechanism of Action

- Group I hormones: These hormones form hormone receptor complex by binding to intracellular receptors and then their biochemical functions are mediated. These are derived from cholesterol and are lipid soluble in nature. *e.g.* estrogen, progesterone, testosterone, T_3, T_4, *etc.*
- Group II hormones: These hormones stimulate the release of certain molecules, known as second messengers, by binding to cell surface receptors which then perform the biological functions. These hormones are lipophobic in nature, with short half-lives. *e.g.* ACTH, FSH, LH, where the second messenger is cAMP; TRH, GnRH, Gastrin, where the second messenger in phospholipid/inositol/Ca^{2+}; STH, LTH, insulin, oxytocin, where the second messenger is unknown [15].

According to the Nature of the Action

- Local hormones: They have local effects like paracrine secretion. *e.g.* testosterone.
- General hormones: These are transported by circulation to organ/tissue. *e.g.* insulin, thyroid hormone, *etc.* [15].

According to the Effect

- Kinetics hormones: These cause muscle contraction, glandular secretion, pigment migration, *etc. e.g.* pinealin, MSH, epinephrine.
- Metabolic hormones: The rate of metabolism is changed by these hormones. *e.g.* insulin, glucagon, PTH, *etc.*
- Morphogenetic hormones: These hormones causes growth and differentiation. *e.g.* STH, LTH, FSH, thyroid hormones, *etc.* [15].

On the Basis of Stimulation of Endocrine Glands

- Tropic hormones: Other endocrine glands are enhanced secretion by these hormones. *e.g.* TSH of the pituitary enhances secretion of the thyroid gland.
- Non-tropic hormones: The non-endocrine target tissues are affected by these hormones. *e.g.* oxygen consumption rate and metabolic activity of cells are increased by thyroid hormone [15].

The function of some of the hormones is given below:

Adrenocorticotropic Hormones (ACTH)

It is a tropic polypeptide hormone produced and secreted by the anterior pituitary gland in response to stress [16]. Its main function is to release cortisol (shown in Fig. **2**) from the cortex of the adrenal gland [17]. (shown in Fig. **3**).

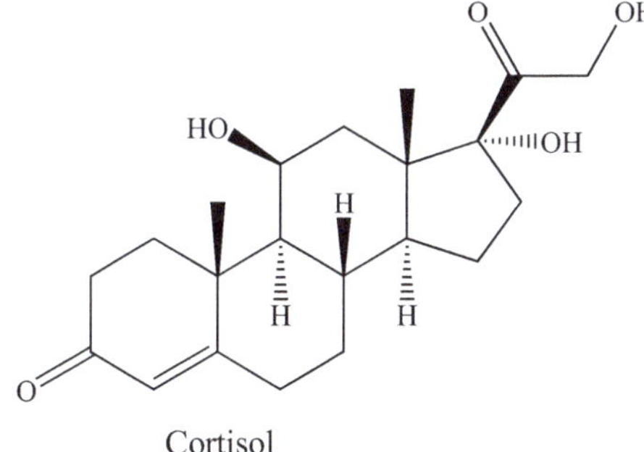

Fig. (2). Structure of Cortisol.

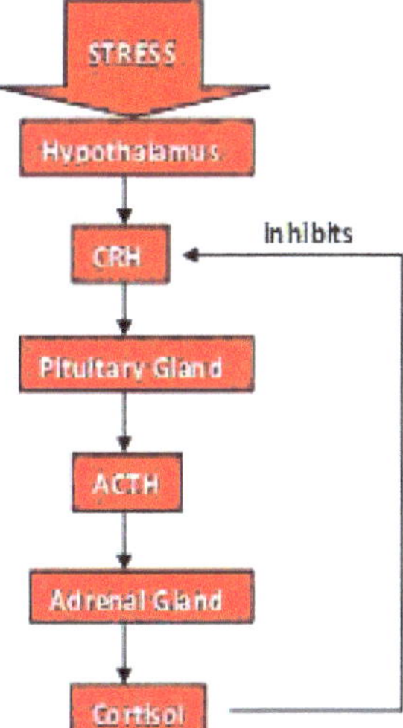

Fig. (3). Function of Cortisol.

Deficiency of ACTH leads to secondary or tertiary adrenal insufficiency while elevated levels of ACTH lead to primary adrenal insufficiency like Addison's disease (shown in Fig. **4**). Its structure consists of 39 amino acids. The first 13 amino acids of ACTH can be cleaved to form α-melanocyte-stimulating hormone (α-MSH). The molecular weight of ACTH in humans is 4540 u and its half-life in human blood is about 10 minutes [18].

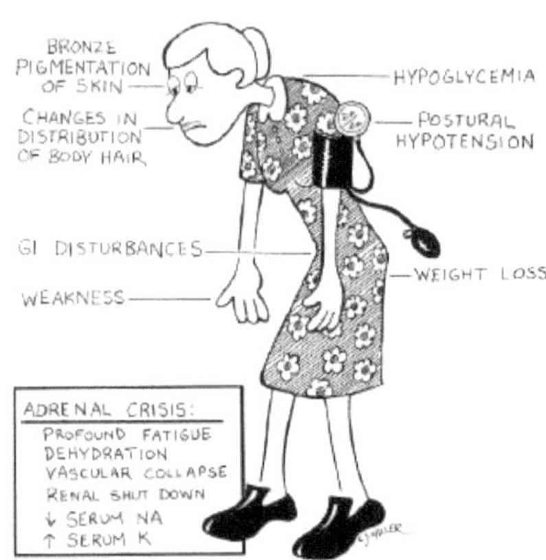

Fig. (4). Deficiency of ACTH [19]. (image courtesy: edu.glogster.com).

Thyroid-Stimulating Hormone (TSH)

It is a pituitary hormone stimulating the thyroid gland for the production of thyroxine (T_4) and triiodothyronine (T_3) (Structures shown in Fig. **5**) [20]. It is secreted by thyrotropic cells present in the anterior pituitary gland.

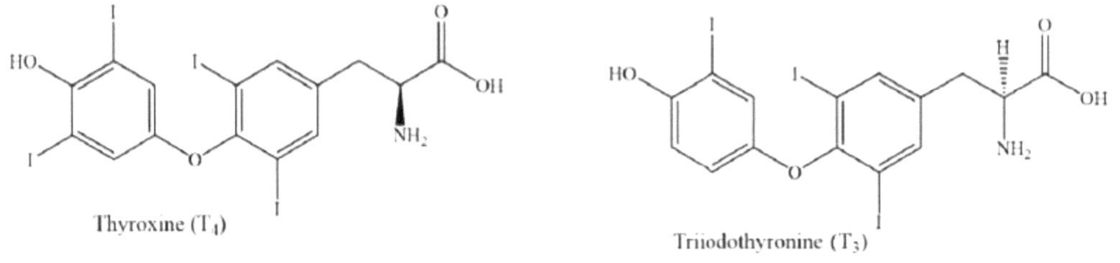

Thyroxine (T_4)

Triiodothyronine (T_3)

Fig. (5). Structure of T4.

T_4 has little effect on metabolism, but when converted to T_3 it stimulates metabolism (shown in Fig. **6**). *e.g.* T_3 and T_4 regulate heart rate and food processing through the intestine, low levels of T_3 and T_4 slow down the heart rate and cause constipation or weight gain, but high levels increase heart rate and cause diarrhea and weight loss. Conversion of T_4 to T_3 mostly occurs in the liver and other organs while only 20% occurs in thyroid itself. The concentration of T_3 and T_4 in blood regulates the production of TSH., a low concentration of T3 and T4, increases the production of TSH, while a high concentration of T_3 and T_4 decreases the production of TSH [20].

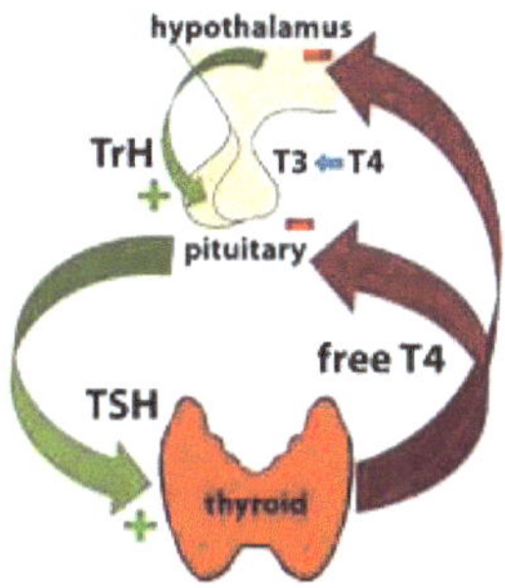

Fig. (6). Function of TSH [21].

Growth Hormones (GH)

It is a single-chain polypeptide with 191-amino acids secreted by somatotropic cells; hence it is also known as somatotropin. Its molecular weight is about 22124 u. This hormone stimulates the growth and cell reproduction in humans and also in other animals. The deficiency of Growth hormone causes growth failure in children and osteoporosis in adults (shown in Fig. 7). Excess growth hormone causes pituitary tumor, which is composed of somatotroph cells. These grow slowly and produce more GH. After becoming large, it causes a headache, deficiency in the release of other pituitary hormones and impaired vision due to pressure on optic nerves [22].

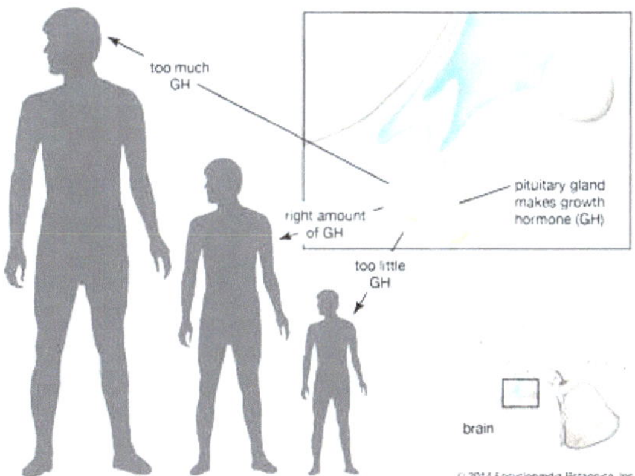

Fig. (7). Growth hormone [23].

Oxytocin

It is a nonapeptide (shown in Fig. **8**) (a peptide with 9 amino acids) with sequence Cys-Tyr-Ile-Gln-Asn-Cys-Pro-Leu-Gly-NH_2 [24]. The cysteine moieties are combined through a disulfide bridge. Its molecular mass is about 1007 dalton [25]. It is released by the posterior pituitary and produced by the paraventricular nucleus of the hypothalamus [26]. It plays an important role during and after childbirth, in bonding, lactation and orgasm [27]. Oxytocin is also known as the 'love hormone' because its level increases while hugging or orgasm.

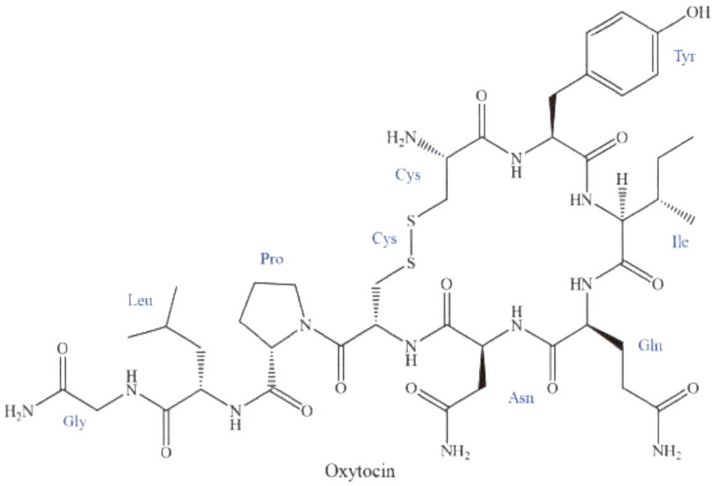

Fig. (8). Structure of oxytocin.

Vasopressin

It is also known as antidiuretic hormone, released from the posterior pituitary (shown in Fig. **9**) [28]. The tonicity of fluid in the body is regulated by vasopressin *i.e.* it maintains the water volume in extracellular fluid [29]. As a result of vasopressin, the kidney reabsorbs solute-free water which returns through the tubules of the nephron to the circulation, thus maintaining the tonicity of the fluid towards normal. The structure of vasopressin (shown in Fig. **10**) is similar to oxytocin, with a difference in the amino acid sequence at the 2nd positions. Vasopressin is also a nonapeptide in the sequence Cys-Tyr-Phe--ln-Asn-Cys-Pro-Arg-Gly-NH_2 with the cysteine moieties binding through a disulfide bond and the C-terminus of the sequence is converted to primary amide [30]. Its deficiency causes diabetes insipidus.

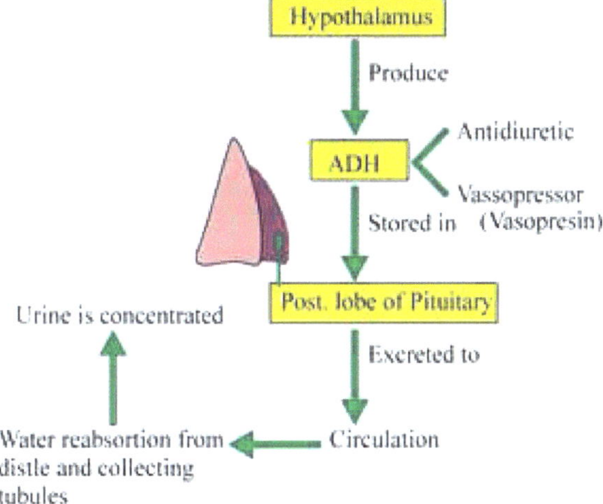

Fig. (9). Function of vasopressin [31].

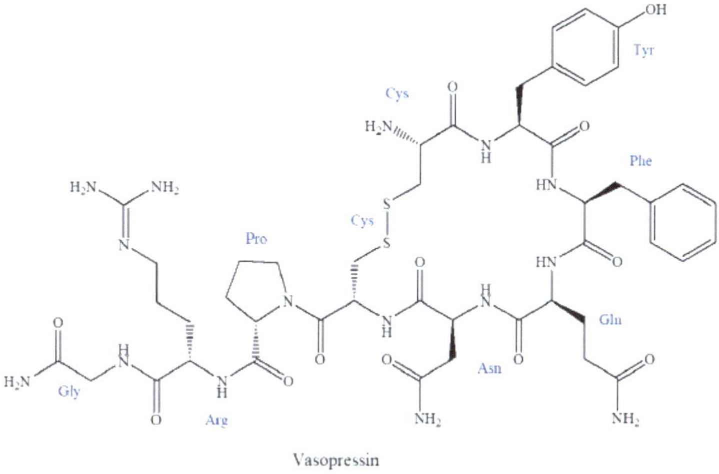

Vasopressin

Fig. (10). Structure of vasopressin.

Aldosterone

It is a steroid hormone and produced by zona glomerulosa in the adrenal cortex. It is the main mineralocorticoid hormone [32]. It plays an important role in the homeostatic regulation of sodium and potassium levels and blood pressure. It acts on the mineralocorticoid receptors present in the distal tubules and collecting ducts of nephron, influencing reabsorption of sodium and potassium excretion (shown in Fig. **11**) [33]. The structure of aldosterone is shown in Fig. (**12**). If dysregulated, aldosterone causes cardiovascular and renal diseases [34].

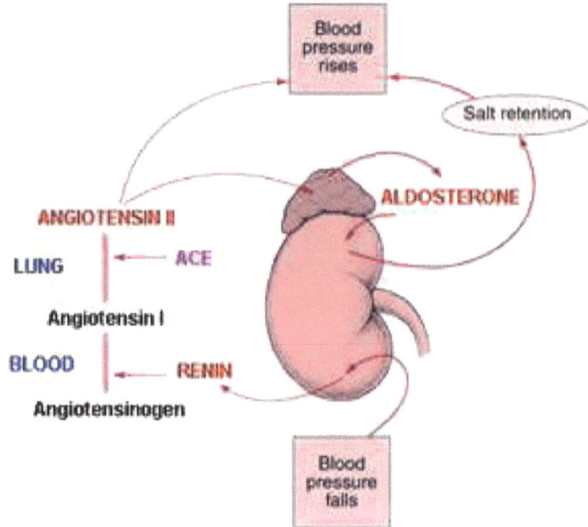

Fig. (11). The function of aldosterone [35].

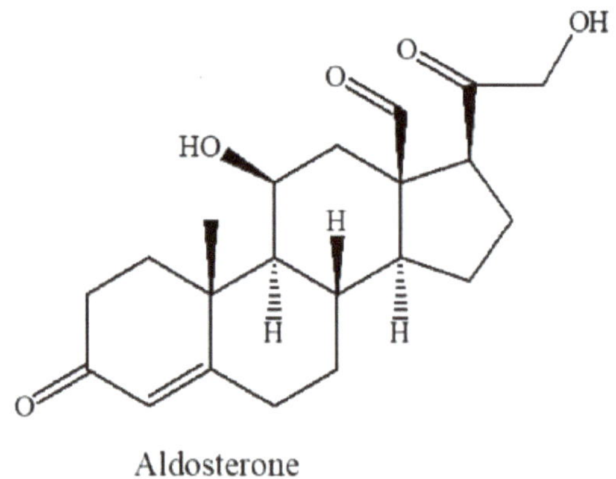

Fig. (12). Structure of aldosterone.

Adrenaline

It is also known as epinephrine (structure shown in Fig. **13**) and is released by adrenal glands [36]. It plays a crucial role in fight and flight response by increasing blood pressure and heart rate, pupil dilation, alter the body's metabolism to increase the blood glucose levels [37] (shown in Fig. **14**). As a medicine, it is used to treat cardiac arrest, superficial bleeding, and anaphylaxis [38].

Adrenaline

Fig. (13). Structure of adrenaline.

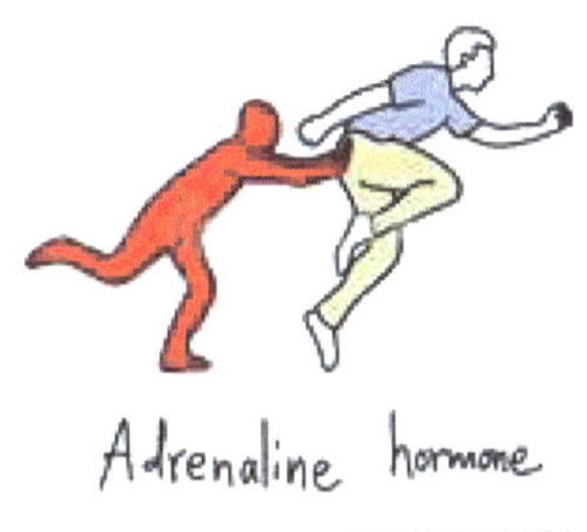

Fig. (14). The function of adrenaline [39].

Noradrenaline

It is also known as Norepinephrine (shown in Fig. **15**). It acts as a hormone and neurotransmitter in the brain and the body. It is synthesized from tyrosine in the adrenal medulla *via* series of enzymatic steps. Release of noradrenaline is high during stress or danger and is low during sleep [40].

Noradrenaline

Fig. (15). Structure of noradrenaline.

Estrogen

Also named as 'oestrogen', it is the primary sex hormone in females. It develops and regulates secondary sex characteristics and the reproductive system of the female. There are three types of estrogens in females having estrogenic hormonal activity *i.e.* estradiol, estrone and estriol, of which estradiol is the most potent (shown in Fig. **16**). Estetrol is also a type of estrogen produced only during pregnancy. All four types of estrogens are synthesized from androgens by the enzyme 'aromatase' [41].

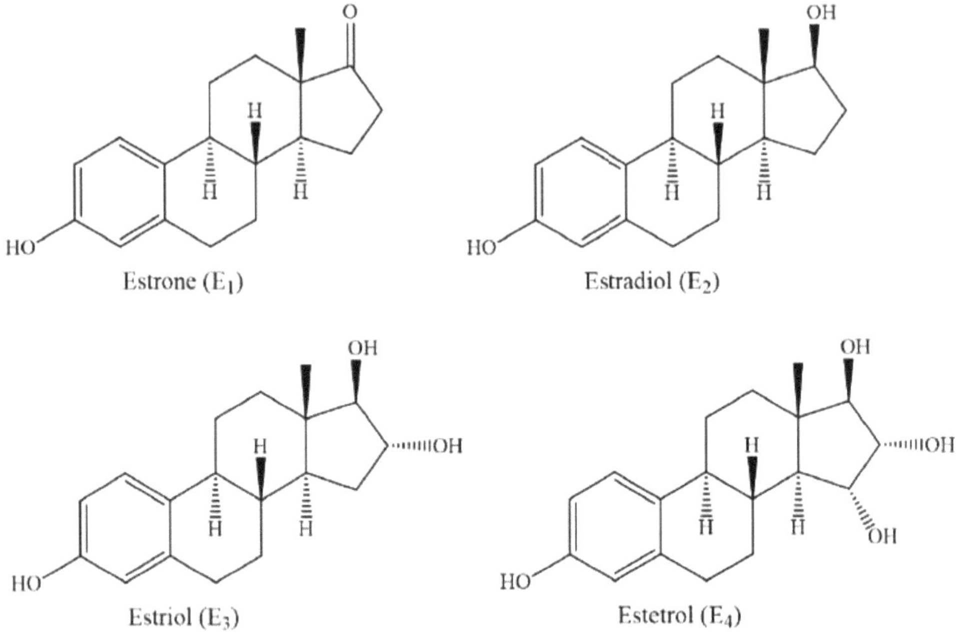

Fig. (16). Structure of estrone, estradiol, estriol and estetrol respectively.

In terms of serum levels, estrone is predominantly present during menopause and estriol is predominant during pregnancy. In non-pregnant females, from menarche to menopause stage, the most important estrogen is estradiol. Estradiol is the strongest estrogen among all, while estriol, despite being plentiful, is the weakest [42].

Progesterone

In the body, the most important progestogen is progesterone (shown in Fig. **17**). It is a sex hormone in the female which is involved in pregnancy, embryogenesis and menstrual cycle. It also acts as a neurosteroid in brain function. It has a role as medication in menopausal hormone therapy. The most important function of progesterone is in thickening the uterus lining every month, to receive a fertilized egg and nourish it. If implantation occurs, more progesterone is produced throughout the pregnancy. During this period, progesterone combines with estrogen and suppresses further ovulation and also stimulates the growth of milk-producing glands. If implantation does not occur then, progesterone level drops, breaking down the endometrium leading to menstruation [43].

Progesterone

Fig. (17). Structure of progesterone.

Testosterone

It is the primary sex hormone in males (structure shown in Fig. **18**). It plays an important role in developing secondary sexual characteristics and reproductive tissues in males [44]. It also prevents osteoporosis [45]. Testosterone is biosynthesized from cholesterol [46]. It is secreted by testicles of males and even by ovaries of females but to a very lesser extent. As a medication, testosterone has been used in the treatment of breast cancer in women and to increase testosterone levels in males [47].

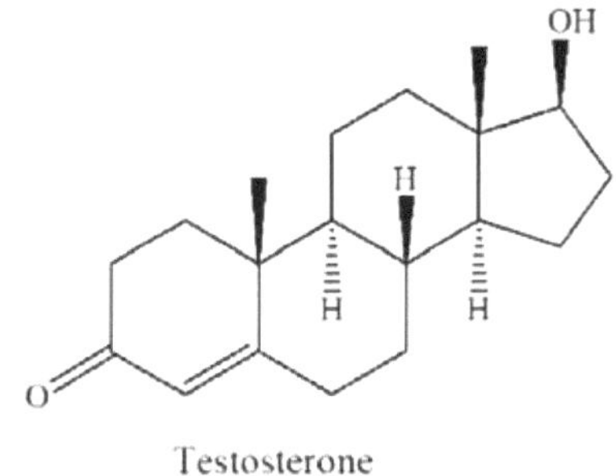

Fig. (18). Structure of testosterone.

Luteinizing Hormone (LH)

It is also known as 'lutrophin' [48]. It is produced in the anterior pituitary gland by gonadotropic cells. In females, it develops the corpus luteum, stimulates the production of progesterone and triggers ovulation, whereas in males it stimulates the production of testosterone [49] (shown in Fig. **19**). As a medication, LH is mixed with FSH and is available in the form of menotropin and urinary gonadotropins, which are commonly used for infertility therapy [49].

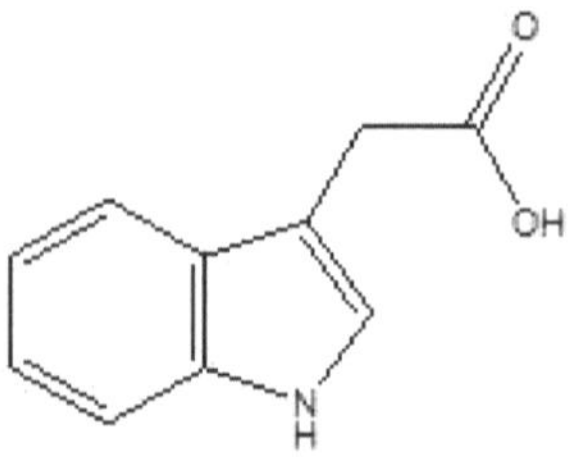

Auxin (Indole-3-acetic acid)

Fig. (19). Production of testosterone [51].

Follicle-stimulating Hormone (FSH)

It is a glycoprotein polypeptide hormone. It is also produced in the anterior pituitary gland by gonadotropic cells. It regulates the growth and development of the reproductive processes of the body. FSH enhances the maturation of primordial germ cells in both males and females [50].

Prolactin (PRL)

It is also known as luteotropic hormone or luteotropin. It is secreted from the pituitary gland during estrogen treatment, mating, nursing, and ovulation. Its main function is to enable mammals, mainly females, to produce milk. It also plays a role in regulating the immune system, pancreatic development and metabolism [52].

REGULATION OF HORMONES

Hormone secretion can be regulated by:

- Other hormones, known as tropic hormones
- Glands and organs
- A negative feedback mechanism

The tropic hormones are secreted by the anterior pituitary in the brain, the hypothalamus and the thyroid gland. The hypothalamus releases thyrotropin-releasing hormone (TRH) which stimulates the release of thyroid-stimulating hormone (TSH) by the pituitary, which in turn (TSH) enhances the thyroid gland to secrete more thyroid hormones. Thus, TRH and TSH act as tropic hormones [53].

Organs and glands regulate the hormones by monitoring blood content. For example, if glucose levels in the blood are too low, then the pancreas releases the hormone glucagon to increase glucose levels and if glucose levels in the blood are too high, then it secretes insulin to lower down the glucose levels [53].

The hormones are also regulated by a negative feedback system. For example, if there is an excess of thyroxine (thyroid hormone) in the blood then the pituitary stops releasing TSH, as TSH enhances the release of thyroxine. Similarly, if the amount of oxygen is too low in the blood then the kidney secretes the hormone erythropoietin (EPO) that stimulates the red bone marrow to produce the red blood cells and when the level of oxygen in blood becomes normal, then the kidney slows down the release of EPO [53].

A feedback system can sometimes also be positive, *e.g.* during childbirth, as the labor contractions increase, the oxytocin hormone begins to release in higher amounts. Also, during the lactation period increase nursing leads to an increase in milk production [53].

HORMONES IN PLANTS

Plant hormones, also known as phytohormones, are produced in the plants and required in extremely low concentrations. These regulate the growth of the plant; without them, plants would be a mass of undifferentiated cells. Unlike animals, plants do not have hormone-producing glands instead every cell in the plants can produce hormones. Plant hormones affect leaf formation and stem growth, time of flowering, fruit development and ripening, longevity and plant death [54].

Type of Plant Hormones

There are generally five types of plant hormones. These are Auxins, Cytokinins, Gibberellins, Ethylene and Abscisic acid. Auxins, Cytokinins and Gibberellins are growth promoters while ethylene and abscisic acid are growth inhibitors.

Auxins

The first plant hormone to be identified and studied is auxin (indole-3-acetic acid) (shown in Fig. **20**). It is formed in shoot tips (mainly in young leaves), in embryos and parts of the flower and seeds [55]. It requires energy for its transport from cell to cell. The movement is polar (moves in one direction only), known as basipetal movement (downward movement in the shoot) and acropetal movement (upward movement in roots) (as shown in Fig. **21**).

Auxin (Indole-3-acetic acid)

Fig. (20). Structure of auxin.

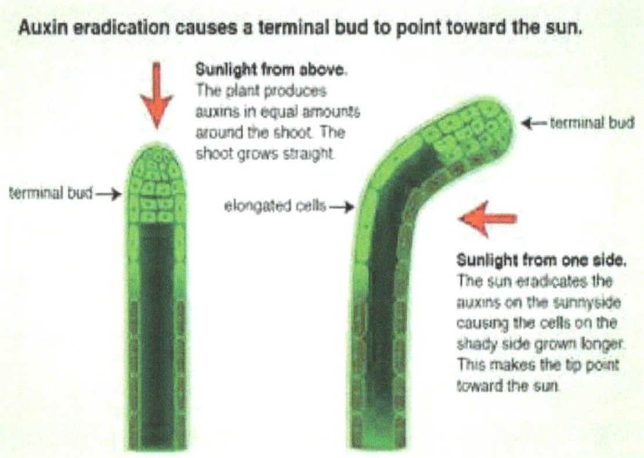

Fig. (21). The function of auxin [57].

The productions of other hormones are also promoted by auxins [55]. These alter the cell wall plasticity and affect cell elongation; differentiate secondary xylem, divides cambium, stimulates ethylene production, thus inhibiting the growth of lateral bud. They also initiate the growth of adventitious roots. The abscission of fruits and leaves is suppressed by auxins. Auxin is produced from amino acid tryptophan and also from carbohydrates known as glycosides. If present in large amounts, then auxins become toxic to plants mostly to dicots than monocots [56]. Therefore, synthetic auxins, like 2,4-D (2,4-dichlorophenoxyacetic), have been used for weed control. Synthetic auxins, like 1-Naphthaleneacetic acid, are also applied to enhance root growth in the cuttings of plants.

Cytokinins

These are a group of chemicals derived from the nitrogen-containing compound (shown in Fig. **22**), adenine, that stimulates cell division and shoot formation. The most active cytokinin is zeatin which is isolated from corn (Zea mays).

The cytokinin Zeatin

Fig. (22). Structure of cytokinin.

Cytokinins are present in root tips, leaves, seeds and fruits, which are active sites of cell division in plants. These are mainly synthesized in roots and transported through the xylem into leaves and fruits. Cytokinins work in a combined form with auxins and promote cell division (shown in Fig. **23**). Equal concentration of auxins and Cytokinins lead to normal cell division. If auxin is greater in concentration than Cytokinins, the roots are formed and if cytokinin concentration is greater than auxin then these lead to the formation of shoots [59]. Cytokinins also retard natural aging or senescence of the plants [58].

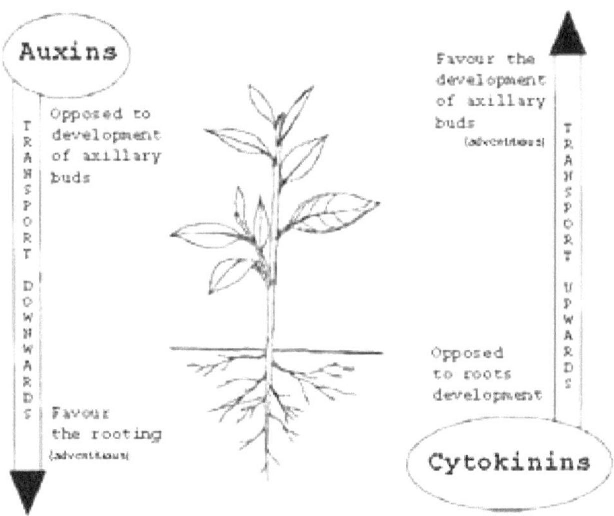

Fig. (23). Promotion of cell division [58]. (image courtesy: autoflower.net).

Gibberellins

Once in Japan it was found that due to a fungus, called Gibberella fujikuroi, infection in the rice plants grew too tall to fall over. It was due to a chemical produced by the fungus which stimulates the growth of rice plants. This chemical was then named Gibberellin [60]. Later it was found that plants themselves produce these chemicals. Till now more than 75 gibberellins have been isolated and they are named as GA1, GA2, GA3, and so on (shown in Fig. **24**).

Gibberellin A1 (GA1) Gibberellic acid (GA3)

Fig. (24). Structure of GA1.

The gibberellins are mainly present in seeds, young shoots and roots and are transferred through the xylem and phloem. These stimulate the growth of stems between their nodes. The growth of dwarf and rosette plants can be promoted by spraying gibberellins on them (shown in Fig. **25**). These also promote flowering and seed growth after germination [61].

Fig. (25). (a) An untreated plant **(b)** plant treated with gibberellins [62].

Ethylene

Ethylene (C_2H_4) (shown in Fig. **26a**) is a plant hormone that exists as a gas. It is produced from methionine in almost every cell of the plant. It is insoluble in water and hence does not accumulate in the cells but diffuses from it into the air and can affect surrounding plants as well. The production of ethylene is increased by auxins and ethylene itself. Large amounts are produced by ripening fruits, senescing flowers, apical meristem of shoots, and roots. This hormone affects the ripening of fruits (shown in Fig. **26b**) and rotting in plants and initiates the abscission of leaves and fruits. Gibberellins and ethylene concentration can determine the sex of flowers in monoecious plants. A high concentration of ethylene in flower buds induces carpellate flowers while a high concentration of gibberellins produces staminate flowers [63].

Fig. (26). (a): Ethylene [64].

Fig. (26). (b): Ripening of fruits [65]. (image courtesy: macmillanhighered.com).

Abscisic Acid

It is synthesized, when plants are mainly in stress, from carotenoids in plastids. It plays an important role in inhibiting cell growth *i.e.* it promotes leaf and seed dormancy and as soon as abscisic acid is dissipated from seeds then growth begins [66]. It also plays a role in the closing of stomata when plants are water-stressed, to minimize the loss of water [67]. (Structure is shown in Fig. **27**).

Abscisic acid

Fig. (27). Structure of abscisic acid.

HORMONES IN INSECTS

Hormones in insects regulate physiological, developmental and behavioral events. These are generated by epithelial glands. These hormones have been utilized to control the population of insects. There are three types of hormones in insects: Brain hormones, Molting hormones and Juvenile hormones (shown in Fig. **28**) [68].

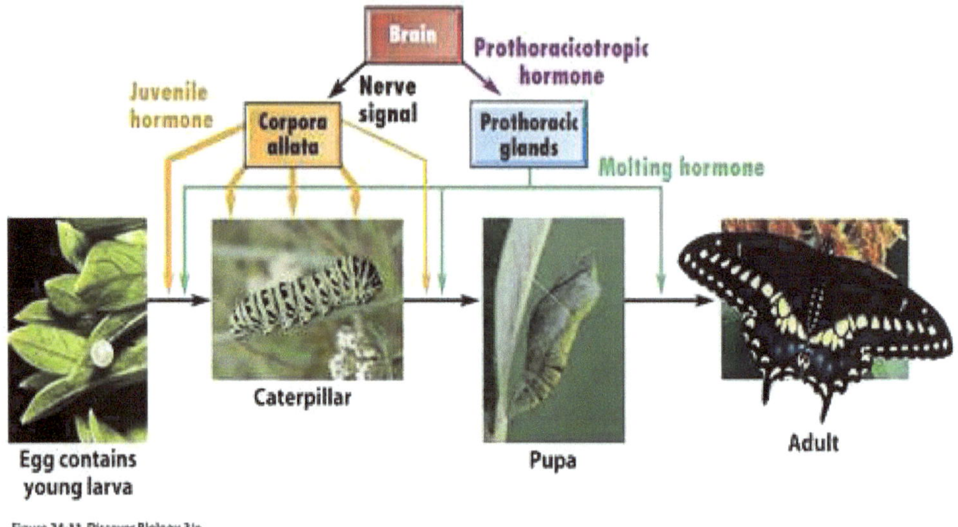

Fig. (28). Insects hormones [69]. (image courtesy: csus.edu).

Brain Hormones

These hormones are also known as neurohormones. These are produced by the central nervous system. These are also known as neuropeptides as these are generally peptides in nature [70]. These hormones play an important role in homeostasis, reproduction, metabolism and development of the insects as well as in the secretion of ecdysteroids and juvenile hormones [71]. The important brain hormones are: Pheromone biosynthesis stimulating hormone, PTTH, Diuretic hormone, Antidiuretic hormone, Allatotrophins and Allatostatins.

The first insect hormone discovered is the Prothoracicotropic hormone (PTTH). It is secreted by the corpus cardiacum (or corpus allatum in some insets), a discrete structure posterior to the brain, also known as the neurohemal organ. It acts on prothoracic glands leading to the release of molting hormone which stimulates the molting process in an insect [72].

The diuretic hormones regulate the water balance in insects. Both diuretic and anti-diuretic hormones control the excretion process in insects. These are produced in neurosecretory cells present in the nervous system, stored and released by neurohaemal sites. Anti-diuretic hormones either inhibit urine production or stimulate reabsorption. Both function as post-eclosion diuresis, postprandial dieresis, clearance of toxic wastes, restricting metabolite loss and excretion of excess metabolic water [73].

Allatostatin hormones are found in insects and crustaceans, inhibiting the generation of juvenile hormones and reducing their food intake. These hormones are of three different types A, B and C identified in different insects, however, all the three are present in the fruitfly, Drosophila. These inhibit the gut motility in insects thus controlling the food intake. These also inhibit the production of juvenile hormones [74].

Pheromones are chemicals which after secretion, acts like hormones outside the body of the individual secreting it and affect the behavior of the individual receiving it. Pheromones are used by basic unicellular prokaryotes as well as by complex multicellular eukaryotes. Even some ciliates, plants and vertebrates also communicate through pheromones [75]. There are many different types of pheromones like aggregation, alarm, epideictic, releaser, signal, primer, territorial, trail, sex, *etc.*

Aggregation pheromones result in a group of individuals at one location, leading to mate selection or defense against predators or overcome host resistance by a mass attack (shown in Fig. **29**). Aggregation pheromones are found in the species of orthoptera, Hemiptera, Coleoptera, Diptera and Dictyoptera [77].

Fig. (29). Aggregation of bugs[76]. (image taken from en.wikipedia.org).

Some species, when attacked by a predator, release alarm pheromones that trigger flight or aggression among the other individuals of the same species. *e.g.* 'Polistes exclamans' uses alarm pheromones to alert others about an incoming threat [78]. This pheromone is also present in some plants; they emit it when they are grazed, resulting in the production of tannin in neighbouring plants [79].

Releaser pheromones can alter the behavior of the individual receiving them. This pheromone evokes a rapid response and is quickly degraded. *e.g.* mammary pheromones released by rabbit (mothers) immediately triggers nursing behavior among their babies [80]. Animals like cats and dogs mark the perimeter of their claimed territory by using territorial pheromones which is present in their urine [80].

Social insects like ants mark their path by using trail pheromones, *i.e.* volatile hydrocarbon, that attract other ants and serves as a guide (shown in Fig. **30**). Species of wasps when finding a new nest then they use this type of pheromone to lead the other members of the colony to the new nest [81].

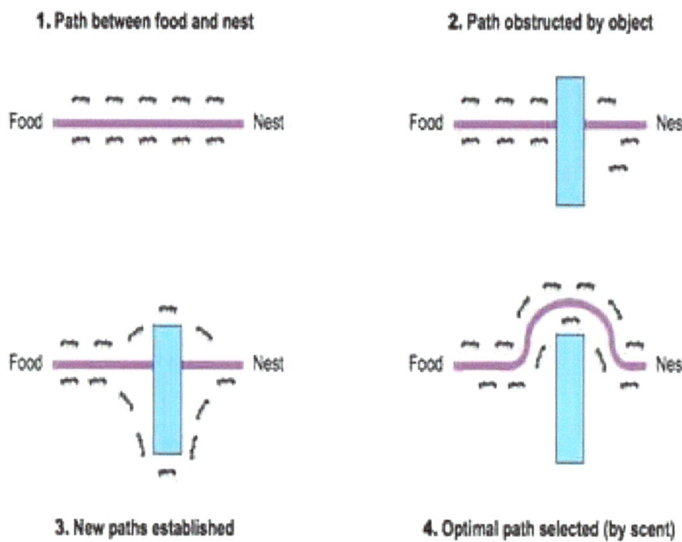

Fig. (30). Mark of the path by insects [82]. (image taken from bioninja.com.au)

Animals also use sex pheromones that indicate the availability of the female for breeding and male animals inform about their species and genotype.

Molting Hormones

These are secreted by prothoracic glands and are responsible for the normal molting, growth and maturation of insects. The types of molting hormones are ecdysone, ecdysterone, and ecdysteroids.

Ecdysteroids are structurally similar to androgens and are involved with reproduction and molting in insects. They derived their name from 'ecdysis' which means 'process of molting in insects. The ecdysteroids are derived from sterols, *e.g.* cholesterol, which the insects must obtain from their diet as they cannot synthesize it. Some ecdysteroids include ecdysterone, turkesterone, ecdysone and 20-hydroxyecdysone (structures are shown in Fig. **31**) [83].

Ecdysone

20-Hydroxyecdysone

Fig. (31). Structure of ecdysone and 20-hydroxyecdysone.

Ecdysteroids, if orally ingested, also have some biological effects in mammals. These also occur in phyla playing different roles like protecting agents against herbivorous insects. The most common molting hormone in insects is 20-Hydroxyecdysone (ecdysterone or 20E).

It is also produced by many plants where it disrupts the reproduction and development of insect pests. Due to the presence of a number of hydroxyl groups molting hormones are hydrophilic in nature. Molting hormones can be used in pest control by blocking their biosynthesis that disrupts the molting process. (shown in Fig. **32**).

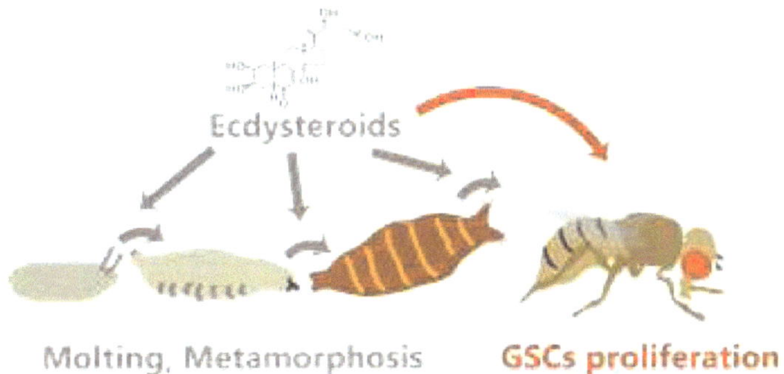

Fig. (32). Biological effect of ecdysteroids[84]. (image courtesy: thenode.biologists.com).

Juvenile Hormones

These are acyclic sesquiterpenoids, first discovered by Vincent Wigglesworth, which regulate reproduction, development, polyphenisms and diapauses. By preventing metamorphosis these hormones ensure the growth of larva. Corpora allata, a pair of endocrine glands present behind the brain, secretes juvenile hormones. These are also important in female insects for producing eggs [85]. Various Juvenile hormones (shown in Fig. **33**) have been discovered till now, like JH 0, JH I, JH II, JH III. JH 0, JH I and JH II are identified only in Lepidoptera while JHB_3 is found in dipteral. JH III is present in most insect species. These hormones are controlled by allatotropins and allatostatins. The allatotropins stimulate the corpora allata to signal the glands for producing juvenile hormones and allatostatins inhibit its production [86].

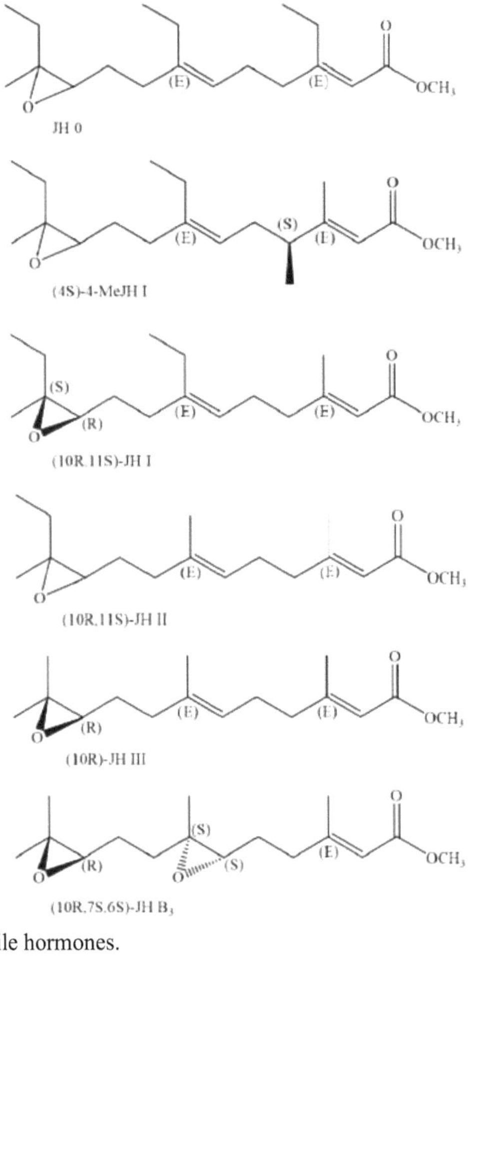

JH 0

(4S)-4-MeJH I

(10R,11S)-JH I

(10R,11S)-JH II

(10R)-JH III

(10R,7S,6S)-JH B_3

Fig. (33). Structures of juvenile hormones.

Juvenile hormone is unstable in light and expensive to synthesize, hence its synthetic analogues are used as insecticides, which prevents the larvae to develop into adult insects. However, at high levels larvae can still molt but only to bigger larvae and not to an adult, thus breaking the reproductive cycle of insects. *e.g.* Methoprene, due to its low toxicity, is used (approved by WHO) to control mosquito larvae in a drinking water cistern.

KEYWORDS

T-cells

A type of lymphocyte that is developed in the thymus gland.

Glucogenesis

Production of glucose.

Glycoprotein

A protein in which a carbohydrates group is attached to a polypeptide chain.

Addison's Disease

Is an endocrine disorder leading to a lack of production of steroid hormones from adrenal glands.

Osteoporosis

A disease in which bone weakens and increases the risk of breaking a bone.

Tonicity

Is a measure of the relative concentration of solutes present in a solution.

The endometrium is the mucous membrane lining inside the uterus.

Rosette Plant

Any plant with a circular arrangement of leaves or leaves-like structures.

SHORT-ANSWER QUESTIONS

1. What do you mean by the term 'hormone'?

2. What is Adrenocorticotropic hormone (ACTH)?

3. What is thyroid-stimulating hormone (TSH)?

4. Write the function of aldosterone?

5. What is the cortisol hormone?

6. How peptide hormones are transported?

7. What is the function of ACTH and LH?

8. What is oxytocin?

9. What are antidiuretic hormones?

10. What are the functions of hormones present in plants?

LONG-ANSWER TYPE QUESTIONS

1. How hormones are classified in animals?

2. Explain the role of thyroxine and triiodothyronine.

3. Differentiate between exocrine and endocrine glands.

4. What is the difference between oxytocin and vasopressin?

5. How hormones are regulated? What do you mean by a negative feedback system?

6. What is the function of auxins and Cytokinins?

7. What are brain hormones? Where are they produced? What is their function?

8. Write a short note on pheromones.

9. Write a short note on juvenile hormones.

10. Explain the function of molting hormones?

MULTIPLE CHOICE QUESTIONS

1. The plant hormone that is responsible for protecting the crops from falling is

a) Auxin

b) Gibberellins

c) Ethylene

d) Cytokinin

2. The hormone that exists in gaseous form is

a) Auxin

b) Ethylene

c) Abscisic acid

d) Florigens

3. The plant hormone responsible for fruit ripening is

a) Auxin

b) Cytokinins

c) Ethylene

d) Traumatic

4. The incorrect statement among the following is

a) Indoleacetic Acid (IAA) is a principle auxin

b) Auxins are the most important plant hormone

c) Auxins are also important in regulating the fall of leaves and fruits

d) Auxins are produced at the region of elongation

5. The plant hormone that breaks the dormancy of plant is

a) Gibberellins

b) Auxin

c) Ethylene

d) Cytokinin

6. The hormone that helps flowers to bloom is

a) Florigens

b) Traumatic

c) Auxin

d) None of the above

7. The hormone that helps in cell division and development in presence of auxin is

a) Florigens

b) Auxins

c) Cytokinins

d) Ethylene

8. In wooden plants, the activity of cambium is increase by

a) Ethylene

b) Gibberellins

c) Cytokinins

d) Auxins

9. Transport of auxin is

a) Symplast

b) Apoplast

c) Polar

d) Non-polar

10. Auxin does not have the function of

a) Enhancing cell division

b) Inducing callus formation

c) Maintaining apical dominance

d) Inducing dormancy

11. Molting process, also known as ecdysis, is controlled by

a) Auxin

b) Ecdysone

c) Cytokinin

d) Juvenile hormone

12. The sentence that best describes hormones is

a) All hormones are lipid-soluble.

b) Hormones are chemical messengers that are released into the environment.

c) Hormones are stable, long-lasting chemicals released from glands.

d) Hormones are relatively unstable and work only in the area adjacent to the gland that produced them.

13. Steroid hormones receptor lies

a) Within the plasma membrane

b) In the cytoplasm

c) In the blood plasma

d) Within the nuclear envelope

14. The hormone that regulates the retention of water in the kidney

a) Oxytocin

b) Thyroxin

c) Vasopressin

d) Prolactin

15. The anterior pituitary does not release the following hormone

a) Growth hormone

b) Thyroid-stimulating hormone

c) Gonadotropin-releasing hormone

d) Melanocyte-stimulating hormone

16. Parathyroid hormone ensures that

a) Sodium levels in urine are constant

b) Potassium levels in the blood don't escalate

c) The concentration of water in the blood is sufficient

d) Calcium levels in the blood never drop too low

17. Which of the following stimulates sodium ion reabsorption by kidneys and is released by the adrenal cortex

a) Cortisol

b) Epinephrine

c) Aldosterone

d) Glucose

18. The function of insulin is to

a) Reduce hyperglycemia

b) Agonistic to glucagon

c) Decrease glycogen storage in liver and muscle

d) All of the above

19. The hormones secreted in the liver are

a) Estrogen

b) Erythropoietin

c) Somatomedin

d) Rennin

20. Hypothalamus is connected to the pituitary through

a) Corpus callosum

b) Cerebral cortex

c) Infundibulum

d) Anterior

21. The main purpose of hormone is

a) To stimulate metabolism

b) To maintain growth

c) To keep the brain functioning

d) None of the above

22. Another name for thyroxine is

a) Thyroid

b) Thymus

c) Triiodothyronine

d) Tetraiodothyronine

23. Which is incorrect among the following?

a) Exocrine glands are not part of the endocrine systems.

b) The mammary glands are part of the endocrine systems.

c) Contents of the endocrine system are released into the bloodstream.

d) The endocrine system is composed of ductless glands

24. Target of ACTH is

a) Mammary glands

b) Adrenal cortex

c) Thyroid gland

d) Most cells

25. Which of the following stimulates and release growth hormones, thyroid

hormones and function of the thyroid gland?

a) Thyroid-stimulating hormone

b) Calcitonin

c) Oxytocin

d) Thyroxine

26. The precursor to T3 is

a) Calcitonin

b) Thyroxine (T4)

c) Melatonin

d) Parathyroid hormone

27. Calcium and phosphorus levels in the blood are regulated by

a) Prolactin

b) Calcitonin

c) Testosterone

d) Estrogen

28. Which of the following regulates cell growth and tissue differentiation?

a) Growth hormone

b) Oxytocin

c) Thyroxine

d) Triiodothyronine

29. Which of the following increases in response to stress?

a) Androgens

b) Adrenaline and Noradrenaline

c) Amylin

d) Insulin

30. Which of the following enhances gluconeogenesis (making new glucose) and glycogenolysis (breaking of glycogen)?

a) Glucagon

b) Oxytocin

c) Amylin

d) Insulin

CONSENT FOR PUBLICATION

Not Applicable.

CONFLICT OF INTEREST

The author confirms that this chapter contents have no conflict of interest.

ACKNOWLEDGEMENTS

Authors are thankful to Department of Chemistry, North Eastern Regional Institute of Science and Technology,Nirjuli, Itanagar for providing facilities to accomplish the work.

REFERENCES

[1] Shuster M. Biology for a Changing World, with Physiology. 2nd ed., New York, NY 2014.

[2] Neave N. Hormones and Behaviour: A Psychological Approach. Cambridge: Cambridge Univ. Press 2008.

[3] Ruhs S, Nolze A, Hübschmann R, Grossmann C. 30 YEARS OF THE MINERALOCORTICOID RECEPTOR: Nongenomic effects *via* the mineralocorticoid receptor. J Endocrinol 2017; 234(1): T107-24.
[http://dx.doi.org/10.1530/JOE-16-0659] [PMID: 28348113]

[4] https://scioly.org/wiki/index.php/Anatomy/Endocrine_System

[5] Saper CB, Scammell TE, Lu J. Hypothalamic regulation of sleep and circadian rhythms. Nature 2005; 437(7063): 1257-63.
[http://dx.doi.org/10.1038/nature04284] [PMID: 16251950]

[6] Spinazzi R, Andreis PG, Rossi GP, Nussdorfer GG. Orexins in the regulation of the hypothalamic-pituitary-adrenal axis. Pharmacol Rev 2006; 58(1): 46-57.
[http://dx.doi.org/10.1124/pr.58.1.4] [PMID: 16507882]

[7] Barbara Y. Heath, John W, Stevens Alan, et al Wheater's functional histology: a text and colour atlas. 5th ed. Edinburgh: Churchill Livingstone/Elsevier 2006; p. 336.

[8] Hall JE. Guyton and Hall Textbook of Medical Physiology. Philadelphia: Elsevier 2016; pp. 466-7.

[9] Philip J. Drawings. Wheater's functional histology: a text and colour atlas. By Deakin, Barbara Young; *et al.* Churchill Livingstone/Elsevier 2006; pp. 342-3.

[10] Colledge NR, Walker BR, Ralston SH, Eds. Davidson's principles and practice of medicine Illustrated by Robert Britton. Edinburgh: Churchill Livingstone/Elsevier 2010; p. 736.

[11] Rosol, T. J.; Yarrington, J. T.; Latendresse, J.; Capen, C. C.; Adrenal Gland: Structure, Function, and Mechanisms of Toxicity. Toxicologic Pathology 2001; 29(1): 41–48. [http://dx.doi.org/10.1080/019262301301418847]

[12] Mancall Elliott L, Brock David G. Cranial Fossae. Gray's Clinical Anatomy. Elsevier Health Sciences 2011; p. 154.

[13] Macchi MM, Bruce JN. Human pineal physiology and functional significance of melatonin. Front Neuroendocrinol 2004; 25(3-4): 177-95. [http://dx.doi.org/10.1016/j.yfrne.2004.08.001] [PMID: 15589268]

[14] Colvin CW, Abdullatif H. Anatomy of female puberty: The clinical relevance of developmental changes in the reproductive system. Clin Anat 2013; 26(1): 115-29. [http://dx.doi.org/10.1002/ca.22164] [PMID: 22996962]

[15] http://www.biologydiscussion.com/hormones/classification-hormones/classification-of-hormones-5-categories/18429

[16] Morton IK, Hall JM. Concise Dictionary of Pharmacological Agents: Properties and Synonyms. Springer Science & Business Media 2012; p. 84.

[17] Dibner C, Schibler U, Albrecht U. The mammalian circadian timing system: organization and coordination of central and peripheral clocks. Annu Rev Physiol 2010; 72: 517-49. [http://dx.doi.org/10.1146/annurev-physiol-021909-135821] [PMID: 20148687]

[18] Yalow RS, Glick SM, Roth J, Berson SA. Radioimmunoassay of human plasma acth. J Clin Endocrinol Metab 1964; 24(11): 1219-25. [http://dx.doi.org/10.1210/jcem-24-11-1219] [PMID: 14230021]

[19] https://edu.glogster.com/glog/addisons-disease-48020875/27im50cwdq6

[20] Korevaar TI, Muetzel R, Medici M, *et al.* Association of maternal thyroid function during early pregnancy with offspring IQ and brain morphology in childhood: a population-based prospective cohort study. Lancet Diabetes Endocrinol 2016; 4(1): 35-43. [http://dx.doi.org/10.1016/S2213-8587(15)00327-7] [PMID: 26497402]

[21] https://www.aacc.org/publications/cln/articles/2013/may/tsh-harmonization

[22] Ranabir S, Reetu K. Stress and hormones. Indian J Endocrinol Metab 2011; 15(1): 18-22. [http://dx.doi.org/10.4103/2230-8210.77573] [PMID: 21584161]

[23] https://www.britannica.com/science/growth-hormone

[24] Yamashita, Kaoru; Kitano, Takashi. Molecular evolution of the oxytocin–oxytocin receptor system in eutherians. Molecular Phylogenetics and Evolution 2013; 67(2): 520–528. [http://dx.doi.org/10.1016/j.ympev.2013.02.017]

[25] https://www.nibsc.org/documents/ifu/76-575

[26] Argiolas, A.; Gessa, G. L.; Central functions of oxytocin 1991; 15(2): 217–231. [http://dx.doi.org/10.1016/s0149-7634(05)80002-8]

[27] Chiras DD. Human Biology. 7th ed. Sudbury, MA: Jones & Bartlett Learning 2012; p. 262.

[28] Anderson DA. Dorland's Illustrated Medical Dictionary. 32nd ed., Elsevier 2012.

[29] Caldwell HK, Young WS III. Oxytocin and vasopressin: genetics and behavioral implications. In: Lajtha A, Lim R, Eds. Handbook of Neurochemistry and Molecular Neurobiology: Neuroactive Proteins and Peptides. 3rd ed. Berlin: Springer 2006; pp. 573-607.

[http://dx.doi.org/10.1007/978-0-387-30381-9_25]

[30] Burtis CA, Ashwood ER, Bruns DE. Tietz Textbook of Clinical Chemistry and Molecular Diagnostics. Elsevier Saunders 5th Ed. 2012; 837-49. ISBN 978-1-4160-6164-9

[31] www.labpedia.net/test/256

[32] Jaisser F, Farman N. Emerging roles of the mineralocorticoid receptor in pathology: toward new paradigms in clinical pharmacology. Pharmacol Rev 2016; 68(1): 49-75. [http://dx.doi.org/10.1124/pr.115.011106] [PMID: 26668301]

[33] Nicpon ME, Katja H. Human Anatomy & Physiology. 9th ed. Boston: Pearson 2013; p. 629.

[34] Gajjala PR, Sanati M, Jankowski J. Cellular and molecular mechanisms of chronic kidney disease with diabetes mellitus and cardiovascular diseases as its comorbidities. Front Immunol 2015; 6: 340. [http://dx.doi.org/10.3389/fimmu.2015.00340] [PMID: 26217336]

[35] http://www.endocrinesurgery.net.au/adrenal-function/

[36] Lieberman M, Marks A, Peet A. Marks' Basic Medical Biochemistry: A Clinical Approach. 4th ed. Philadelphia: Wolters Kluwer Health/Lippincott Williams & Wilkins 2013; p. 175.

[37] Khurana. Essentials of Medical Physiology. Elsevier India. 2008; 460..

[38] Everard ML. Acute bronchiolitis and croup. Pediatr Clin North Am 2009; 56(1): 119-133, x-xi. [x–xi.]. [http://dx.doi.org/10.1016/j.pcl.2008.10.007] [PMID: 19135584]

[39] https://www.pixtastock.com/illustration/38052285

[40] Aronson JK. "Where name and image meet"--the argument for "adrenaline". BMJ 2000; 320(7233): 506-9. [http://dx.doi.org/10.1136/bmj.320.7233.506] [PMID: 10678871]

[41] Burger HG. Androgen production in women. Fertil Steril 2002; 77 (Suppl. 4): S3-5. [http://dx.doi.org/10.1016/S0015-0282(02)02985-0] [PMID: 12007895]

[42] Labhart A. Clinical Endocrinology: Theory and Practice. Springer Science & Business Media 2012; p. 548.

[43] Jameson JL, De Groot LJ. Endocrinology: Adult and Pediatric E-Book. Elsevier Health Sciences 2015; p. 2179.

[44] Mooradian AD, Morley JE, Korenman SG. Biological actions of androgens. Endocr Rev 1987; 8(1): 1-28. [http://dx.doi.org/10.1210/edrv-8-1-1] [PMID: 3549275]

[45] Tuck S, Francis R. Testosterone, bone and osteoporosis. Front Horm Res 2009; 37: 123-32. [http://dx.doi.org/10.1159/000176049] [PMID: 19011293]

[46] Luetjens CM, Weinbauer GF. Testosterone: Biosynthesis, Transport, Metabolism and (Non-genomic) Actions. In: Nieschlag E, Behre HM, Nieschlag S, Eds. Testosterone: Action, Deficiency, Substitution. 4th ed. Cambridge: Cambridge University Press 2012; pp. 15-32. [http://dx.doi.org/10.1017/CBO9781139003353.003]

[47] Testosterone. Drugs.com. American Society of Health-System Pharmacists. 2015..

[48] Ujihara M, Yamamoto K, Nomura K, *et al.* Subunit-specific sulphation of oligosaccharides relating to charge-heterogeneity in porcine lutrophin isoforms. Glycobiology 1992; 2(3): 225-31. [http://dx.doi.org/10.1093/glycob/2.3.225] [PMID: 1498420]

[49] Ezcurra D, Humaidan P. A review of luteinising hormone and human chorionic gonadotropin when used in assisted reproductive technology. Reprod Biol Endocrinol 2014; 12(1): 95. [http://dx.doi.org/10.1186/1477-7827-12-95] [PMID: 25280580]

[50] https://www.webmd.com/women/fsh-test#1

[51] http://www.vivo.colostate.edu/hbooks/pathphys/endocrine/hypopit/lhfsh.html

[52] Bole-Feysot C, Goffin V, Edery M, Binart N, Kelly PA. Prolactin (PRL) and its receptor: actions, signal transduction pathways and phenotypes observed in PRL receptor knockout mice. Endocr Rev 1998; 19(3): 225-68.
[http://dx.doi.org/10.1210/edrv.19.3.0334] [PMID: 9626554]

[53] https://www.thoughtco.com/hormones-373559

[54] https://biology.tutorvista.com/biomolecules/plant-hormones.html

[55] Osborne DJ, McManus MT. Hormones, Signals and Target Cells in Plant Development. Cambridge University Press 2005; p. 158.
[http://dx.doi.org/10.1017/CBO9780511546228]

[56] Walz A, Park S, Slovin JP, Ludwig-Müller J, Momonoki YS, Cohen JD. A gene encoding a protein modified by the phytohormone indoleacetic acid. Proc Natl Acad Sci USA 2002; 99(3): 1718-23.
[http://dx.doi.org/10.1073/pnas.032450399] [PMID: 11830675]

[57] https://betterbonsai.co.za/index.php/plant-biology-auxin/

[58] https://www.autoflower.net/forums/threads/cytokinins-auxins-good-read.40635/

[59] Sipes DL, Einset JW. Cytokinin stimulation of abscission in lemon pistil explants. J Plant Growth Regul 1983; 2(1-3): 73-80.
[http://dx.doi.org/10.1007/BF02042235]

[60] Grennan AK. Gibberellin metabolism enzymes in rice. Plant Physiol 2006; 141(2): 524-6.
[http://dx.doi.org/10.1104/pp.104.900192] [PMID: 16760495]

[61] Tsai FY, Lin CC, Kao CH. A comparative study of the effects of abscisic acid and methyl jasmonate on seedling growth of rice. Plant Growth Regul 1997; 21(1): 37-42.
[http://dx.doi.org/10.1023/A:1005761804191]

[62] https://ramneetkaur.com/gibberellins/

[63] Wang Y, Liu C, Li K, *et al.* Arabidopsis EIN2 modulates stress response through abscisic acid response pathway. Plant Mol Biol 2007; 64(6): 633-44.
[http://dx.doi.org/10.1007/s11103-007-9182-7] [PMID: 17533512]

[64] https://www.askiitians.com/blog/bet-didnt-know-plant-hormones/

[65] https://www.macmillanhighered.com/BrainHoney/Resource/6716/digital_first_content/trunk/test/hillis2e/hillis2e_ch26_4.html

[66] Feurtado JA, Ambrose SJ, Cutler AJ, Ross AR, Abrams SR, Kermode AR. Dormancy termination of western white pine (Pinus monticola Dougl. Ex D. Don) seeds is associated with changes in abscisic acid metabolism. Planta 2004; 218(4): 630-9.
[http://dx.doi.org/10.1007/s00425-003-1139-8] [PMID: 14663585]

[67] Else MA, Coupland D, Dutton L, *et al.* Decreased root hydraulic conductivity reduces leaf water potential, initiates stomatal closure, and slows leaf expansion in flooded plants of castor oil (Ricinus communis) despite diminished delivery of ABA from the roots to shoots in xylem sap. Physiol Plant 2001; 111(1): 46-54.
[http://dx.doi.org/10.1034/j.1399-3054.2001.1110107.x]

[68] https://www.nature.com/subjects/insect-hormones

[69] https://www.csus.edu/indiv/l/loom/preview19.htm

[70] Williams CM. Physiology of Insect Diapause. X.. An endocrine mechanism for the influence of temperature on the diapausing pupa of the cecropia silkworm. Biological Bulletin 1956; 110(2): 201-18.
[http://dx.doi.org/10.2307/1538982]

[71] Williams, Carroll M.; The juvenile hormone of insects 1956; 178(4526): 212–213. [http://dx.doi.org/10.1038/178212b0]

[72] Klowden MJ. Physiological Systems in Insects. 2nd ed., Academic Press 2007.

[73] Coast GM, Orchard I, Phillips JE, *et al.* Insect diuretic and antidiuretic hormones. Adv In Insect Phys 2002; 29: 279-409. [http://dx.doi.org/10.1016/S0065-2806(02)29004-9]

[74] Stay B, Tobe SS. The role of allatostatins in juvenile hormone synthesis in insects and crustaceans. Annu Rev Entomol 2007; 52: 277-99. [http://dx.doi.org/10.1146/annurev.ento.51.110104.151050] [PMID: 16968202]

[75] Definition of pheromone. MedicineNet Inc. 2012..

[76] https://en.wikipedia.org/wiki/Pheromone#/media/File:Bug_aggregation.jpg

[77] www.msu.edu2018.

[78] Post DC, Downing HA, Jeanne RL. Alarm response to venom by social wasps Polistes exclamans and P. fuscatus. J Chem Ecol 1984; 10(10): 1425-33. [http://dx.doi.org/10.1007/BF00990313] [PMID: 24318343]

[79] Bothma J du P. Game Ranch Management. 4th ed., Van Schaik 2002.

[80] Kimball JW. Pheromones. Kimball's Biology Pages 2008.

[81] Excited ants follow pheromone trail of the same chemical they will use to paralyze their prey. 2006..

[82] http://www.vce.bioninja.com.au/aos-2-detecting-and-respond/coordination--regulation/pheromones.html

[83] Hussain SM, Brandon M, Doe CQ. Steroid hormone induction of temporal gene expression in Drosophila brain neuroblasts generates neuronal and glial diversity. eLife 2007; 6.

[84] https://thenode.biologists.com/new-role-insect-steroid-hormone-link-mating-germl-ne-stem-cells/highlights/

[85] Nijhout HF. Insect Hormones. Princeton: Princeton University Press 1994.

[86] Schooley DA. Schooley D.A. Identification of an allatostatin from the tobacco hornworm Manduca sexta. Proc Natl Acad Sci USA 88. 1991; 9458-9462.

CHAPTER 9

Fundamentals of Thermodynamics: Principle Applicable To Biological Processes

Nene Takio[1,*], **Pratibha Yadav**[2] and **Anindita Hazarika**[1]

[1] *Department of Chemistry, NERIST, Nirjuli, Itanagar-791109 (AP), India*

[2] *Department of Chemistry, IIT Delhi, Hauz Khas New Delhi-110016, India*

Abstract: All living organisms require energy from the surrounding and utilized by biomolecules that mediates the flow of energy from exergonic reactions to the energy requiring processes of life. This unit deals with the energy behind life in the form of three laws of thermodynamics. Derived the mathematical equations for the laws. Comparison of thermodynamic parameters like ΔH or ΔS and ΔG have been done to provide meaning ful insights about a process and its use in calculating the relative contributions of molecular phenomena to an overall process.

Keywords: Biomolecules, Calorimeter, Denaturation, Glycolysis, Phototrophs, Pyruvate.

INTRODUCTION

Living organisms need energy for various life processes like movement, biomolecular synthesis, transportation of ions and molecules across membranes, *etc.* All species gain energy from their environment and use it to carry out their life process. Studying certain energy transformation phenomena requires awareness of thermodynamics processes. Thermodynamics gives us an analysis of energy transformations occurring in the cell. Biological processes also obey the law of thermodynamics. The principles of thermodynamics help us to understand the relationship between heat, energy, and matter in the biological system. This chapter presents several basic thermodynamics principles that give us a broader understanding of various biological processes [1]. Sun is the ultimate source of all forms of energy. Living organisms rely directly on sunlight like plants (Fig. **1**). Some bacteria live in extreme conditions, such as Antarctica, also some blue-green algae grow in the lakes under dense layers of ice, and these also require energy in those remote areas. All living organisms require energy to synthesize

* **Corresponding author Nene Takio:** Department of Chemistry, NERIST, Nirjuli, Itanagar-791109 (AP), India. E-mail: nentakio@ymail.com

Meera Yadav & Prof. Hardeo Singh Yadav (Eds.)

food, to reproduce and store and carry out metabolic activities for survival. The energy of sunlight and its wavelength λ or frequency ν is interdependent, which is shown by the relation, $E = h\,c/\lambda = h\,\nu$, where

h = Planck's constant (6.634×10^{-34} Js)

c =speed of light in vacuum (2.99×10^{8} m/s)

Fig. (1). The sun is the source of energy of all virtue of life. We can harvest its energy in the form of electricity by using windmills driven by air heated by the sun.

What is the Basic Concept of Thermodynamics?

In every thermodynamic analysis, a distinction between system and surrounding is necessary. A system is the part of the universe we're concerned with. It may be a combination of Chemicals in a test tube, or a single cell, or a whole organism. The surrounding environment includes everything else other than the system. These systems are categorized into isolated, closed, and open types (as shown in Fig. **2**).

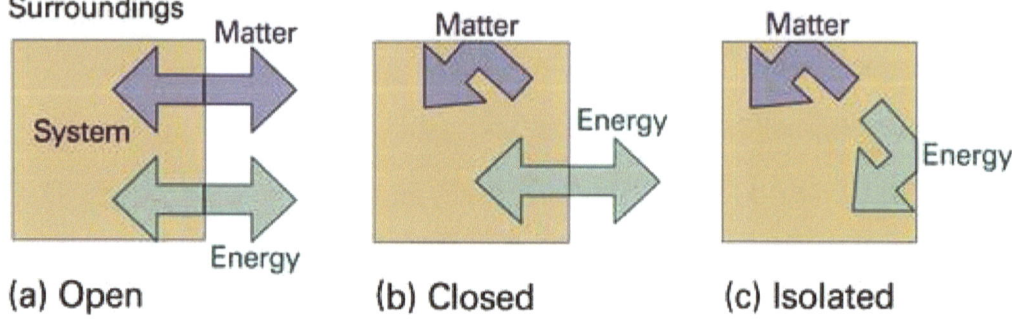

Figure 2-1
Atkins Physical Chemistry, Eighth Edition

Fig. (2). The characteristics of isolated, closed, and open systems. Isolated systems exchange neither matter nor energy with their surroundings. Closed systems may exchange energy, but not matter, with their surroundings. Open system may exchangeeither energy or matter with surrounding.

An isolated system doesn't share matter or energy with the external environment.

A closed system doesn't share any matter with the surroundings.

An open system has external interaction with both system and surroundings.

All living beings are usually open systems that can exchange matter and energy with the surrounding [1].

Even when the system is closed, it can still interact with its surroundings. One such kind of interaction is called work. If the surroundings exert a force, F, on the system and this force is exerted along a path of length l; the surrounding does work on the system:

$W = Fl$

If W = +ve, then it indicates work is done on the system.

If W = -ve, it indicates work is done by the system.

STATE VARIABLES

State variables that characterize the condition of a complex system are known as state functions.

Every State variable corresponds to one of the underlying state space coordinates. Such variables include macroscopic properties, including pressure, volume, temperature, density, composition, surface area, *etc.* For example, an ideal gas system. To define this system, we only have to specify the three variables p, v, and t. State variables have intensive and extensive properties. In intensive properties, the object does not change its size, whereas, in extensive properties, the object size changes. The extensive properties depend on the size of the system, *i.e.* when a system's size doubles, the extensive property also gets double. On the other hand, intensive properties will remain unchanged irrespective of changes in the size of the system. (a) Intensive properties: Temperature, pressure, concentration, density, dipole moment, refractive index, viscosity, surface tension, molar, volume, gas constant, specific heat capacity, vapour pressure, specific gravity, dielectric constant, emf of a dry cell, *etc.* (b) Extensive properties: Volume, energy, heat capacity, enthalpy, entropy, free energy, length, and mass.

ENERGY TRANSFER INTO HEAT & WORK

Every system has a fixed quantity of energy called internal energy. System internal energy can be transferred in two ways. i) one is the transfer of heat in the form of work. Whenever the system and surroundings have a difference in temperature ii) if two are in thermal contact, the energy in the form of heat is exchanged.

The flow of heat occurs from the body with a higher temperature to the body having a lower temperature. The internal energy of the body receiving heat is increased, whereas the internal energy of the body is decreased on losing heat. The second one is the pressure-volume work. Let's check an example of a system containing gas in a cylinder (as shown in Fig. **3**). Now the gas is compressed by a moving piston with constant pressure P. In moving the piston; the system does some amount of work. Thus, a decrease in its internal energy takes place. The force the piston exerts on the gas is expressed by equation F = PA, where A is the piston region, where force is exerted along with a distance 1. The area between the initial and the final volume is A, times the distance l, through which the piston moves:

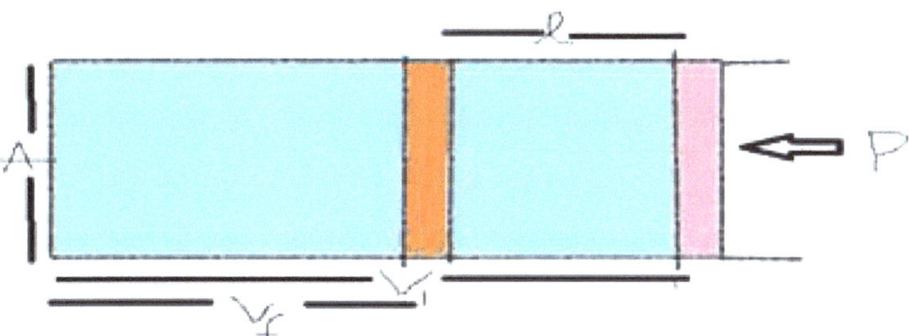

Fig. (3). An example of pressure- volume work of gas being compressed by an external pressure P.

$V_{initial} - V_{final} = Al$

However, the volume is decreasing and its change, ΔV is hence negative:

$\Delta V = V_{initial} - V_{final} = Al$

Then W = Fl =PΔV

Since Al = -ΔV W = -PΔV

Here –ΔV is negative W = +ve

Since work is done by the surrounding. (Fig. **3**). An example of pressure-volume work of gas being compressed by an external pressure P.

THE FIRST LAW: LAW OF CONSERVATION OF TOTAL ENERGY

At the beginning of thermodynamic understanding, it was discovered that heat is converted from one energy form to another, which signifies a change in total energy transfer is always conserved. The first law of thermodynamics states that the isolated system's total energy is retained. It is clearly a point about energy conservation. When work and/or heat are changed, the internal energy shifts. The internal energy depends only on a system's present state, hence classified as state functions. It does not depend on the direction of the reaction; hence it is not a path function. Internal energy, E of any system will only change in the form of heat or work when energy flows into or out of the system. The shift in internal energy for any phase that transforms one state (state 1) into another (state 2),

Change in internal energy is given by $\Delta E = E_2 - E_1 = q + w$,

'q' = heat absorbed from the surrounding, 'w' = work is done by the surroundings.

We typically deal with constant pressure in biological systems; hence the enthalpy, 'H' can be applied. It is termed as the heat transferred at constant pressure. The heat transferred (H) can be calculated by a van't Hoff plot of R (ln Keq) versus 1/T is plotted or by using a calorimeter. Examples of work done in biological systems include the flight of insects and birds, the circulation of blood by a pumping heart, the transmission of an impulse along a nerve, and the lifting of weight by someone who is exercising. However, if a person cannot shift the weight during exercising then we can say no work has been done in the thermodynamic sense. Work in chemical and biochemical systems is interdependent to change in pressure and volume [2].

The work done on the system is denoted as w = PΔV, where P is the pressure, V is change in volume given by

$\Delta V = V_2 - V_1$.

Work may be done in many ways like mechanical, electrical, magnetic as well as chemical work (reaction) where ΔE, w, q all should have the same units.

ENTHALPY IS A USEFUL FEATURE FOR BIOLOGICAL SYSTEMS

When no mechanical work is done at constant volume.

The work done is W = -PΔV.

Then ΔE = q.

Therefore, ΔE is a useful quantity at a constant volume process. Biological processes take place under constant pressure. At constant pressure, ΔE is not exactly the same as the heat transferred. For this purpose, chemists gave a function that is particularly suited to processes under constant pressure called as

Enthalpy, H and is defined as H = E + PV

At constant Pressure

ΔH=ΔE

+PΔV= q + w +PΔV= q –PΔV + PΔV= q

ΔE = heat transferred at constant volume

ΔH = heat transferred at constant pressure

Since biochemical reactions generally occur in liquids or solids rather than gases, so the volume variation is usually very small, and the enthalpy and internal energy are almost identical. The standard state is useful in comparing thermodynamic parameters of different reactions. The normal state of the solvent in a solution is the activity of the unit. Enthalpy, internal energy radiation, and other thermodynamic quantities are often given, calculated, or marked with standard conditions and then with the superscript symbol ("o"), such as ΔH °, ΔE ° and so on. Enthalpy changes can be measured experimentally for biochemical processes by measuring the heat absorbed (or given off) by the process in a calorimeter. Alternatively, the normal state enthalpy shift for the process can be calculated for any process AB at equilibrium from the temperature dependence of the equilibrium constant:

ΔH= -R d (lnkeq)-d(1/T)

Here, R is the gas constant, defined as R= 8. 314J /mol ⁻

K.A plot of R (lnKeq) versus 1/ Tis called a van' tHolt plot. The example below Figs. (**4** and **5**) demonstrates how a van't Hoff plot is constructed and determines enthalpy changes for a reaction.

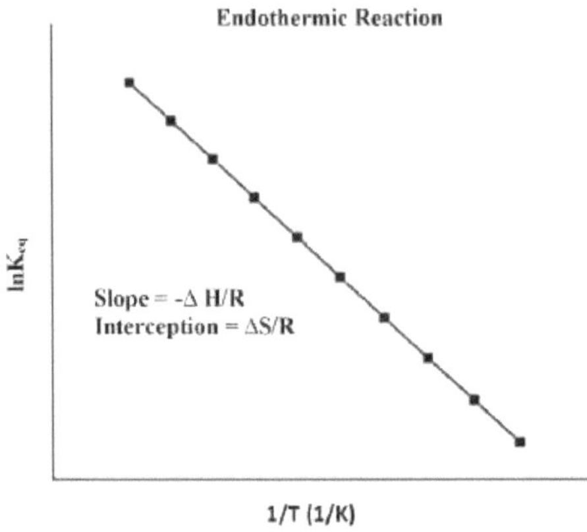

Fig. (4). ln Keq *vs.* 1/T (endothermic reaction).

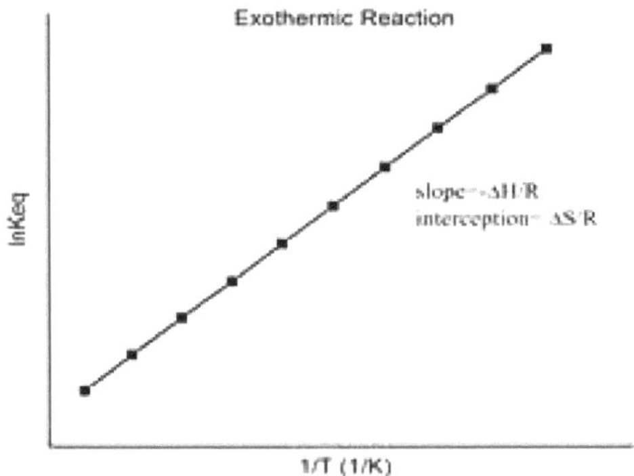

Fig. (5). ln Keq *vs.* 1/T (exothermic reaction).

THE SECOND LAW: SYSTEM TENDS TOWARD DISORDER AND RANDOMNESS

The second law of thermodynamics was defined in a number of ways such as;

The system appears to move from ordered states (low entropy) to disordered states (high entropy), entropy and surroundings of a system remain unchanged in a reversible process, system entropy and surroundings of a system increase irreversibly. All processes that occur naturally proceed towards equilibrium to achieve minimum potential energy. The criteria of the spontaneity of an isolated system are that there must occur an increase in the entropy of the system, and it is positive in nature. For example, a hot coffee continues to cool till its temperature is the same as that of the surroundings. This final stage is called the equilibrium stage. As long as the spontaneous process in a system continues, its entropy keeps increasing until it reaches its equilibrium stage, and thus entropy attains its maximum. So the condition to attain maximum entropy in an isolated system is to be in equilibrium with the system.

For spontaneity: $\Delta S > 0$ for equilibrium $\Delta S = 0$

From there, we can conclude, "The universe energy is constant. The entropy of the universe always leads to the maximum. Many of these second law statements cited the concept of entropy S, which is a quantity to estimate the disorderliness and randomness of the system. A state with low entropy is organized and in order whereas a state with high entropy represents a disordered state. (As shown in Fig. **6**). Entropy is definable in a variety of quantitative ways.

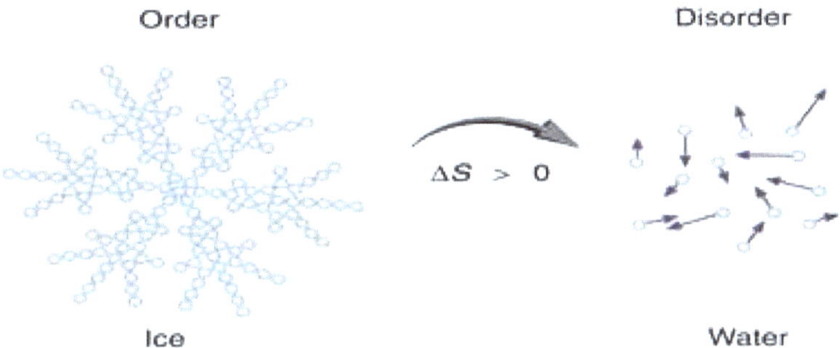

Fig. (6). Entropy change during transformation of ice to water.

If W is the number of ways the system components can be arranged with no entropy change, then entropy can be written as $S = k \ln W$,

Where k is the Boltzmann's constant ($k=1.38\text{x}10^{-23}$J/K). The entropy of heat transferred in the reversible process is given as dS reversible= dq/T. Here dS in the reversible process is the entropy change in the system, q is the transfer of heat at temperature T.

Attributes of the Entropy Function

The important characteristics of the entropy functions can be highlighted again as follows:

1. The value of ΔS for a cyclic change of state is always zero.

2. The change in the value of the entropy function in going from one state to another is independent of the path.
3. It is an extensive property.
4. It is a function of independent variables which are used to define the state of a system.
5. The unit of entropy is the unit of heat divided by kelvin temperature, *i.e.* $J\ K^{-1}$.

The processes which are not reversible are called irreversible process. A system that is going through an irreversible phase may still be able to return to its initial state. However, the uncertainty exists in restoring its environment to its own initial state. The irreversible mechanism increases universe entropy. Since entropy is a function of a state, the shift in system entropy is the same, whether the process is reversible or irreversible. Previously it was believed that reversible biological processes are a combination of two irreversible processes. Although a single enzyme was once thought to catalyze both the forward and reverse chemical changes, researchers found that usually two different enzymes of similar structure are required to perform this change, resulting in a pair of thermodynamically irreversible processes. Energy transfer is never 100 percent efficient in biological processes. For example, in photosynthesis, not all light energy is absorbed by the plant. It reflects some energy, and some is lost as heat. The loss of energy to the natural environment causes disorder or entropy to increase. Animals cannot take energy directly from the sunlight as compared to plants and other photosynthetic species. Most of this energy is lost during the metabolic processes by the producers and primary consumers that are predated. For species at higher trophic stages, much less energy is available. The lower the energy available, the greater number of species that can be backed up. For this reason, an ecosystem includes more producers than consumers. Living systems need a constant supply of energy to sustain this highly ordered state. For starters, the cells are highly ordered and have low entropy. The energy is lost or converted to the environment in the course of holding this order. Thus, while cells are organized, the processes carried out to preserve energy result in an increase in entropy in the surroundings of the cell/organism. The energy transfer causes entropy to increase in the universe [2, 3].

THE THIRD LAW: ENTROPY CHANGE AT ABSOLUTE ZERO

The third law of thermodynamics states that when the temperature reaches 0 K, the entropy of any crystalline, ideal ordered substance must reach zero, and at t = 0 K, the entropy is zero. Based on that, a quantitative, absolute zero scale can be defined for any substance as given below

$$S = Cp \ln(T_2/T_1)$$

Where C heat capacity at constant pressure. Any substance's heat capacity is the amount of heat of 1 mole required to increase the temperature of that substance by 1 degree. This is defined mathematically,

At constant pressure $Cp=dH/ dt$,

Entropy modifications are more useful for biological processes than total entropies. The changes in entropy for a phase can be measured if the enthalpy changes and free energy changes are known [1, 2].

GIBBS FREE ENERGY

The Gibbs free energy, G, is defined as $G = H- TS$

For any process AB at constant P and T, the free energy change is given by $\Delta G=\Delta H- T \Delta S$

At equilibrium $\Delta G = 0$,

There is no net flow of energy in the reverse or forward direction. When $\Delta G=0$, $\Delta S=\Delta H/T$, all thermodynamic properties are balanced. If no changes in ΔG, the reactions move spontaneously to a final state with lower free energy. If ΔG= -ve, the reaction moves towards spontaneity and is exothermic and if ΔG = +ve, it proceeds in the reverse direction and is endothermic.

THE STANDARD STATE FREE ENERGY CHANGE

A reaction condition is influenced by the nature of reactants and products as well as temperature, pressure, pH, concentrations. Let's take an example of a reaction between reactants R_1 and R_2 which convert into products P_1 and P_2

$R_1 + R_2 \rightarrow P_1 + P_2$

The free energy change is given by, $\Delta G = \Delta G^{\circ} + RT \ln [P_1][P_2] [R_1][R_2]$

ΔG^{o} = standard free energy change R = gas constant = $1.98x10^{-3}$ kcal mol^{-1} deg^{-1}

T = usually room temperature = 298 K $[P_1][P_2]$ K = $[R_1][R_2]$

At equilibrium where $\Delta G = 0$ is given as $=\Delta G^{o} + RT \ln [P_1][P_2] [R_1][R_2]$

$\Delta G^{o}= - RT \ln [P_1][P_2] [R_1][R_2]$

But $[R_1][R_2] = K_1eq\ [P_1][P_2]$

$\Delta G^{o}= - RT \ln K1eq$

This relationship allows the normal state-free energy shift. Any process can be calculated in any of these forms if the equilibrium constant is known. More specifically, the equilibrium point for a solution reaction is a function of the normal Free State energy change for the process.

EXAMPLE:

The equilibrium constants calculated by Brandt for denaturing chymotrypsinogen at several temperatures measure denaturation process. For example, the equilibrium constant is 0.27 at 54.5 °C so $\Delta G°$ = -(8. 314J /mol.K) (327.5 K) ln (0.27)

ΔG °= -(2. 72kJ /mol) ln(0.27)

ΔG °=3.56 kJ/mol.

The positive sign of ΔG ° means the process of unfolding is unfavorable. Having calculated both $\Delta G°$ and $\Delta H°$ for the denaturation of chymotrypsinogen, we can also use the given equation to calculate $\Delta S°$

$$\Delta S = \frac{\Delta G - \Delta H}{T}$$

At 54.5°C (327.5 K), $\Delta S°$ = -(3560- 533, 000J /mol)/327.5 K

$\Delta S°$ =1620 J/mol/ K

Fig. (**7**) indicates the dependence on chymotrypsinogen denaturation temperature at pH 3 of $\Delta S°$. A positive $\Delta S°$ suggests that the unfolding of protein occurs in a disorderly manner. Comparison of the value of 1.62kJ/molK with the values of $\Delta S°$ in Table **1**. concludes that the present value is very high for chymotrypsinogen.

Calculated parameters for different proteins and conditions ΔH °kJ/mol $\Delta S°$ kJ/mol ΔG °kJ/mol ΔC kJ/mol ß-Lactoglobulin Chymotrysinogen pH 3, 25°C 164 0.440 31.0 10.95 M urea, pH 3, 25°C -88 -0.300 2.5 9.0 Myoglobin pH 9, 25°C 180 0.400 57.0 5.9 Ribonuclease pH 2.5, 30°C 240 0.780 3.8 8.4 (Table **1**).

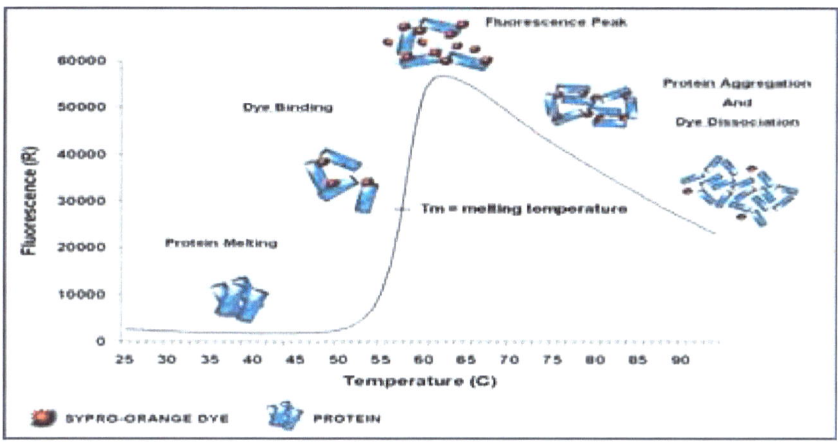

Fig. (7). Thermal shift assay of protein denaturation [5].

Table 1. Calculated parameters for different proteins.

Protein (and conditions)	ΔH°kJ/mol	ΔS° kJ/mol	ΔG°kJ/mol	ΔC kJ/mol
Chymotrysinogen pH 3, 25°C	164	0.440	31.0	10.9
ß-Lactoglobulin 5 M urea, pH 3, 25°C	-88	-0.300	2.5	9.0
Myoglobin pH 9, 25°C	180	0.400	57.0	5.9
Ribonuclease pH 2.5, 30°C	240	0.780	3.8	8.4

THERMODYNAMIC PARAMETERS AFFECTING BIOCHEMICAL EVENTS

A single parameter (like ΔH or ΔS,) cannot define the whole reaction. A positive ΔH° of the protein unfolding may reflect hydrogen bond breakage in protein molecules and also the effect of water on hydrophobic groups.

The comparison of different thermodynamic parameters can give imperative data about the method. For illustration, the exchange of Na^+ and Cl^- particles from the gas stage to an aqueous solution indicate a negative ΔH° (favorable ionic stabilization) and a moderately little ΔS° (Table **2**).The term with negative entropy alludes to the arrangement of the water atoms within the hydration locale of the Na and Cl particles. The unfavorable commitment of –TΔS is more than compensated by increase heat of hydration, which in common makes particle hydration a beneficiation process.

Negative changes in the entropy of acetic acid dissociation in negative water represent the arrangement of water molecules in ion hydration shells. Consequently, ΔG dissolution of acetic acid in water is positive and therefore, acetic acid is a weak acid [6]. Transferring pure nonpolar hydrocarbon particles to water are good examples of exposing hydrophobic protein groups to solvent for protein unfolding. Transferring of toluene to water state includes a-ve ΔS°, +ve ΔG° and a small change in ΔH° in compared to ΔG°.

ΔC is negative for acetic acid dissociation and positive for toluene transfer to the water. The theory is that both polar, as well as non-polar molecules, drive neighbouring water molecules into organisations, but in different ways. The molecules of water close to a polar solvent are organized but not stable. Water forms hydrogen bonds close to non-polar solutes to regroup faster than pure water hydrogen bonding. The hydrogen bonding between water molecules formed close to an ion, on the other hand, are less labile compare to bonding form in pure water indicating that in the case of acetic acid, ΔC should be negative for ion dissociation in solution.

EFFECT OF pH ON STANDARD FREE ENERGIES CHANGE

The normal description of the standard state for biochemical reactions where hydrogen ions (H) are formed is an awkward one. The standard condition for the H ions is 1 M, corresponding to pH 0. At this pH, almost all enzymes would be denatured, and there will be no biochemical reactions. Using free energies and equilibrium constant calculated at a pH 7 makes more sense. Thus, biochemists use a modified form of the standard condition, denoted with prime (') symbols, as in ΔG°', Keq', ΔH°' and so on [4]. The values defined indicate a standard condition of 10- 7M^+ and unit activity (1 M for standard solutions, 1atm for gases and pure solids specified as unit activity) is assumed for all other components at pH 7 in the ionic forms. The two standard conditions can easily be connected.

For any reaction where H is produced as product, $A \rightarrow B^- + H^+$

The equilibrium constant relation for any two standard states is Keq =Keq[H^+] And ΔG° is calculated by ΔG°=ΔG° + RT ln [H^+]. For the reaction in which H^+ is absorbed $A + H^+ \rightarrow B$ The equilibrium constant is related by Keq=Keq[H^+] and Δ G°' is given by ΔG°' = ΔG°+ RTln[1/H^+] = Δ G°'- RTln[H^+]

REACTION OF CONCENTRATION AND FREE ENERGY CHANGE

$\Delta G = \Delta G^o + RT \ln [P_1][P_2] [R_1][R_2]$

ΔG^o = standard free energy change, R = gas constant = 1.98 x10^{-3} kcal mol^{-1} deg^{-1},

T = usually room temperature = 298 K $[P_1][P_2]$ K = $[R_1][R_2]$

The above equation indicates that changes in the free energy of the reaction can differ significantly from standard state values if the concentrations of the reactants and products differ notably from the unit activity (1 M for solution) and results can be drastic. Take Phosphocreatine hydrolysis:

Phosphocreatine + H_2O → creatine + Pi

The reaction is extremely exothermic, where ΔG° at 37°C is -42.8 kJ/mol. Phosphocreatine, creatine, and inorganic phosphate concentrations usually vary between 1-10 mM.

By taking concentrations of 1mM and using the above equation, phosphocreatine hydrolysis gives ΔG° value as ΔG = - 42. 5 kJ/mol + (8.314 J/mol K) (310 K) ln (0.001)(0.001) (0.001), ΔG = -60.5 kJ/mol

The difference in standard state and concentration of 1 mM for such a reaction is, therefore, approximately-17,7 kJ/mol at 37 ° C.

COUPLED PROCESS

The biochemical reactions in living organisms have the thermodynamic potential of positive G°, which include adenosine triphosphate (ATP) synthesis, high-energy molecule productions and the formation of cellular ion gradients. Such processes are directed towards unfavorable direction thermodynamically *via* the coupling process. In intermediate metabolism, oxidative phosphorylation, and membrane transport, are highly crucial.

By summarizing the free energy changes for each reaction, we can predict whether pairs of coupled reactions will occur spontaneously. Consider for example, in the glycolysis process the phosphor (enol) pyruvate (PEP) is converted into pyruvate (as shown in Fig. **8**). The hydrolysis process is favourable and helps in the conversion of adenosine diphosphate (ADP) to ATP, an energetically unfavourable process.

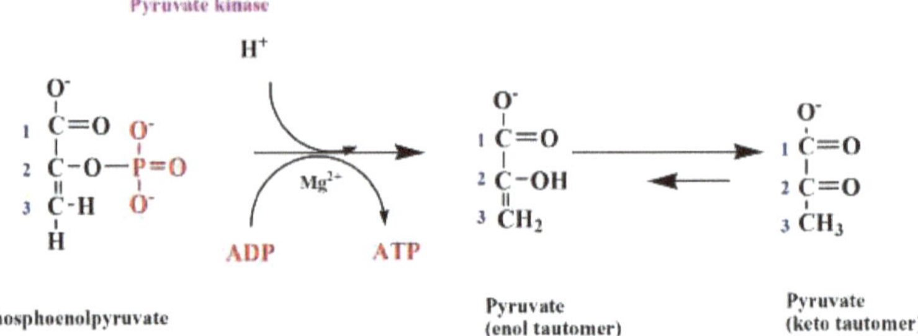

Fig. (8). Conversion of phosphoenolpyruvate to pyruvate.

Using values of ΔG° of human erythrocyte: PEP + H_2O → pyruvate + Pi ΔG= -78 kJ/mol

ADP + P →ATP+ H_2O ΔG= +55kJ /mol

PEP+ADP → pyruvate + ATP, Total ΔG =-23 kJ/mol

The net reaction catalyzed by this enzyme depends on the coupling of the two above-mentioned reactions to produce the net reaction with a net negative part of G. There are various examples of coupling reactions that include reactions and processes with +Δ G °values that are powered by coupling reaction with -ΔG ° [4, 7].

Changes in the spontaneity of the reaction due to the ΔH and ΔS signs (a sign of ΔG). ΔH ΔS ΔG = ΔH- TΔS - +. The reaction is both enthalpy-favorable (exergonic) and entropy-favorable, spontaneous. – – Enthalpically reaction is favored but opposed entropically. Spontaneous only at temperatures below T = ΔH/ΔS. + + The reaction is unfavored enthalpically(endo), but entropy is favored. It reacts spontaneously at temperatures higher than T = ΔH/ΔS + - where the reaction is both unflavored and non-spontaneous (Table **2**).

Table 2. Variation of reaction spontaneity (sign of ΔG) with the signs of ΔH and ΔS.

ΔH	ΔS	ΔG= ΔH- TΔS
–	+	The reaction is both enthalpically favored(exo) and entropically favored.It is spontaneous(exergonic) at all temperatures.
–	–	The reaction is enthalpically favored but entropically opposed. It is spontaneous only at temperatures below T=ΔH/ΔS.

+	+	The reaction is enthalpically opposed(endo) but entropically favored. It is spontaneous only at temperatures above T=ΔH/ΔS
+	−	The reaction is both enthalpically and entropically opposed. It is unspontaneous(endergonic) at all temperatures.

ATP (ADENOSINE TRIPHOSPHATE)

A cell does three types of work: 1. Mechanical work: Cilia beating, muscle contraction, *etc.* 2. Transport: Movement of substances across membranes 3. Chemical work: Allowing spontaneous and non-spontaneous reactions, *e.g.* Protein synthesis. ATP is the agent that drives several forms of the cell. ATP is considered as a chemically active organic fuel involved in various biochemical mechanisms (as shown in Fig. **9**). Present in all types of life form, they are also termed as 'unit of currency' of biomolecules for cellular energy production. When food is ingested initially, through cellular respiration, mainly ADP and AMP are formed and finally, ATP is synthesized.

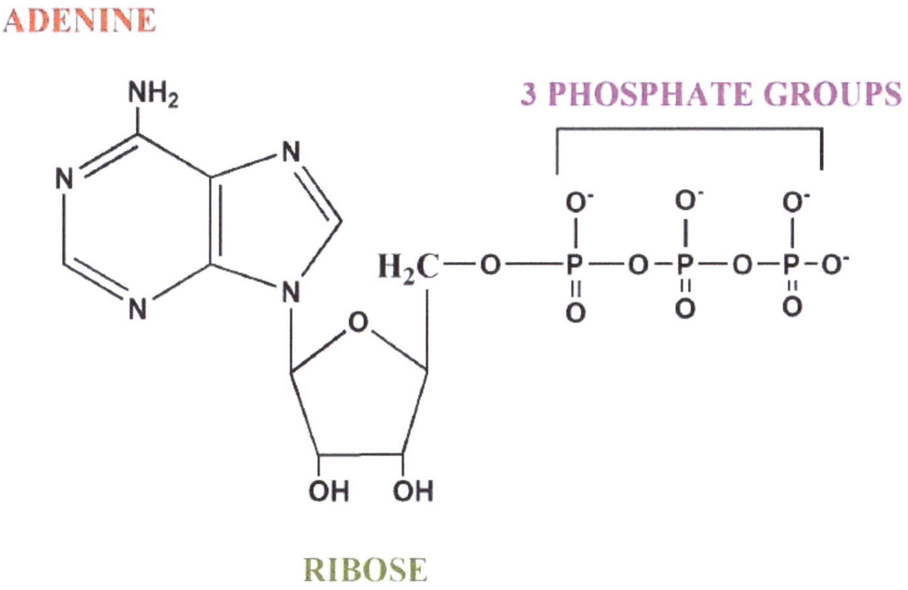

Fig. (9). Structure of ATP.

They act as a precursor to DNA and RNA. ATP is abundant at a normal intracellular concentration of 1–10 mM. ATP dephosphorylation and ADP and AMP rephosphorylation occur frequently during aerobic metabolism.

Many different cellular processes can produce ATP; the three dominant pathways in eukaryotes species are (1) glycolysis (2) citric acid cycle or oxidative phosphorylation and (3) β-oxidation. The net process of conversion of glucose to CO2, which is a combination of pathways 1 and 2, is called cellular respiration, which creates approximately 30 ATP from each molecule of glucose. The ATP production *via* a non-photosynthetic aerobic pathway occurs primarily in mitochondria, covering 25% of the total volume of any typical cell.

At pH 6.8-7.4, ATP exists in stable form but rapidly hydrolyzes at high pH to ADP and phosphate. Living cells retain the equilibrium ratio of ATP to ADP at a point ten, with ATP concentrations five times higher than ADP concentrations [4, 8].

THERMODYNAMIC HYDROLYSIS OF ATP

ATP hydrolysis represents one of the main biochemical mechanisms through which the free energy acquired during catabolic cellular metabolism is liberated to cellular power activities. ATP's hydrolysis into ADP and inorganic phosphate yield 30.5 kJ/mol of enthalpy, with a free energy transition of 3. 4kJ/mol. The actual change in free energy for ATP hydrolysis, symbolized by ΔG, which varies according to cell type and prevailing physiological conditions. When phosphate groups are hydrolyzed, energy is released [8].

ATP + H_2O→ ADP + Pi($\Delta G°$ = -35.7 kJ/mol) $\Delta G°$ = RT lnKeq [ADP] [Pi] Keq = [ATP] [H_2O] Fig. 9.9.

Structure of ATP

The activation energy for the phosphoryl group transfer (200 to 400 kJ/mol) reaction was significantly higher than the free energy of ATP hydrolysis (-30.5 kJ/mol).

$\Delta G'°$=-30.5 kJ/mole = -7.3 kcal/mole

ΔG= -52 kJ/mole = -12.4 kcal/mole

Cellular conditions: [ADP][Pi]/[ATP] = 1/850

ATP and its sequential hydrolysis products, adenosine diphosphate (ADP) and adenosine monophosphate (AMP) are nucleotides composed respectively of adenine, ribose and 3, 2, or 1 group (s) phosphate (as shown in Fig. **10**).

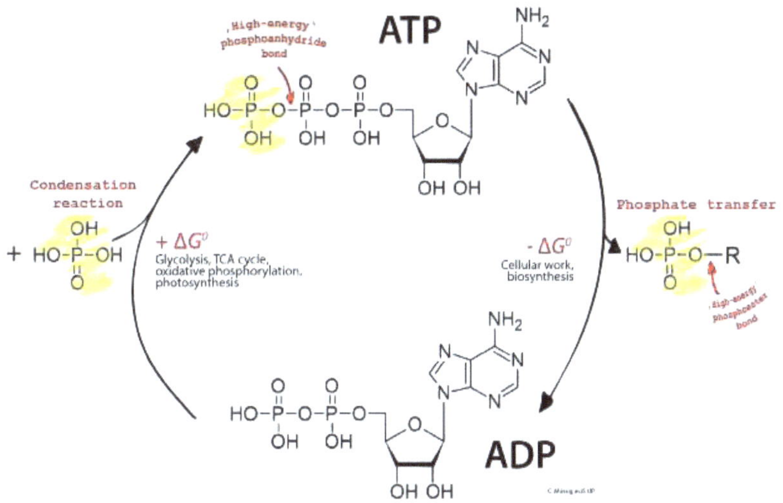

Fig. (10). ATP to ADP conversion cycle [9].

ATP, ADP and AMP are found in the cytosol as well as in the nucleus and mitochondria. ATP accounts for about 75 percent or more of the sum of all 3 adenine ribonucleotides in normal respiratory cells. At pH 7.0, ATP, ADP, and AMP phosphate groups are ionised in such a way to form several charged anions like ATP^{4-}, ADP^{3-}, AMP^{2-}. Cellular fluid hold high concentrations of Mg^{2+} ions, both ATP and ADP primarily exist as complexes of $MgATP^{2-}$ and $MgADP^{-}$. In fact, ATP participates as its complex form in the phosphate transfer reactions. But with Mn^{+}, ATP can also form a complex.

In biological systems, energy is stored in the form of ATP and hence it acts as the main free energy donor for metabolic activity.

A cell can absorb an ATP molecule within a minute of its formation. They have a high turnover rate. For instance, a human in resting condition can consume about 40 kg of ATP per day. The ATP is consumed at a rate of even 0.5 kg per minute during strenuous labour. Endergonic processes such as biosynthesis, active transportation, *etc.* can only occur if ATP is regenerated from ADP continuously. Phototrophs use the free energy to regenerate ATP in light, while chemotrophs extract ATP *via* food oxidation (Fig. **10**). ATP to ADP conversion cycle [9].

FACTORS AFFECTING HYDROLYSIS OF FREE ENERGY OF ATP

There are several factors that affect the hydrolysis of ATP, which makes the true reactions quite complicated. Some of them have been discussed below:

1.ATP and ADP have inorganic phosphate groups which are weakly acidic, have low pK values and actions of the various biological activities depend on the pH of the medium of reaction, which can overall affect and change the value of ΔG ° (as shown in Fig. **11**).

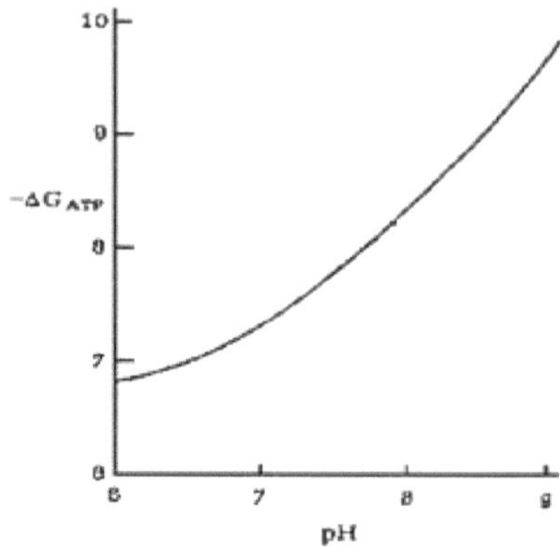

Fig. (11). Effect of pH on the free energy of hydrolysis of ATP (ΔGATP) at 10 mM Mg^{2+} [8].

$HATP^{3-}$<==> ATP^{4-} + H^+ ; $pK_{1'}$ = 6.95 $HADP^{2-}$<==> ADP^{3-} + H^+ ; $pK_{2'}$ = 6.88 H_2PO_{4-}<==> HPO_4^{2-} + H^+, $pK_{3'}$ = 7.20

The resulting conditions affect the pK values by releasing H^+. The released H^+ can be useful for ATP hydrolysis assay.

2. The value of ΔG' changes in the presence of Mg^{2+} due to the reaction of different reactants with bivalent cation to form complexes (as shown in Fig. **12**).

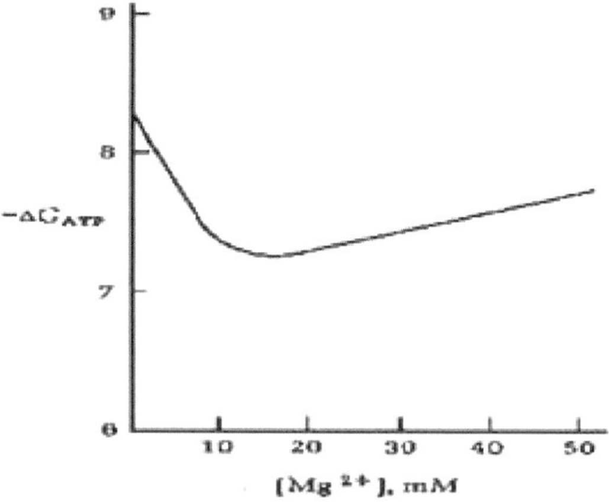

Fig. (12). Effect of Mg^{2+} concentration on the free energy of hydrolysis of ATP (ΔGATP) at pH 7.0 [8].

For Mg^{2+} the following equilibria are important:

Mg^{2+} + ATP^{4-}<==> $MgATP^{2-}$ Mg^{2+} + ADP^{3-}<==>$MgADP^{-}$ Mg^{2+} + HPO_4^{2-}<==> $MgHPO_4$

HYDROLYSIS OF ATP AND EQUILIBRIA OF COUPLED REACTIONS

To understand the importance of ATP's role in energy coupling, the following reaction is taken as an example with a positive + ΔG ° (say + 4kcal/mol)

A=B, G°= + 4 kcal/mol

This reaction is unfavorable in thermodynamic terms without free energy input. The equilibrium constant, K′eq at 25 ° C according to the equation:

$$\frac{[B]_{eq}}{[A]_{eq}} = K'_{eq} = 10 - \Delta G^{o'}/1.36 = 1.15 \times 10^{-3}$$

Therefore, if the molar ratio of B/A is greater than 1.15 x 10^{-3}, then A will not spontaneously become B. However, if the reaction is combined with the hydrolysis of ATP (when the value of ΔG ° is -7.3 kcal/mol) and when the ratio of [B]/[A] is 1, then A is converted to B. The overall new reaction can be written as:

$A + ATP + H_2O \rightarrow B + ADP + Pi + H^+$

Since the sequential reaction values are additive, the normal free energy change value of the above reaction must be

$\Delta G^{o\prime} = \Delta G^{o\prime} + \Delta G^{o\prime} = + 4 + (-7.3) = -3.3$ kcal/mol.

The Keq of this new coupled reaction will be written as

$$K_{eq}' = \frac{[B]_{eq}}{[A]_{eq}} \times \frac{[ADP]_{eq}[Pi]_{eq}}{[ATP]_{eq}}$$

At equilibrium, the ratio [B]/[A] is:

$$\frac{[B]_{eq}}{[A]_{eq}} = \frac{[ATP]_{eq}}{[ADP]_{eq}[Pi]_{eq}}$$

The [ATP]/[ADP][Pi] ratio of the ATP-producing cells is held at a high value, usually in the order of 500. Thus,

$$\frac{[B]_{eq}}{[A]_{eq}} = 2.67 \times 10^2 \times 500 = 1.34\ 10^5 = 10^8 \text{ approx.}$$

This concludes that ATP's hydrolysis allows A to be transformed to B until the ratio of [B]/[A] reaches a value of 1.34 x 105. This equilibrium is significantly different from the value of 1.15×10^{-3} that excludes ATP hydrolysis In other words, ATP's coupled hydrolysis modified the equilibrium ratio between B and A by a factor of about 108. Thus, by using light or substrates as free energy sources, the cells retain a high level of ATP. The one molecule ATP hydrolysis then increases the product equilibrium ratio to the reactants of a coupled reaction by a very huge factor of about 108. More precisely, an ATP molecule's hydrolysis alters an equilibrium ratio of a coupled reaction (or reaction sequence) by a factor of 108. For example, in a coupled reaction, the hydrolysis of 3ATP molecules alters the balance ratio by a factor of 1024. Thus, a series of thermodynamically unfavourable reactions can be transformed into an advantageous one by combining it with a large number of hydrolyzed ATP molecules [4, 6, 8].

DAILY HUMAN REQUIREMENT FOR ATP

In most biological environment, high-energy radiation molecules like ATP are used and recycled rapidly. Surprising and impressive results are found in rough calculations. Suppose the average adult consumes approximately 11,700 kJ (2,800 calories) per day and ATP synthesis work with about 50% thermodynamic efficiency. Approximately 5860 kJ of the 11,700 kJ a person consumes as food ends up in the form of synthesized ATP. As discussed earlier, under cellular conditions, around 50kJ of free energy is produced on hydrolysis of 1mole of ATP, suggesting that each day the body produces 5860/50=117 moles of ATP. The molecular weight of the disodium salt of ATP is 551 g/mol, and the average person hydrolyzes about (117) 551 g = 64, 467 g/mol of ATP per day. The average adult weighs around 70 kg and absorbs around 65 kg of ATP per day. It is approximately the same as its own weight. Fortunately, we have a very efficient ATP/ADP recycling system. The energy released from food is stored as ATP and used to release ADP and phosphate; again our body recycles it into ATP *via* intermediate metabolism for reuse. The average 70 kg body contains only about 50 grams of total ATP/ADP. Therefore, each ATP molecule in our body must be recycled almost 1,300 times a day [9]. If this were a fact, our habit of dependence on ATP would cost more than a million dollars a day at current commercial prices of around \$ 20 per gram. Hence, the understanding of biochemistry to maintain the extraordinary activity and vitality of biology deserves our respect and attention.

SUMMARY

All living beings require energy to perform activities like, movement, growth, biomolecular synthesis, transportation of ions and molecules across cellular membranes. All species must obtain energy from their environment and must make efficient use of that energy to conduct life processes.

• The first thermodynamic rule states that the total energy of an isolated system is always conserved. Enthalpy denotes as H = E+PV. The heat transmitted in a constant pressure phase is denoted as ΔH. Volume variations are usually minimal for biochemical reactions in liquids, and enthalpy and internal energy are also equal.

• Second law of thermodynamics can be stated in the following ways: Systems move from a state of low disorderliness to a state of high disorders. Entropy remains unchanged in reversible processes and increases during irreversible conditions. • According to the third law of thermodynamics, entropy must be zero for any crystalline, perfectly ordered substance at T= 0K.

• Comparison of many thermodynamic parameters may give useful insights into whether a reaction is feasible. It can be helpful in the determination if a reaction will proceed forward as written and to measure the relative overall contributions.

• Biomolecules mediates energy flow from exergonic reactions necessary for the life process. They form complex molecules and take part in various metabolic activities.

KEYWORDS

Pyruvate

It is a mixture of salt, ester, the anion of pyruvic acid. Pyruvate is generated at the end of the glycolysis process and can be converted into lactate or acetyl CoA.

Denaturation

Modification of protein structure due to external stress (heat acid or alkali), which can lead to structural deformation and cannot perform the cellular function.

Calorimeter

It is a device to measure the heat generated during mechanical, electrical, or chemical reactions. It also calculates the heat capacity of materials.

Glycolysis

It is a biological process that converts glucose into pyruvate and releases ATP as a form of energy. It is also called a citric acid cycle.

Phototrophs

An organism that uses light energy like sun and CO_2 to produce its own food. *e.g.* Algae, cyanobacteria, plants.

SHORT-ANSWER TYPE QUESTION

1. Define a state function.

2. Explain the expression, $S = k \ln W$.

3. How many ways the internal energy can be changed according to the first law of thermodynamics?

4. At what condition $\Delta H = q$ is achieved for any system?

5. The body performs many processes that depend upon energy and could not occur without the supply of ATP. How does energy from ATP power cellular reactions?

6. Explain the structure of ATP.

7. What is the third law of thermodynamics of absolute zero entropy?

8. How energy produced is stored and utilized in our body?

9. What are irreversible and reversible processes? Give examples for both.

10. What is the significance of entropy?

11. Define the spontaneous process with examples.

12. What are the shortcomings of the first law of thermodynamics?

13. How many forms of energy are present in the Universe?

14. Explain the various thermodynamic processes and their types.

15. Explain the first law of thermodynamics?

16. What are coupled reactions in the biological system?

17. At what condition, $Cp = Cv + R$?

18. What is the difference between entropy & enthalpy?

19. What are extensive and intensive properties?

20. What do you mean by thermodynamics hydrolysis of ATP?

LONG-ANSWER TYPE QUESTIONS

1. What are the conditions required for the system to undergo spontaneity?

2. Derive the equation, $H = U + P\Delta V$.

3. What thermodynamics changes do you observe during protein denaturation?

4. Explain Hydrolysis of ATP?

5. What is the daily requirement for ATP?

6. Compare the structure of ATP and ADP mentioning its activation energy difference.

7. How pH and concentration affect the standard free energy change?

8. What is a coupled reaction and why it is important?

9. Explain the third law of thermodynamics by giving an example.

10. Differentiate the entropy of real gas and ideal gas.

11. In a certain condition, work done on a system of 600 J produces 250 J of heat. How do you get change in the internal energy of a system?

MULTIPLE CHOICE QUESTIONS

1. How many ATP is generated in an aerobic respiration:

(a) 2 (b) 34

(c) 36 (d) 16

2. ATP is synthesized inside

(a) Golgi bodies (b) nucleus

(c) Mitochondria (d) all cell organelles

3. For work done at maximum in an open system, the process should in

(a) Adiabatic in nature (b) irreversible in nature

(c) Isobaric in nature (d) reversible

4. The condition of absolute zero temperature is discussed in

(a) Adiabatic process

(b) Third law of thermodynamics

(c) Second law

(d) Isothermal process

5. Protein unfolding causes

(a) No entropy change (c) less entropy decrease

(b) Entropy is decreased significantly (d) entropy increases significantly

6. Which of these is not true about ATP hydrolysis?

(a) Exothermic process (b) ADP generation

(c) $\Delta G = -30.5$ kJ/mol (d) $\Delta G = 30.5$ kJ/mol

7. Homogeneous system

a) consists of anyone phase among solid, liquid and gas

b) consists only solid phase, not liquid and gas

c) consists of all three phases at a particular temperature

d) none of the above

8. At constant volume, the value of Work done of a process is

a) -ve

b) 0

c) +ve

d) none of these

9. Which of these is a state function?

a) Heat

b) Mass

c) Volume

d) Density

10. All spontaneous process follows

a) quasi-static condition

b) reversible condition

c) irreversible condition

d) none of these

11. A process becomes irreversible when there is

a) no equilibrium condition

b) involvement of dissipative effects

c) either a. or b. or both

d) none of these

12. During heat transfer, if two bodies temperature differences in which heat increases the irreversibility of the process

a) remain same

b) increase

c) decrease

d) the process proceeds towards reversibility

13. The energy of the universe

a) always constant

b) decreasing

c) either increases or decreases

d) increasing

14. At equilibrium condition,

a) entropy is maximum for a system

b) entropy is minimum for a system

c) entropy is the same for both system and surrounding

d) none of the these

15. How are the entropy increase of the universe during the irreversible process and its irreversibility related?

a) The higher the entropy increase of the universe, the lower will be the irreversibility

b) The higher the entropy increase of the universe, the higher will be the irreversibility

c) change in entropy increase of the universe does not change the irreversibility

d) none of the above

16. When system and surrounding temperature differences decreases, the energy also remains

a) constant

b) increasing

c) none of these

d) decreasing

17. Entropy of the Universe is

a) expanding

b) constant

c) shrinking

d) none of these

18. How do you define an ideal gas?

a) PV = RT

b) P = VR/T

c) PV = R/T

d) none of these

19. Which one is the example of an ideal gas?

a) nitrogen

b) air

c) water vapour

d) hydrogen

20. Which common refrigerant is used in fridges?

a) sulphur dioxide

b) ethyl chloride

c) propane

d) Ammonia

21. What point should be we be looking for in a good refrigerant?

a) Not toxic

b) non-corrosive

c) enthalpy of vaporization is at minimum

d) all of these

22. When there is an increase in air humidity ratio, the temperature

a) increase

b) decrease

c) constant

d) unpredictable

23. Substance whose chemical composition is homogenous with no change in mass is called

a) good substance

b) pure substance

c) solid substance

d) ideal substance

24. In which of the following process there is no transfer of heat between the system and surrounding?

a) isobaric

b) adiabatic

c) isochoric

d) isothermal

25. Heat is a

a) extensive property

b) intensive property

c) transfer function

d) none of these

26. Choose the correct statements.

(1) Extensive property depends on the quantity of matter

(2) Volume is an extensive property

(3) Energy is a state function

(4) Heat is a state function

a) (1), (2) and (4) are correct

b) (1), (3) and (4) are correct

c) All of them is correct

d) (1), (2), (3) are correct

27. At what process energy is not conserved but destroyed?

a) reversible condition

b) irreversible condition

c) both reversible and irreversible condition

d) none of the above

28. What happens when we use electric energy to light our homes?

a) Energy is destroyed

b) energy is created

c) Energy is converted from more energy to less energy value

d) Energy is converted from less energy to more energy value

29. What is the humidity of the air at zero rates of evaporation of water?

a) 0%

b) 55%

c) 100%

d) Unpredictable

30. The pH of boiling water is

a) 0

b) 7

c) > 7

d) < 7

31. Which substance given below cannot be converted into a simpler form?

a) compounds

b) molecules

c) atoms

d) elements

32. During the isothermal condition

a) Temperature is constant

b) heat is constant

c) internal energy is constant

d) all of these

33. How many kg of CO is produced by 1 kg of carbon?

a) 3/7

b) 11/7

c) 11/3

d) 4/11

34. An equilibrium relation is achieved between all three bodies when two bodies are thermally at equilibrium with the third body is stated in

a) Kelvin Planck's law

b) Isobaric process

c) Zeroth law of thermodynamics

d) The second law of thermodynamics

35. During the irreversible process, which of the following occur?

a) loss of heat

b) no gain of heat

c) no loss of heat

d) gain of heat

36. The Second law of thermodynamics is explained in which of the following?

a) The construction of a working engine is not possible in a cyclic process that only converts heat energy to work.

b) Body heat cannot be transfer from low to high temperature without any help from an outside source.

c) Mechanical energy can be converted into heat energy

d) all of the above

37. Thermodynamic law where the temperature is a thermodynamic property is given by

a) Zeroth law

b) Second law

c) Third law

d) First law

38. The equation $\Delta U + P\Delta V$ is given by

a) enthalpy

b) entropy

c) none of these

d) Work

39. A sequence of operations that take place in a certain order while restoring the initial conditions at end of the operation is called

a) irreversible process

b) reversible cycle

c) adiabatic process

d) none of these

40. 4/3 kg oxygen needs 1kg carbon to generate __________ kg of CO gas.

a) 8/3

b) 11/7

c) 7/3

d) 11/3

41. Which law states that total energy is distributed equally by the degree of freedom of a substance?

a) law of equipartition of energy

b) law of conservation

c) law of degradation

d) none of these

42. In which of the following expansion or compression processes, the temperature remains constant?

a) isolobal

b) isothermal

c) adiabatic

d) isobaric

43. Gas with the highest molecular mass is

a) Nitrogen

b) Hydrogen

c) Oxygen

d) Methane

44. At 0°C entropy of H_2O is

a) 1 b) -1 c) 0 d) 5

45. Which of the following represents an adiabatic process?

a) no heat transfer

b) the temperature fluctuates

c) Work done = change in internal energy

d) all of the above

46. Which of the following has the highest specific heat capacity value?

a) Alcohol

b) Water

c) Oil

d) Hydrogen

47. Which of the following open system process allow the transfer of mass in and out of the system?

a) Low

b) non-flow

c) none of these

d) adiabatic process

48. Gas used for street lighting and domestic heating in town and villages is

a) Mond gas

b) Producer gas

c) Coal gas

d) Coke oven gas

49. At what temperature absolute zero degrees is achieved?

a) 0°C

b) -237 °C

c) -273°C

d) 273°K

50. Substance with the highest calorific value is

a) Bituminous coal

b) Peat

c) Lignite

d) Anthracite coal

51. Which one of them given below is an intensive property?

a) Volume

b) Mass

c) Energy

d) Temperature

52. One Joule (J) is same as

a) 1 kilo Newton metre

b) 1 Newton metre

c) 10 Nm/s

d) 10 kNm/s

53. Which of the following is one of the main reasons for the irreversible reaction?

(a) all of them

a) limitless expansion

b) the difference in heat transfer

c) mechanical friction

54. The condition when a substance evaporates from its liquid state is complete, is called

a) steam

b) perfect gas

c) vapor

d) air

55. Which of the following changes occur when gas is heated?

a) Volume

b) Temperature

c) all of them

d) Pressure

56. Which of the following is not considered biofuel?

a) ethanol

b) methane

c) fossil fuel

d) ammonia

57. Which of the following will allow a reversible cycle?

a) The temperature and pressure of the system must remain equal with the surrounding

b) All process must be very slow in a reaction

c) all of them

d) No friction in parts of the engine

58. Molecular oxygen consists of how many numbers of atoms?

a) 4 b) 2 c) 16 d) 8

59. Which of these given below to define a spontaneous chemical reaction?

a) The entropy keeps on increasing

b) The change in entropy and enthalpy decides which direction the reaction will proceed

c) The enthalpy keep on decreasing

d) Both enthalpy and entropy is constant

60. Maximum entropy is observed in which of the following reaction?

a) mixing of acid and base

b) water condensation

c) sublimation process of camphor

d) ice melting

CONSENT FOR PUBLICATION

Not Applicable.

CONFLICT OF INTEREST

The authors confirm that this chapter contents have no conflict of interest.

ACKNOWLEDGEMENTS

Authors are thankful to Department of Chemistry, North Eastern Regional Institute of Science and Technology,Nirjuli, Itanagar for providing facilities to accomplish the work.

REFERENCES

[1] Bray HG, White K. Thermodynamics in Biochemistry. 2nd Ed., UK: J & A Churchill Ltd. 1966.

[2] Cantor CR, Schimmel PR. Biophysical Chemistry. San Francisco: W.H. Freeman 1980.

[3] Atkins PW. The Second Law. 1st Ed., New York: W. H. Freeman & Co. 1984.

[4] Alberty RA. Standard Gibbs free energy, enthalpy, and entropy changes as a function of pH and pMg for several reactions involving adenosine phosphates. J Biol Chem 1969; 244(12): 3290-302. [Internet]. [http://dx.doi.org/10.1016/S0021-9258(18)93127-3] [PMID: 4307313]

[5] Brandts JF. The thermodynamics of protein denaturation. I. The denaturation of chymotrypsinogen. JACS 1964; 86(20): 4291-01. [http://dx.doi.org/10.1021/ja01074a013]

[6] Edsall JT, Gutfreund H. Biothermodynamics: The study of Biochemical process at Equilibrium. New York: John Wiley 1983.

[7] Becker WM. Energy and the Living Cell; An Introduction to Bioenergetics. New York: Lippincott Williams and Wilkins 1977.

[8] Phillips RC, George P, Rutman RJ. Thermodynamic data for the hydrolysis of adenosine triphosphate as a function of pH, Mg2+ ion concentration, and ionic strength. J Biol Chem 1969; 244(12): 3330-42.http://www.jbc.org/content/244/12/3330 [Internet]. [http://dx.doi.org/10.1016/S0021-9258(18)93131-5] [PMID: 5792663]

[9] Available from. https://en.wikipedia.org/wiki/Adenosine_triphosphate

CHAPTER 10

Bioenergetics, Metabolism of Biomolecules, Photosynthesis and Respiration, Transcription and Translation, Recombinant DNA Technology

Saroj Yadav[1,*], **Kamlesh Singh Yadav**[2] and **Pratibha Yadav**[3]

[1] *Department of Chemistry, Delhi University, India*

[2] *Department of Chemistry, DDU Gorakhpur University Gorakhpur-273009 Uttar Pradesh, India*

[3] *Department of Chemistry, IIT Delhi Hauz Khas New Delhi-110016, India*

Abstract: This unit describes how living organisms are procuring their life with energy transform in order to perform biological work. This chapter explores the in-depth metabolism of carbohydrates, lipids and fats, proteins, nucleic acids and nucleotides. The relationship between photosynthesis and cellular respiration, the structure of DNA and the technology of recombinant DNA have been described in detail.

Keywords: Catabolism, Citric acid cycle, Glycolysis, Oxidative phosphorylation, Photorespiration, Photosynthesis, Recombinant DNA.

INTRODUCTION

Bioenergetics is a branch of biochemistry. A study about the transformation of energy in the living systems is bioenergetics. This type of energy is created by living organisms. All living organisms use energy, but they use it in different ways. Consider plants and animals, the plants use energy in such a manner that causes them to release oxygen as a byproduct, while animals require oxygen for living. All living organism converts foods, sunlight, minerals, water, oxygen, *etc.* into beneficial energy. The application of bioenergetics helps us to understand how living cells transform energy often through the production, retention or consumption of adenosine triphosphate (ATP). Progressions of the bioenergetics, cellular respiration or photosynthesis, are essential to most aspects of cellular breakdown; therefore to life itself [1, 2].

Metabolism is a chemical transformation occurring within the living organisms for the sustenance of life. Some enzymes are produced during the metabolism of

* **Corresponding author Saroj Yadav:** Department of Chemistry, Delhi University, India; E-mail: saroj123@rediffmail.com

Meera Yadav & Prof. Hardeo Singh Yadav (Eds.)

cells. These enzyme-catalyzed reactions are responsible for the growth and replication of living organisms as well as their response to various stimuli in their surroundings. The explanation of digestion is about all chemical reactions that occur in maintaining the living cells and the organisms. This type of chemical reaction keeps our body alive and functioning [3]. Metabolism can be suitably assigned into two kinds:

a. Catabolism –the energy is obtained by the breaking of molecules called catabolism.

b. Anabolism – It is a constructive part of metabolism. The synthesis of more complex substances from smaller units in living cells is called anabolism.

Anabolism uses energy kept within the system of adenosine triphosphate (ATP) to form larger molecules from smaller molecules. Catabolic reactions are destroying larger molecules into smaller molecules by producing ATP and raw materials for anabolic reactions. The production of ATP is necessary for the proper routine of energetic functions like living cell reactions and procedures of the body that entail energy is delivered from the exchange of ATP to ADP. This energy is used in the transmission of nerve impulses, muscle contraction, active transport through plasma membranes, protein synthesis and cell division. Since all digestible forms of carbohydrates are finally converted into glucose, it is significant to consider how glucose is competent to energy supply in the form of Adenosine triphosphate (ATP) to numerous cells and tissues.

Lipids are macromolecules which dissolve in a non-polar solvent and it is an essential part of living organisms. Sometimes scientists define lipids as hydrophobic or amphipathic small molecules. Triglycerides (TGs) and cholesterol store excess energy from the diet. These are often considered a contributing factor in heart diseases and obesity in humans, while lipids are also required for physiological activity. The main functions of TGs are storing energy in adipocytes and muscle cells, dietary fat is a collective constituent of the cell membranes, steroids, bile acids, and building blocks of the molecule. Vertebrates and humans use their excessive fats as sources of vitality for body parts such as the heart to function. Lipid metabolism in plants takes place in different ways as compared to that taking place in animals. Hydrolysis is the ingestion of unsaturated fats into the epithelial cells of the intestinal divider. In the epithelial cells, unsaturated fats are packaged and shipped to the rest of the body [4].

Proteins are made by amino acids and include many essential biological compounds such as enzymes, hormones or antibodies. Each protein contains 2,000 amino acids which are associated with peptide bonds. Protein is a very important material for the growth of the human body. Proteins are excessively enormous to be consumed into the circulatory system, so they are separated into amino acids and little pieces with a few amino acids before absorption. The split of the proteins into amino acids is involved by several enzymes (As shown in Fig. **1**).

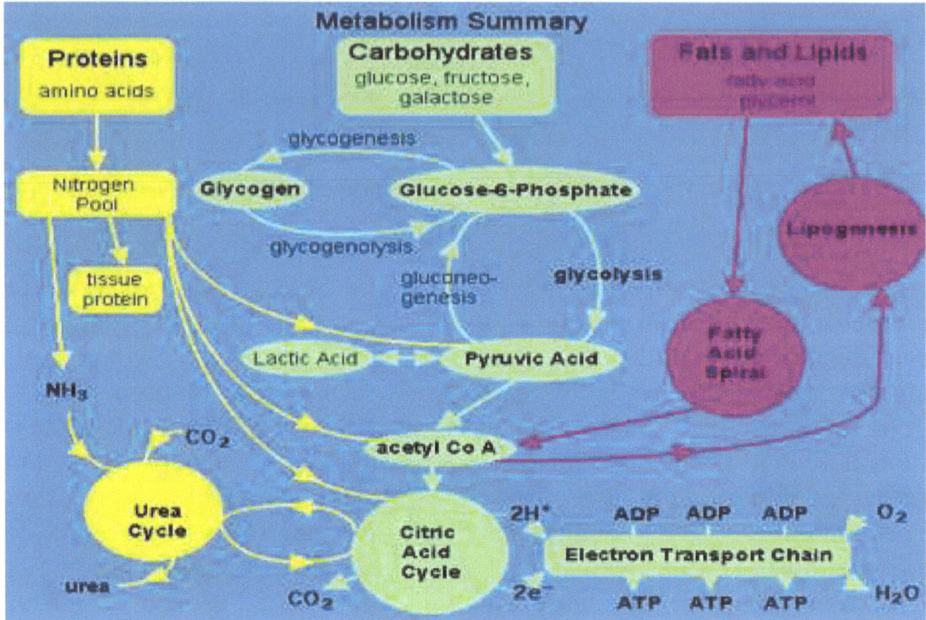

Fig. (1). Living Organisms Metabolism. (Chemistry.elmhurst.edu).

Photosynthesis and cellular respiration which enables us to live on earth, are interconnected through an imperative relationship.

Oxidative phosphorylation is the last metabolic process of cellular respiration that takes place following the breaking of glucose and the citric acid cycle. During cellular respiration, twenty-six out of thirty ATP units are produced by one glucose unit from oxidative phosphorylation. Within metabolism, the replications of DNA is a biological activity that gives two new DNA strains from one original DNA. This process occurs in all living organisms and is the basis for biological inheritance. Transcription is the process of making an RNA duplicate and directs the synthesis of the protein. The translation is the process of making a sequence of amino acids or a protein as per the information carried in the molecule of mRNA.

The genetic code describes the link between the sequence of base pairs in a gene and the corresponding amino acid sequence. The ribosome reads the sequence of the mRNA in groups of three bases to assemble the protein in the cell cytoplasm [5].

BIOENERGETICS

Albert Lehninger, in 1948 contributed to the present-day understanding of bioenergetics at a molecular level. He was an American biochemist who worked in the field of bioenergetics. He discovered, along with Eugene P. Kennedy, that mitochondria are the site where oxidative phosphorylation takes place. Bioenergetics is a subfield of biochemistry that compacts with miscellaneous aspects of energy transactions and transformations. The synthesis of ATP from redox energy is an active conveyance to the metabolism [6].

A continuous energy exchange takes place between the living organisms and their surroundings. The autotrophs can get vitality from sunlight (during photosynthesis), prepare nutrients and break down them. The heterotrophs take energy by utilizing these nutrients in the food by breaking down the complex bonds into simpler ones. The production of ATP takes place in autotrophs by consuming sunlight in the processes of photosynthesis, however heterotrophs by eating nutrients, generally proteins, carbohydrates, and fats utilize this energy in the form of ATP. Energy is obtained by living organisms from organic or inorganic materials which are taken in the form of food [7, 8].

Theory of Bioenergetics

Living cells and life forms require energy to develop and duplicate. This process is called cell development and this is energy-dependent. The transformation of energy from one form to another is a remarkable feature in living cells. The study of energy transfer inside the living cells takes place under some principles. There are two principles of bioenergetics. Energy cannot be created or destroyed but energy can be transfer from one form to another form. Green plants and algae are autotrophs; they make their own food using sunlight [2].

Laws of Thermodynamics in Bioenergetics

The overall quantity of energy within the universe remains constant for any physical or chemical change; energy can be changed from one form or transported from one form into another, but there is no creation of energy or its destruction, according to the first law of thermodynamics. In all living organisms, chemical energy can be changed into heat or electrical or mechanical energy. For a closed system, the change in internal energy is equal to the difference between energy used in work done by the system on its surroundings and heat supplied to the system. This relation is shown by the following equation:

$\Delta U = Q - W$

According to the second law of thermodynamics, the total entropy of a system will be increased in case of a chemical process spontaneously. Entropy is a state function of the system, so any change in the entropy of the system depends upon the initial and final states of the system. The entropy does not change in the reversible process but in the case of an irreversible process, the total entropy always increases. The entropy is expressed by an equation given below which shows the relation between the free energy change (ΔG) and the change of entropy (ΔS) of a system at constant pressure and temperature. It combines the first and second laws of thermodynamics.

$\Delta G = \Delta H - T\,\Delta S$

(ΔH change of entropy and T is RT)

Subsequently, the total enthalpy change equals the entropy change in the metabolic reactions. The above equation can be written as follows:

$\Delta G = \Delta E - T\Delta S$

If the value of ΔG is negative, the reaction will be spontaneous.

The total energy in an isolated or closed system remains the same. This system can be completed by chemical or physical processes. Living organisms are a part of an open system in which there is a conversation of both material and energy with their surroundings and this system never attains equilibrium with their surroundings. In the living system, we talk about the energy changes happening in a chemical reaction. The measure of vitality fit for accomplishing work during a response at steady temperature and weight communicates as the Gibbs free vitality (G). When free energy is gained by the system then the value of ΔG is positive is called endergonic reactions.

In exothermic reactions, during a chemical reaction heat is released; so the value of ΔH has negative and the heat of the reactant is more than the product. When a reaction system gets heat from their surroundings is called an endothermic reaction, and in this case, the value of ΔH is positive. We can calculate entropy for that system that contains disorder or order fewer molecules.

ΔG and ΔH are expressed in J/mol or cal/mol units and entropy are joules/mole degree Kelvin (J/molK), respectively. The entropy of the universe increases during all chemical or physical activity in the surrounding, which can be explained by the second law of thermodynamics. So, living beings protect their internal order by taking energy from the environmental factors *i.e*, free energy as supplements or daylight and getting back to their environmental factors as an equivalent measure of vitality as warmth and entropy [1, 2, 6].

Equilibrium Constant and Free Energy

The total energy needed for the reaction is done during any work at fixed temperature as well as environmental pressure; that energy is called Gibbs free energy (G).

It is general molecular reaction formula.

$$aA + bB \leftrightarrow cC+dD \tag{1}$$

In this equation, small letters a, b, c and d are present the number of molecules of reactant (A, B) and product (C, D).

The equilibrium constant of equation 1 is shown below.

$$K_{eq} = \frac{[C]^c[D]^d}{[A]^a[B]^b} \tag{2}$$

Where [A], [B],[C] and [D] molar concentration.

The relation of free energy change is expressed in bellow reaction equation 3.

$$\Delta G = \Delta G° + RT\ loge\ [C]\ [D]/\ [A]\ [B] \tag{3}$$

Where $\Delta G°$ is the change in free energy, R and T are gas constant and room temperature, respectively [A, B, C, D] molar concentrations.

We can calculate standard free energy change by free energy change at standard conditions. Just a physical constant is individual designed for each reaction as K eq, therefore $\Delta G°'$ is persistent. A simple association equation is shown below:

$$\Delta G°' = -RT\ lnK'eq \tag{4}$$

As shown in Table **1**, if the equilibrium constant of any given reaction is one, then the value of $\Delta G°'$ is zero. The value of standard free energy $\Delta G°'$ is negative due to the value of K'_{eq} of a reacting system is larger than one and if the value of K'_{eq} is < 1.0, then $\Delta G°'$ is up. Under standard conditions, the different values of standard free energy of product and reactant. While the reactants have more free energy than products, this reacting system will be spontaneous and $\Delta G^{o'}$ has a negative value in standard conditions. If the value of $\Delta G^{o'}$ is positive, in this case, the reactant has less energy than the product so the reaction cannot proceed spontaneously. The reactions are allowed or not allowed depending upon the relationship between standard equilibrium constant and standard free energy, which is shown in below Table **1**.

Table 1. Relationship between change in standard free energy and the Equilibrium constant for a given reaction.

When K'_{eq} is	$\Delta G^{o'}$ is	Reaction with 1M components as the reactant
>1.0	-ve	Reaction go to forward
1.0	0	Reaction go to an equilibrium
<1.0	+	Reaction go to backward

For example, let us calculate the standard free energy changes of the reaction catalyzed by the protein phosphoglucomutase: Glucose-1-phosphate and glucose-6-phosphate. The compound assessment shows that whether we start with 20 mM glucose-1-phosphate (no glucose-6-phosphate) in the presence of phosphoglucomutase or with 20 mM glucose-6-phosphate, the final equilibrium mixture, in either case, will contain 1 mM glucose-1-phosphate and 19 mM glucose-6-phosphate at 25 °C and pH 7.0. (Recollecting that enzyme doesn't influence the equilibrium position). From this information, we can calculate the equilibrium constant.

$$K'_{eq} = \frac{[\text{glucose-6-phosphate}]}{[\text{glucose-1-phosphate}]} = 19\text{ mM}/\ 1\text{ mM}$$

If the value of K_{eq} is 19 mM/1mM then we can calculate the value of $\Delta G^{o'}$.

$$\Delta G^{o'} = -RT\ \ln K'_{eq} = -7{,}296\text{ J/mol} = -7.3\text{ KJ/mol}$$

Endergonic and Exergonic Reactions

Exergonic Reactions

Exergonic reactions discharge energy from a spontaneous chemical reaction without any simultaneous usage of energy (as shown in Fig. **2**). The reactions are significant in terms of biology as those reactions have the functionality to create work, for example, cellular respiration. Maximum reactions involved in the cleavage of bonds during the formation of reaction intermediates are unmistakably observed during respiratory pathways. The bonds which can be created at some stage in the formation of metabolites are more potent than the cleaved bonds of the substrate.

$\Delta G = G_{products} - G_{reactants} < 0$. In this case $Q < K_{eq}$

Endergonic Reactions

Endergonic is the opposite of exergonic in being non-spontaneous and needs an input of free energy (as shown in Fig. **2**). Maximum of the anabolic reactions like photosynthesis and DNA and protein synthesis are endergonic in nature.

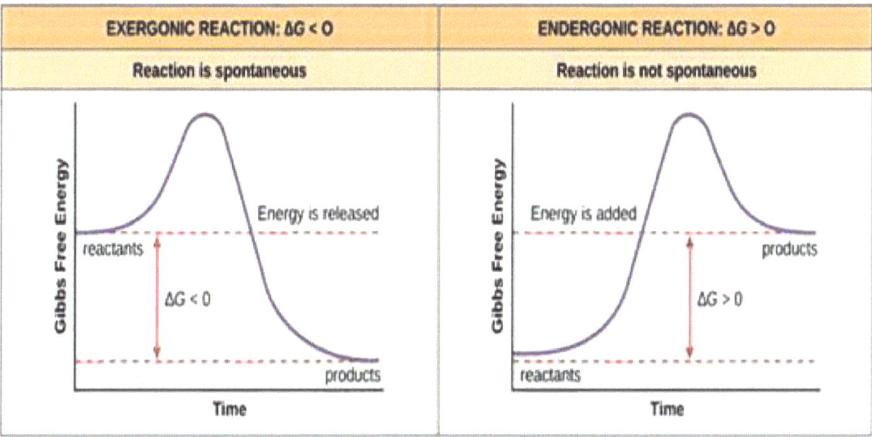

Fig. (2). Exergonic *vs* Endergonic reactions. (Bio.libretexts.org).

$\Delta G = G_{products} - G_{reactants} > 0$

In this case $Q > K_{eq}$ (as shown in Fig. **3**).

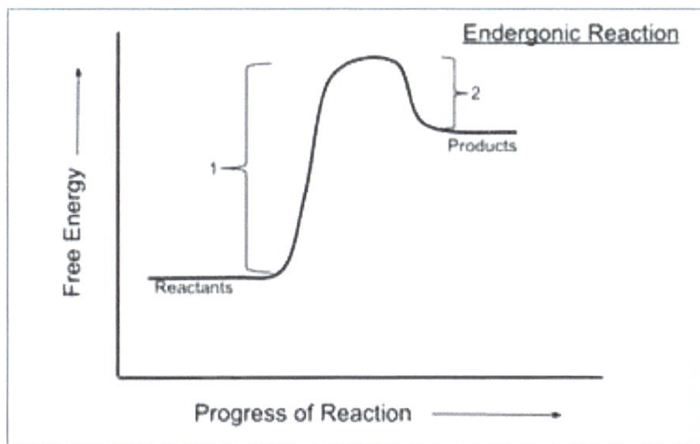

Fig. (3). Free energy *vs* Progress of reaction in case of endergonic reactions. (en.wikipedia.org/).

Endergonic reactions are capable to proceed in nature only when the thermodynamic conditions of the possibility of a reaction (-ΔG and +ΔS) are accomplished. The reaction is drawn or dragged when there is a sharp application of the products of the endergonic reactions by an ensuing exergonic reaction, thereby certifying that the concentration of the products of the endergonic reaction is every time low. The reaction can also be shoved or pushed by coupling them to strong exergonic reactions like ATP hydrolysis through mutual intermediates which provides the excess free energy required for the transition to occur. The relative strength of the bonds being formed due to the endergonic reaction is generally weaker than the bonds that were initially present in the substrate. In Table 10.2 are summarized the properties of various reactions.

Table 2. Properties of various reactions.

Category of Reaction	Spontaneity of Reaction	Variation in Free Energy (ΔG)	Variation in Entropy ($T\Delta S$)	Variation in Enthalpy (ΔH)
Exergonic	Yes	Negative and <0	Positive, or increasing, Equal to 0	Absorb/ liberate heat, so ΔH >/< 0 with $\Delta H < T\Delta S$
Exergonic	Yes	Negative, < 0.	Negative, or decreasing, < 0.	Liberate heat, so $\Delta H < 0$, and $\|\Delta H > \| T\Delta S\|$
Endergonic	No	Positive, and > 0.	Positive, or increasing, and > 0.	Absorb heat, so $\Delta H > 0$, and $\Delta H > T\Delta S$
Endergonic	No	Positive, and > 0.	Negative, or decreasing, and < 0.	Absorb/ Liberate heat, so ΔH >/< 0 with $\|\Delta H< \| T\Delta S\|$

Activation Energy

Activation energy is the insignificant quantity of energy required by the exergonic reactions to begin the energy-releasing reactions. Generally called activation energy (E_A), this energy is usually provided as heat from the surroundings when a much-needed transition state before the uncomplicated reaction is initiated. These values determine the rate of a given reaction and the higher the value of E_A the slower the rate of the reaction. E_A is greatly lowered during catalysis especially by enzymes as shown in Fig. (4).

This is indeed interesting that E_A is required even for the reactions which have a negative ΔG value. This is explained by the fact for the covalent bonds to break and release energy in catabolic exergonic reactions it is necessary that they should be contorted with a little energy. With respect to the reactants and products, this contortion is highly transient and highest energy requiring in nature. Thus E_A is always positive in value. The fate of the reaction in terms of being exergonic or endergonic in nature is determined by the energy levels of the reactants and the products. The cumulative bond energy of the reactants and the products in the form of heat/temperature or pressure is known to accelerate the collision of molecules so that the activation energy barrier is crossed and the transition state is reached. In the biological system, the presence of an activation energy barrier especially for exergonic catabolic reactions of essential components of the cell protects them from self-disintegration at room temperature [1].

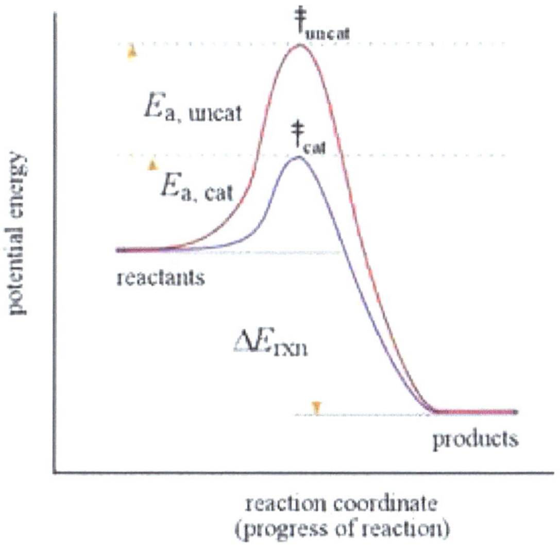

Fig. (4). Reaction coordinate diagrams for catalyzed reactions.

Coupled Reaction in Bioenergetics

This is a common feature in biological systems where some enzyme-based reactions are explainable as two coupled half-reactions, one spontaneous and other non-spontaneous. The hydrolysis of **ATP** (adenosine triphosphate) is to generate **ADP** (adenosine diphosphate) as the spontaneous coupling reaction (as shown in Figs. **4** and **5**).

$$ATP + H_2O \rightleftharpoons ADP + Pi$$

Where Pi, is an inorganic phosphate ion.

The phosphor anhydride bonds framed by catapulting water between two phosphate gatherings of ATP display a huge negative ΔG of hydrolysis and are hence frequently named "high vitality" bonds. In any case, likewise, with all bonds, vitality is needed to break these bonds, yet the thermodynamic Gibbs vitality distinction is "vitality delivering" while including the solvation thermodynamics of the phosphate particles; ΔG for this response is - 31 kJ/mol.

ATP is the major 'vitality' particle created by living digestion, and it fills in as a class of 'vitality source' in the cell: ATP is dispatched to any place as a non-unconstrained response needed so the two responses are coupled, so the general response is thermodynamically preferred.

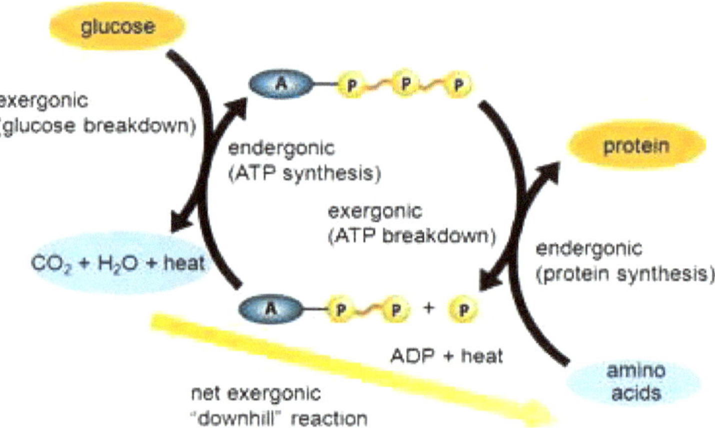

Fig. (5). Coupled reaction.

The formations of carboxylic acids by oxidation of aldehydes RCHORCHO is an organic compound and all living cells contain NAD (nicotinamide adenine dinucleotide is a coenzyme). In the condensed form, NAD+ acts as an oxidizing agent that can receive electrons from other molecules. The NAD+-linked oxidation of an aldehyde is virtually irreversible with an equilibrium that strongly favours the products (ΔG>>0ΔG>>0:

$$RCHO + NAD^{+} + H_2O \rightleftharpoons RCOOH + NADH + H^{+} \text{ (Spontaneous)}$$

The position of equilibrium for phosphorylating carboxylic acids lies very much to the left:

RCOOH + Pi ⇌RC(=O)(O−Pi) + H_2O (Nonspontaneous)

(Pi) is an inorganic phosphate ion.

METABOLISM OF CARBOHYDRATES

A vital component for various living metabolism is carbohydrates. In processes of photosynthesis, all green plants and algae synthesize carbohydrates by consuming CO_2 and H_2O, by the energy absorbed from sunlight on the inside. Cellular respiration is found in all animals and humans. The carbohydrates are broken down through cellular respiration when all living organisms consume plants as their food [9]. All animals and plants provisionally accumulate the released energy in the form of high energetic molecules, ATP, for use in numerous cellular processes.

Carbohydrates are made by simple unites of monomers that contain carbon, oxygen and hydrogen. Even though humans consume different types of carbohydrates, digestion cleaves complex carbohydrates into some small units of monosaccharides for cell metabolism: glucose, fructose and galactose, and the main constituent of products are glucose approximately 80%. The majority of the fructose and galactose move to the liver, where it changes to glucose.

Glycolysis

Glycolysis can be described as a metabolic pathway, where the breakdown of glucose into two molecules of pyruvic acid takes place under aerobic conditions; or the production of a small amount of energy in the case of lactate under anaerobic conditions. Embden, Meyerhof and Parnas have exposed this pathway. The site of glycolysis is the cytoplasm of virtually all the cells of the body [10, 11].

Classification of Glycolysis

This is classified into two kinds on the basis of the existence of oxygen.

1. **Aerobic glycolysis**: At the point when the event of oxygen is the abundant sum, the completion item is pyruvate sideways with the creation of 8ATP atoms.
2. **Anaerobic glycolysis:** At the point when oxygen is restricted, the last item is lactate close to the creation of 2ATP particles.

Glycolysis Pathway in Steps

It is an extramitochondrial pathway and eleven kinds of enzymes are involved. In the process of glycolysis, there is the conversion of glucose into pyruvate through 10 steps of metabolic reactions. The study of the glycolytic pathway can be separated into two phases. The first phase is the activation of glucose and the second is the payoff phase. It is displayed in Fig. (**6**).

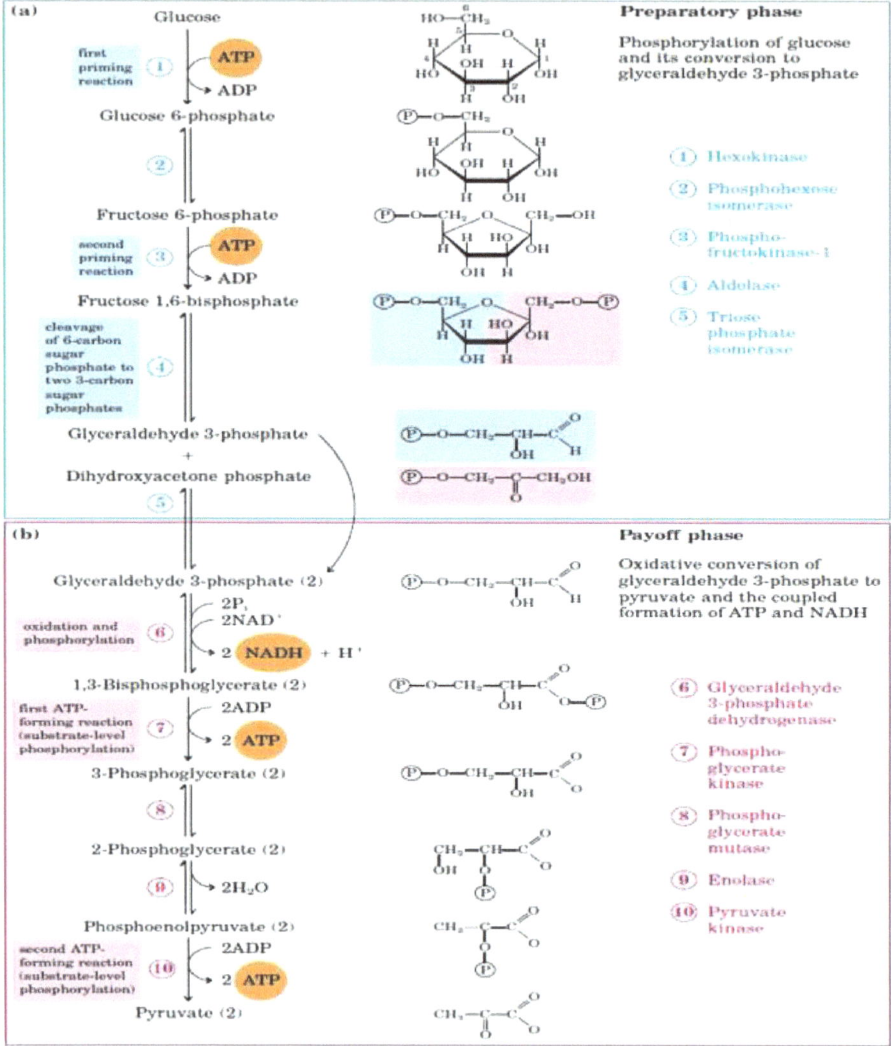

Fig. (6). Two phases of glycolysis.

Glucose Activation Phase

It is called an introductory phase of the pathway. In this phase, two molecules of ATP are invested and there is the formation of two triose phosphate by the breaking of the hexose chain. This phase has five steps of cellular metabolism. Phosphorylation of glucose is the conversion into glyceraldehyde 3-phosphate. The five types of the enzyme are involved in this pathway; namely hexokinase, phosphohexoseisomerase, phosphofructokinase-I, aldolase and triosephosphate isomerase.

Step-1

In the initial step of glycolysis, the glucose is prepared for the resulting schemes by phosphorylation at the C6 carbon. The cycle includes the exchange of phosphate from the ATP to glucose shaping Glucose-6-phosphate within the presence of the protein hexokinase and glucokinase (in creatures and microorganisms).

Step-2

In this step is the conversion of glucose-6-phosphate into fructose-6-phosphate due to isomerization and this reaction is catalyzed by the aldose-ketoseisomerase enzyme. There is no conversion of ATP or ADP. There is an arrangement of the furanose ring in the fructose-6-phosphate before the introductory of the glucopyranose ring of the glucose-6-phosphate to a straight arrangement.

Step-3

This step reaction is irreversible and utilization of one ATP for phosphorylation. There is phosphorylation of fructose-6-phosphate to fructose-1,6-biphosphate with help of the enzyme phosphofructokinase-1. It catalyses the transfer of a phosphate group from ATP to the fructose-6-phosphate. Phosphofructokinase-1enzyme regulates the breakdown of glucose and it is an important enzyme in glycolysis.

Step-4

The enzyme aldolase takes part by converting fructose 1, 6-biphosphate into 3 carbon units; one unit is glyceraldehyde-3-phosphate (GAP) and another one is DHAP; both are isomer of one another. This progression of digestion response is reversible.

Step-5

In this progression, dihydroxyacetone phosphate is isomerized into glyceraldehyde 3-phosphate within the sight of the chemical triose phosphate isomerase. This response finishes the main period of glycolysis. This reaction is additionally fast and reversible.

Payoff Phase

This stage is additionally called the energy extraction stage. In this stage, the transformation of glyceraldehyde-3-phosphate to pyruvate and ATP takes place. This means after 5 steps there is a payoff stage now.

Step-6

In this stage is the formation of 1, 3-Biphospholycerate by the glyceraldehyde---phosphate reaction with inorganic phosphate in the presence of glyceraldehyde 3-phosphate dehydrogenase enzyme. The enzyme acts as a catalyst. The group of aldehyde is oxidized to acid with a lot of energy. Through this response, NAD+ is condensed to NADH. Fig. (7) describes this reversible reaction step.

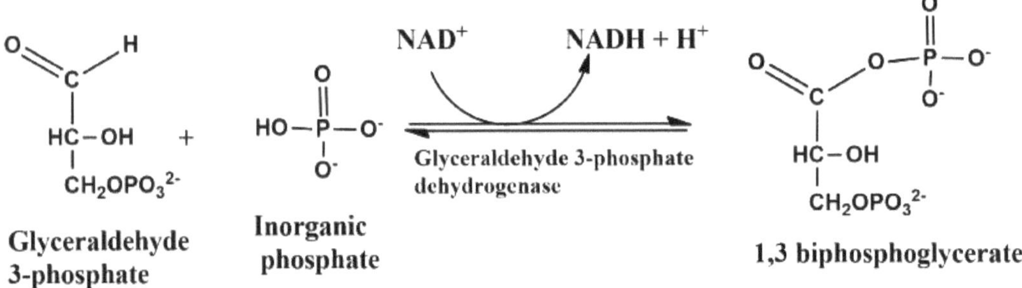

Fig. (7). The initial step of payoff phase of glycolysis.

Step-7

The protein phosphoglycerate kinase transfers the high-energy group of phosphoryl commencing the acid group of 1,3-bisphosphoglycerate to ADP, framing ATP and 3-phosphoglycerate. This kind of response where ATP is framed at substrate level is called substrate level phosphorylation (as shown in Fig. **8**).

Fig. (8). Conversion of 1,3-Bisphosphoglycerate to 3-Phosphoglycerate.

Step-8

2-phosphoglycerate to 3-phosphoglycerate is isomerized by moving the phosphate bunch from third to second carbon particle. The compound is phosphoglucomutase. It is a promptly reversible response and Mg^{2+} is basic for this response (as shown in Fig. **9**).

Step-9

The enolase enzyme is involved in the changing of 2-phosphoglycerate into phosphoenolpyruvate (as appeared in Fig. **10**) and one water molecule is taken out. A high energetic phosphate bond is delivered. This conversion is reversible and the enzyme has an Mg^{++} metal ion.

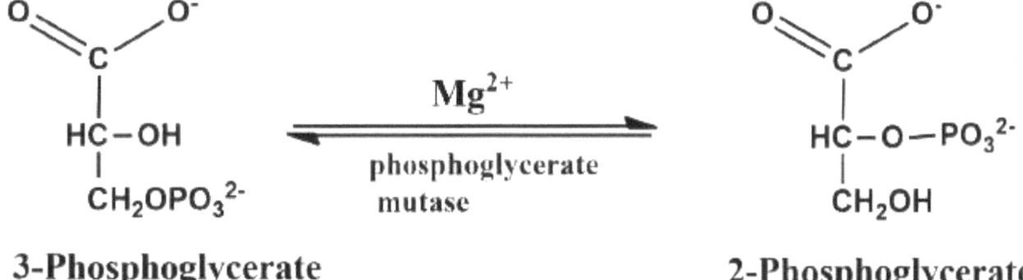

Fig. (9). Conversion of phosphoglycerate.

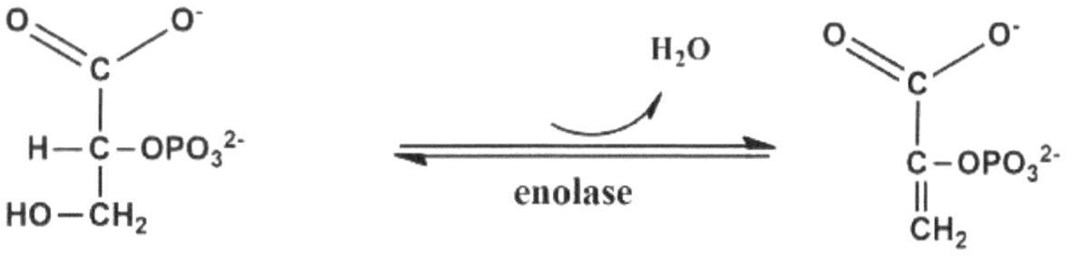

Fig. (10). Conversion of glycerate to enol pyruvate.

Step-10:

The transformation of a phosphoryl group from the phosphoenolpyruvate into ATP in the presence of metallic enzyme pyruvate kinase. The transition phase of this reaction is enol and pyruvate ions with isomerized into keto pyruvate. This enzyme is key to glycolysis and this is irreversible transformation [8, 11].

Anaerobic Condition

At the point when living tissues can't be provided with adequate oxygen to help vigorous reduction of the pyruvate and NADH delivered in glycolysis, NAD+ is recovered from NADH by the decrease of pyruvate to lactate. The decrease of pyruvate is catalyzed by lactate dehydrogenase (as shown in Fig. **12**).

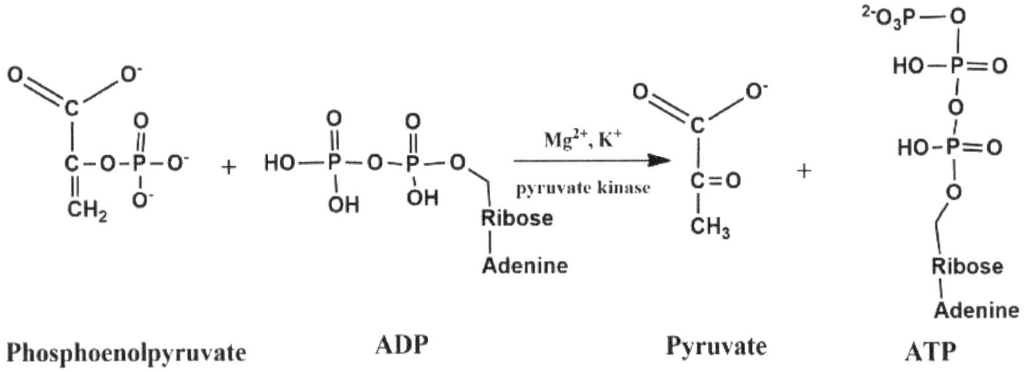

Fig. (11). Dephosphorylation of phosphoenol to pyruvate by pyruvate kinase. www.bioinfo.org.cn.

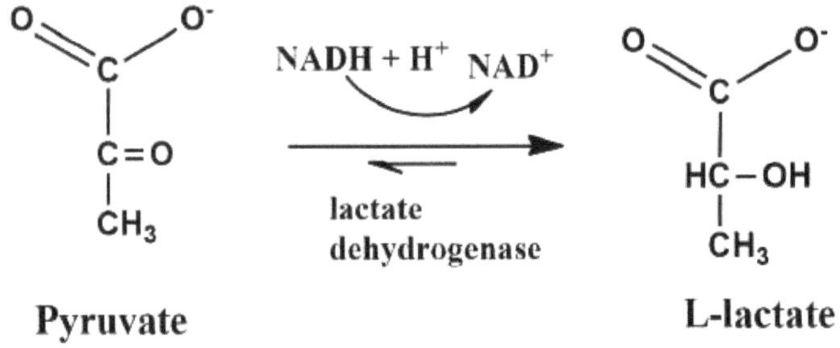

Fig. (12). Formation L-lactate. https://proteopedia.org.

Significance of the Glycolysis Pathway

- Glycolysis is a practically widespread focal pathway of glucose catabolism happening in the cytoplasm of the apparent multitude of tissues of organic frameworks prompting age of vitality as ATP for indispensable exercises.
- It is the pathway through which the biggest transition of carbon happens in many

cells.

- Some plant tissues which are adjusted for the capacity of starch, for example, potato tubers and a few plants adjusted to development in immersed water, for example, watercress determine the greater part of their vitality from glycolysis.
- In plants, glycolysis is the key metabolic segment of the respiratory cycle, which produces vitality as ATP.
- Numerous kinds of anaerobic microorganisms are totally subject to glycolysis.

LIPIDS METABOLISM

In 1815, Henry Braconnot grouped lipids into two classifications, suifs (solid oils or fat) and huiles (liquid oils). In 1823, Michel Eugène Chevreul built up a more definite order, including oils, fat, waxes, soaps, resins and unstable oils (or fundamental oils). According to T. P. Hilditch in 1947, lipids are divided into "simple lipids", with lubricates and waxes (true waxes, sterols, alcohols), and "complex lipids", with phosphorous-containing lipids and glycolipids. Various applications of lipids are found, out of them one is the very important application is for energy storage in a compact form. Lipid digestion is the union and debasement of lipids in cells, including the separate or capacity of fats for vitality. These fats are received from expanding food and engrossing them or these are blended by a living being's liver. Lipogenesis is the way toward orchestrating these fats. The dominant part of lipids is found in the human body from ingesting food ib the form of fatty substances and cholesterol. Other kinds of lipids found in the body are unsaturated fats and film Lipids. Lipid digestion is frequently viewed as the assimilation and ingestion cycle of dietary fat; be that as it may, there are two different ways life forms can utilize fats to acquire vitality: devoured dietary fats and capacity fat.

Vertebrates and humans utilize the two techniques for fat utilization as their wellsprings of vitality for organs, for example, the heart to work. Since lipids are hydrophobic particles, they should be solubilized before their digestion starts. Lipid digestion frequently starts with hydrolysis, which happens with the assistance of different catalysts in the stomach-related framework. Lipid digestion exists in plants; however, the cycle is different from that of animals. The second step after the hydrolysis is the retention of the unsaturated fats into the epithelial cells of the intestinal divider. In the epithelial cells, unsaturated fats are bundled and shipped to the remainder of the body. The significant dietary lipids for people and different living beings are creature and plant fatty substances, steroids, alcohols, and phospholipids membrane. The cycle of lipid digestion combines and debases the lipid stores and delivers the auxiliary and utilitarian lipids normal for singular tissues [4, 13].

Biosynthesis of Lipid

It occurs principally in the liver and adipocytes (vertebrates). Fatty acid combination and corruption happen by two totally isolated pathways. At the point when glucose is abundant; a lot of acetyl CoA is delivered by glycolysis and can be utilized for unsaturated fat synthesis. Glucose oxidation in the pentose phosphate pathway gives NADPH to fatty acid synthesis. The fatty substances are generated in living being due to an overconsumption of dietary starch. The starch is sugar; the extra sugar is converted into fatty substances. This includes the blend of unsaturated fats from acetyl-CoA then the esterification of unsaturated fats in the creation of fatty oils, a cycle called lipogenesis. The unsaturated fats might be hence changed over to fatty oils that are bundled in lipoproteins and discharged from the liver cells. The union of unsaturated fats includes a desaturation response, where twofold security is brought into the greasy acyl chain. Desaturation of steric acid by stearoyl-CoA desaturase-1 produces oleic acid.

Terpenes and isoprenoids, the carotenoids, are made by the adjustment of isoprene units got from the reactive precursors' isopentenyl pyrophosphate and dimethylallylpyrophosphate. These precursors can be made in various manners. In animals and archaea, the mevalonate pathway (shown in Fig. **13**) creates these mixes from acetyl-CoA, while in plants and microorganisms the non-mevalonate pathway utilizes pyruvate and glyceraldehyde 3-phosphate as substrates. One significant response that utilizes these enacted isoprene contributors is steroid biosynthesis. At this point, the isoprene units are consolidated to make squalene and shaped into a lot of rings to make lanosterol. Lanosterol would then be able to be changed over into different steroids, for example, cholesterol and ergosterol.

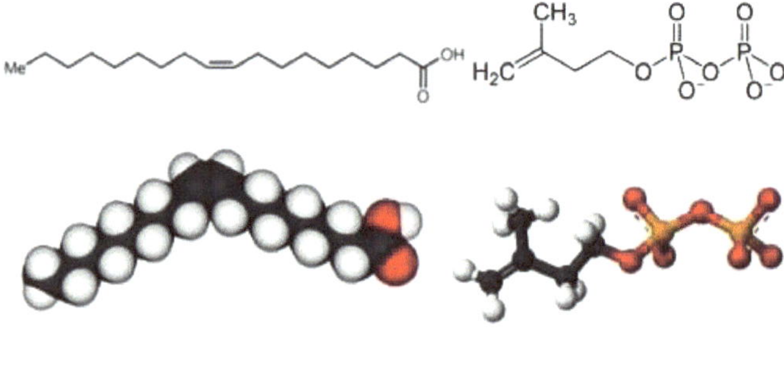

Oleic acid **Isopentenyl pyrophosphate**

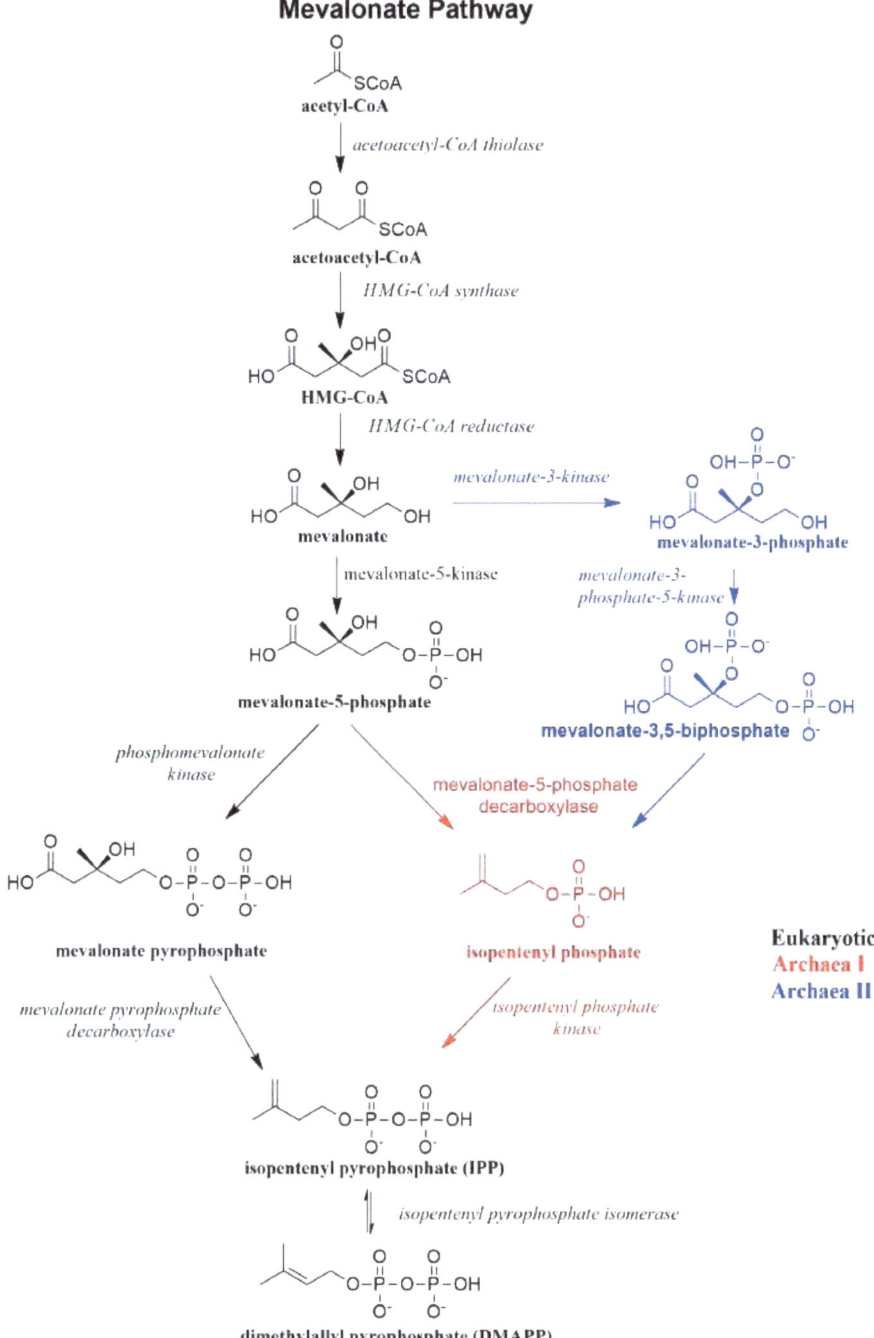

Fig. (13). Mevalonate pathway. Image credit: wikimedia commons. https://images.app.goo.gl/vDBZkJRpn3mKHSzW9.

Degradation of Lipid

Triacylglycerol (TAG) contains an enormous quantity of chemical vitality. Complete oxidation of 1 g of TAG yields 38 kJ, 1g of saccharides or proteins just 17 kJ. Man, that gauges 70 kg has 400 000 kJ in his TAG (that weight around 10,5 kg). This makes him ready to endure starving in weeks. TAG amasses transcendently in adipocyte cytoplasm. Living cells can't change unsaturated fats to glucose. Gluconeogenesis requires other things (1) vitality, (2) carbon buildups. Unsaturated fats are a rich wellspring of vitality yet they are not the wellspring of carbon buildups (there is any way one significant special case, for example, odd-numbered unsaturated fats). Cells cannot able to change over AcCoA to neither pyruvate nor OAA. Plant cells are prepared for the change of AcCoA to OAA in the glyoxylate cycle. B-oxidation is the metabolic cycle by which unsaturated fats are separated in the mitochondria or in peroxisomes to create acetyl-CoA.

METABOLISMS OF PROTEINS

Protein digestion indicates the different biochemical cycles liable for the combination of proteins and amino acids, and the breakdown of proteins (and other enormous particles) by catabolism. Dietary proteins are first separated into singular amino acids by different catalysts and hydrochloric corrosive present in the gastrointestinal tract. These amino acids are additionally separated to α-keto acids which can be reused in the body for the age of vitality, and creation of glucose or fat or other amino acids. This separation of amino acids to α-keto acids happens in the liver by a cycle known as transamination, which follows a bimolecular ping pong mechanism.

Protein Synthesis

Proteins are composed of amino acids utilizing data encoded in genes. Every protein has its own novel amino acid grouping that is indicated by the nucleotide arrangement of the gene encoding this protein. The hereditary code is a three-nucleotide set called codons and every three-nucleotide mix assigns an amino acid, for instance, AUG (adenine-uracil-guanine) is the code for methionine. Since DNA contains four nucleotides, the absolute number of potential codons is 64; hence, there is some repetition in the hereditary code, with some amino acids indicated by more than one codon. Genes encoded in DNA are translated into pre-messenger RNA (mRNA) by proteins, for example, RNA polymerase. Most living beings at that point cycle the pre-mRNA (otherwise called an essential record) by utilizing different types of Post-transcriptional modification to frame the mRNA, which is then utilized as a layout for protein blend by the ribosome. In prokaryotes, the mRNA may either be utilized when it is delivered or be limited by a ribosome. Interestingly, eukaryotes create mRNA in the cell core and afterward move it into the cytoplasm, where protein amalgamation at that point happens. The speed of protein production is higher in prokaryotes than eukaryotes and can spread up to 20 amino acids every second [13, 15]. The way toward blending a protein from an mRNA layout is shown in Fig. (**14**). The mRNA is stacked onto the ribosome and is examined as three nucleotides one after another by coordinating every codon to its base matching anticodon situated on an exchange RNA molecule, which conveys the amino acid comparing to the codon it perceives. The catalyst aminoacyl-tRNA synthetase "charges" the tRNA particles with the right amino acids. Proteins are consistently biosynthesized from N-end to C-end.

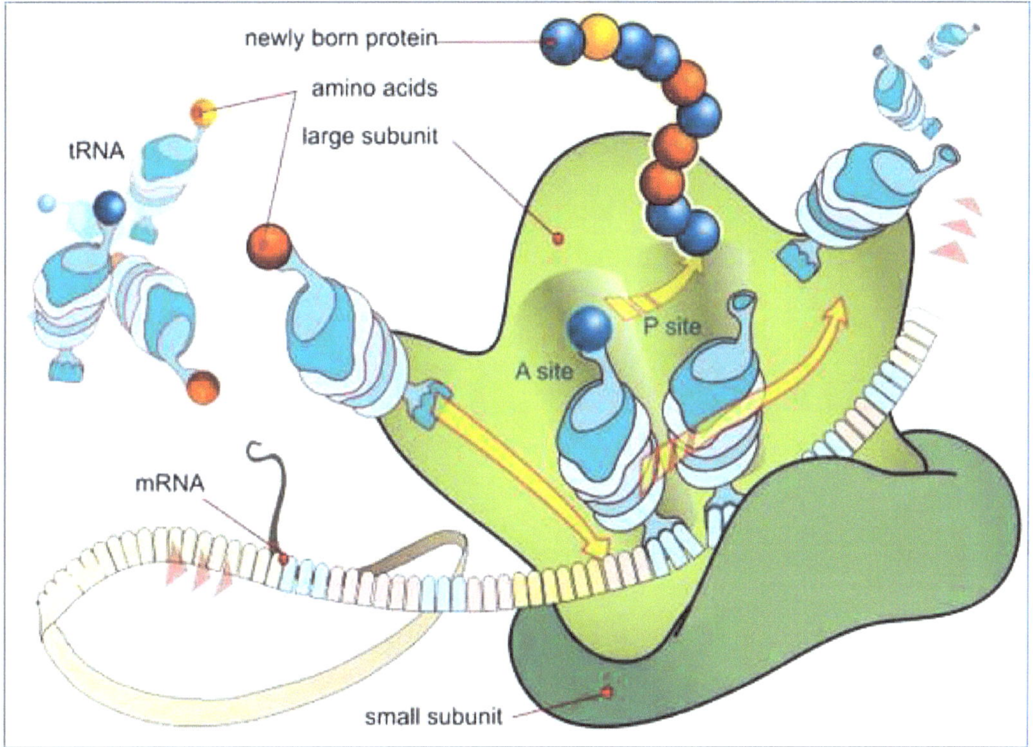

Fig. (14). A ribosome produces a protein [21]. Image credit: https://commons.wikimedia.org/wiki/File:Ribosome_mRNA_translation_en.svg#/media/File:Ribosome_mRNA_translation_en.svg.

Chemical Synthesis of Protein

Short proteins can likewise be combined synthetically by a series of strategies known as peptide combination, which depends on natural synthesis techniques, for example, concoction ligation to create peptides in high yield. Chemical synthesis considers the presence of non-common amino acids into polypeptide chains, for example, the connection of fluorescent tests to amino acid side chains. These techniques are valuable in the research centre of organic chemistry and cell science, however by and large not for business applications. Chemical synthesis is inefficient for polypeptides longer than around 300 amino acids, and the orchestrated proteins may not promptly expect their local or native tertiary structure. Most compound combination strategies continue from C-end to N-end and inverse the natural response.

Protein Degradation

Proteolysis is the breakdown of proteins into smaller polypeptides or amino acids. Uncatalysed, the hydrolysis of peptide bonds is extremely slow, taking hundreds of years. Proteolysis is typically catalysed by cellular enzymes called proteases, but may also occur by intra-molecular digestion. In all tissues, the majority of intracellular proteins are degraded by the ubiquitin (Ub)–proteasome pathway (UPP) (2). However, extracellular proteins and some cell surface proteins are taken up by endocytosis and degraded within lysosomes. Denaturation disrupts the normal alpha-helix and beta sheets in a protein and uncoils it into a random shape. Denaturation occurs because the bonding interactions responsible for the secondary structure (hydrogen bonds to amides) and tertiary structure are disrupted. The protein degradation is summarised in Scheme **1**.

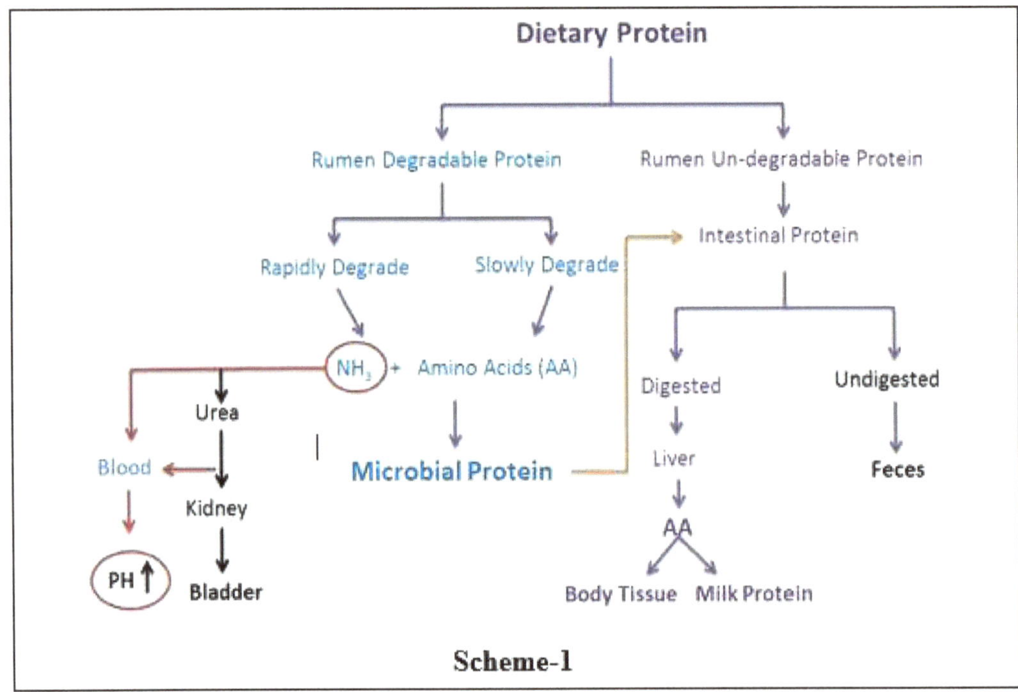

Scheme-1

METABOLISM OF NUCLEIC ACID

Nucleic acids are polymers of nucleotides. Nucleotides are prepared by an anabolic system that consists of phosphate, a pentose sugar and a nitrogenous base.

Synthesis of Nucleic Acids

Nucleotides (as appeared in Fig. **15**) can be divided into purines and pyrimidines. These both are fundamentally delivered in the liver. Nucleic acid metabolism is the process by which nucleic acids (DNA and RNA) are synthesized and degraded. Nucleic acids are the polymers of nucleotides. Nucleotide synthesis is an anabolic mechanism generally involving the chemical reaction of phosphate, a pentose sugar, and a nitrogenous base. The syntheses of nucleic acid are three types on the basis of chemical compositions.

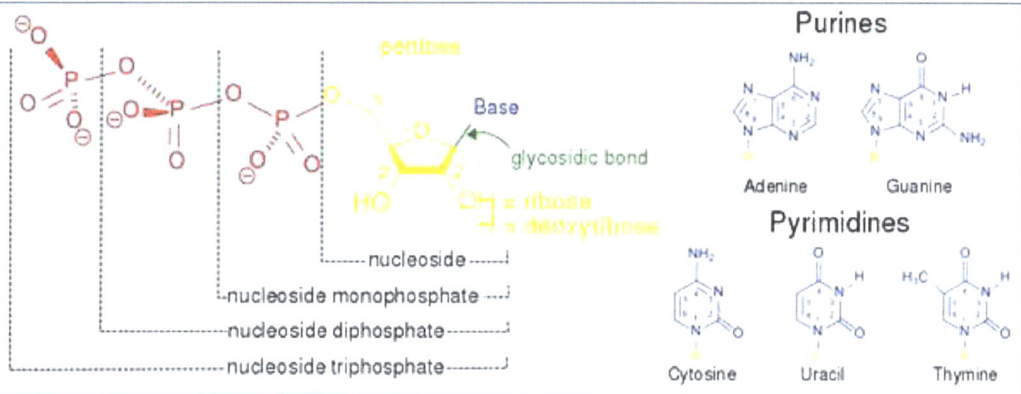

Fig. (15). Composition of nucleotides that make up nucleic acid [22].

(A) Purine synthesis

(B) Pyrimidine synthesis

(C) Converting nucleotides to deoxynucleotide

Adenine and guanine (as shown in Figs. **16** and **17**) are the two nucleotides delegated purines. In purine amalgamation, PRPP is transformed into inosine monophosphate or IMP. The creation of IMP from PRPP requires glutamine, glycine, aspartate, and 6 ATP, in addition to other things. Demon is then changed over to AMP (adenosine monophosphate) utilizing GTP and aspartate, which is changed into fumarate. While IMP can be legitimately changed to AMP, a blend of GMP (guanosine monophosphate) requires a middle advance, where NAD+ is utilized to shape the moderate xanthosine monophosphate or XMP. XMP is then changed over into GMP by utilizing the hydrolysis of 1 ATP and the transformation of glutamine to glutamate. AMP and GMP would then be able to be changed over into ATP and GTP, separately, by kinases that include extra phosphates.

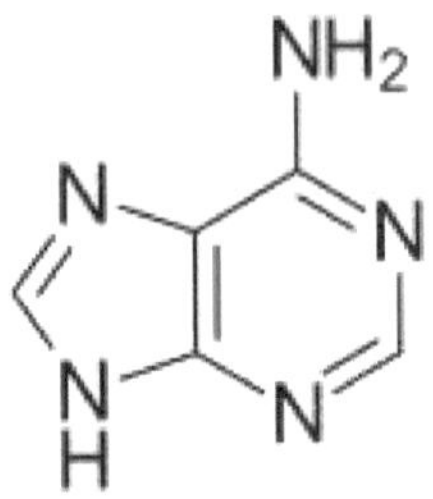

NH_2, N, N, N, N, H

Fig. (16). Adenine.

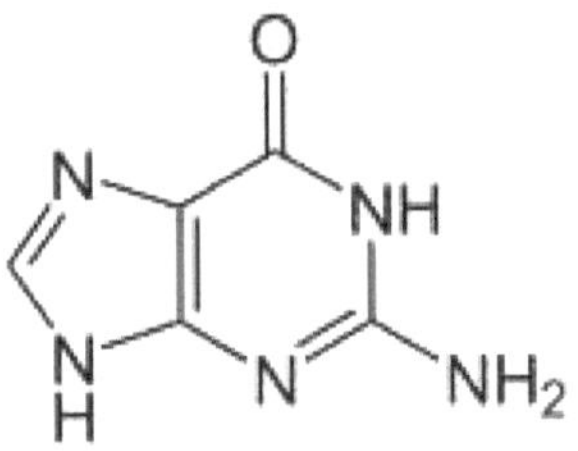

O, N, NH, N, N, NH_2, H

Fig. (17). Guanine.

ATP animates the creation of GTP, while GTP invigorates the creation of ATP. This cross guideline keeps the general measures of ATP and GTP the equivalent. The abundance of either nucleotide could improve the probability of DNA changes, where an inappropriate purine nucleotide is embedded.

Pyrimidine nucleotides include cytidine, uridine, and thymidine as shown in Figs. (**18**, **19** and **20**). Uridine is the starting precursor for the synthesis of pyrimidine nucleotides. This response requires aspartate, glutamine, bicarbonate, and 2 ATP atoms (to give vitality), just as PRPP which gives the ribose-monophosphate. Not at all like in purine blend is sugar/phosphate bunch from PRPP not added to the nitrogenous base until towards the finish of the cycle. After uridine-monophosphate is incorporated, it can respond with 2ATP to shape uridine-triphosphate or UTP. UTP can be changed over to CTP (cytidine-triphosphate) in a response catalyzed by CTP synthetase. Thymidine amalgamation initially requires a decrease of the uridine to deoxyuridine, before the base can be methylated to create thymidine. ATP, a purine nucleotide, is an activator of pyrimidine blend, while CTP, a pyrimidine nucleotide, is an inhibitor of pyrimidine combination. This guideline assists with keeping the purine/pyrimidine sums comparable, which is advantageous because equivalent measures of purines and pyrimidines are required for DNA blend. Lacks of compounds engaged with pyrimidine amalgamation can prompt the hereditary sickness oroticaciduria which causes an extreme discharge of orotic corrosive in the pee.

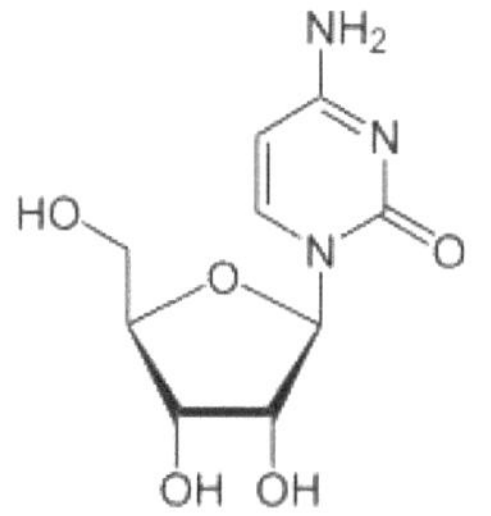

Fig. (18). Cytosine.

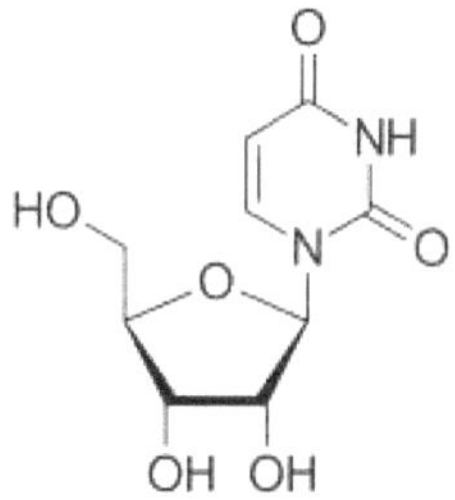

Fig. (19). Uracil.

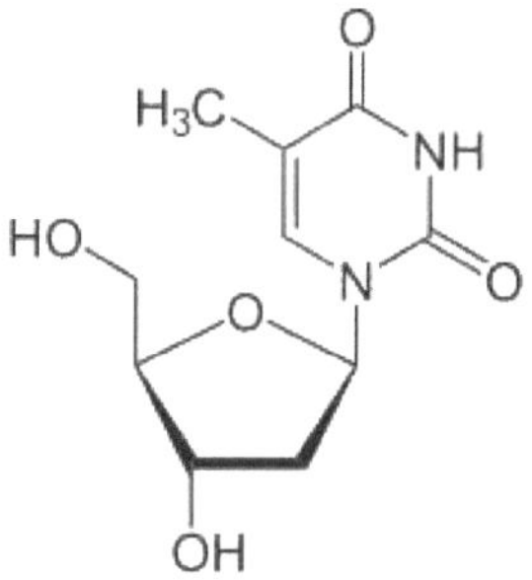

Fig. (20). Thymine.

Ribose is a carbohydrate and the main constituent of RNA. First uses the ribose in the synthesis of nucleotides. Nucleotides are organic compound and it is made up of nucleosides and phosphate units. Livings being have two types of nucleotides one is DNA and another is RNA. DNA exists with deoxyribose; it is made by the loss of a 2'-hydroxy group of the ribose in the presence of catalytic enzyme ribonucleotidereductase. The nucleotides must be in the diphosphate structure for the response to happen. To include thymidine, a part of DNA that just exists in the deoxy structure, uridine is changed over to deoxyuridine (by ribonucleoti dereductase), and afterward is methylated by thymidylate synthase to make thymidine.

Decomposition of Nucleic Acid (DNA and RNA)

The nucleic acids are polymers with molecular weights as high as 100,000,000 grams per mole. They can be broken down or digested, to form monomers known as nucleotides. Each nucleotide contains three units: a sugar, an amine, and a phosphate. The nucleotides are broken down into **uric acid** and this is the second major organic waste product that we excrete in our urine. There are some practical gatherings or invention structures that can alter the DNA. These atoms have sufficient reactivity to make changes and break covalent bonds inside DNA. Practically all the DNA responses fall into only two general classes: 1) the response of a DNA nucleophile with an electrophile) the response of a DNA pi bond or C-H bond with a revolutionary (As shown in Fig. **21**). An investigation was made into the disintegration of nucleic acids, uric corrosive and urea by various gatherings of soil microorganisms including bacilli, non-coryneform bars, corynebacteria (arthrobacters and non-arthrobacters), streptomycetes, growths and yeasts. Hydrolysis of nucleic acids was discovered to be a typical wonder.

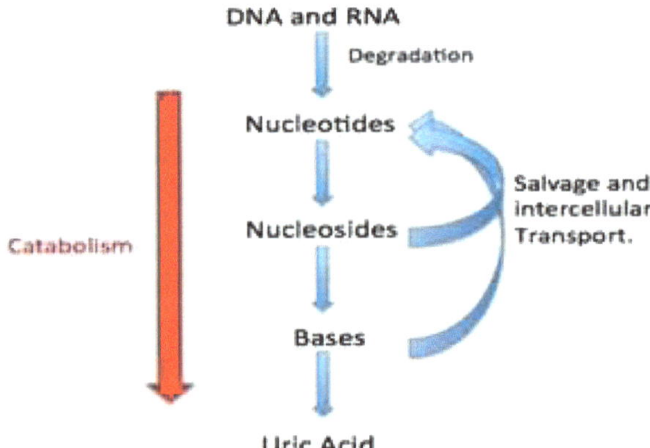

Fig. (21). Nucleic acid degradation.

Catabolism of Pyrimidine

Base cytosine and uracil are changed over into beta-alanine and later to malonyl-CoAwhich are required for unsaturated fat union, in addition to other things. Thymine, then again, is changed over into β-aminoisobutyric corrosive which is then used to shape methylmalonyl-CoA. The extra carbon skeletons, for example, acetyl-CoA and Succinyl-CoA would then be able to be oxidized by the citrus extract cycle. At the end of the degradation of pyrimidine are formed ammonia, water and carbon dioxide. The product ammonium formed enters into the urea cycle that takes place in the cytosol and the mitochondrial cell.

Purine Catabolism

Purines are degraded to uric acid and this takes place in the liver cells of the human body in presence of certain enzymes. Initially, nucleotide loses its own phosphate through 5'-nucleotidase. Then deamination of nucleoside and adenosine takes place and hydrolyzes to give hypoxanthine through adenosine deaminase and nucleosidase respectively. Oxidation of hypoxanthine gives xanthine and in turn, uric acid is formed by the reaction of xanthine oxidase. Dissociation of another purine nucleoside and guanosine takes place to give guanine. Deamination of guanine leads to the formation of xanthine *via* guanine deaminase which in turn is converted into uric acid. The oxygen finally accepts the electron during purine disintegration. Uric acid formed is finally released out of the animal body. Free purine and pyrimidine bases formed are transported within and across the cell membrane and restored to make more nucleotides through nucleotide salvage.

PHOTOSYNTHESIS AND RESPIRATION

The correlation between photosynthesis and cellular respiration is that the by-product of one framework is the reactants of the other. Photosynthesis includes the utilization of energy from daylight, water and carbon dioxide to generate glucose and oxygen. Cellular respiration utilizes glucose as well as oxygen to deliver carbon dioxide and water. The conditions for photosynthesis are the inverse of cellular respiration.

In order to survive, human beings, animals and plants are dependent upon the cycle of cellular respiration and photosynthesis as related in Fig. (**22**). The oxygen released by plants as a result of photosynthesis is taken by human beings and several other animals. The blood carries the oxygen and is transported to the cells for cellular respiration. Carbon dioxide is released during respiration from the body of animals and is trapped by plants. It gives vitality to the development and advancement of the plants. This is the cycle that will never end and thus sustains life on earth.

The process of photosynthesis is major utilized by plants and other microorganisms to release energy. Cellular respiration helps in the breaks down of the energy for doing useful work. Despite the contrasts between these two forms, there are some similarities also. Such as, both the processes built and make use of ATP, the energy currency [7, 8, 16].

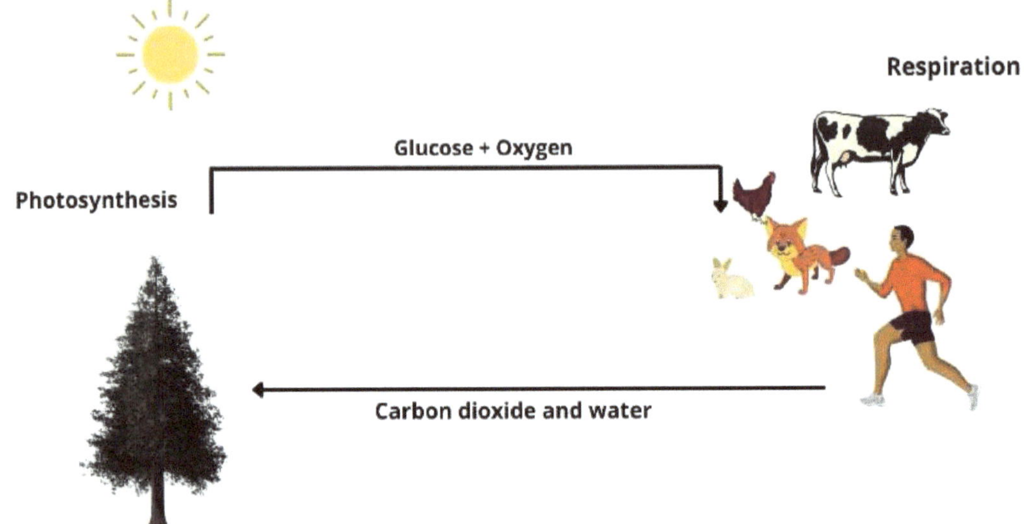

Fig. (22). Photosynthesis and respiration. photosynthesiseducation.com.

Differences between Photosynthesis and respiration are given in following Table 3

Table 3. Difference between photosynthesis and respiration.

S. N.	Photosynthesis	Respiration
1.	Occurs in green plants only	Occurs in plants and animals
2.	$CO_2 + H_2O$ = Sugar + O_2	Sugar + O_2 = $CO_2 + H_2O$
3.	Requires the energy from the sunlight rays	Acquire energy from chemical reaction alone within the body
4.	Make their own food	Depends on the by-products of photosynthesis.
5.	Oxygen is the gaseous by-product	Carbon dioxide is the gaseous by-product
6.	Needs energy	Produces energy
7.	Location is in chloroplast	Location is in mitochondria
8.	Includes synthesis of carbohydrate	Includes break down of carbohydrate

Similarities between Photosynthesis and Respiration

- Both the processes are involved in the manufacture of energy.

- Both mechanisms are located in the cell organelle which was considered as an endosymbiotic organism. It is chloroplast in plants and mitochondria in animals.
- In critical conditions, both can have an alternate pathway to achieve.

OXIDATIVE PHOSPHORYLATION

Oxidative phosphorylation is an operational method to synthesize ATP in both green plants as well as in animals. It includes the chemiosmotic coupling of electron transport and ATP synthesis. Oxidative phosphorylation takes place within the mitochondrial cell. The mitochondrial cell is bounded by the inner and outer membrane. The area inside the internal membrane is the matrix and the intermembrane area is the distance between the two membranes. NADH and $FADH_2$, produced during glycolysis and the citric acid cycle undergo oxidation in mitochondria. The electron released during oxidation is accepted by protein complexes implanted within the mitochondrion inner film which are consisted of numerous polypeptides able to receive and donate electrons.

These protein complexes are the principal additives of the respiration chain. Electrons launched from NADH and $FADH_2$ are added alongside the respiration chain and ultimately donated to oxygen to generate water. Redox potential associated with each electron carrier determines the direction of electron flow. Three predominant accountable protein complexes are protected withinside the respiratory chain which allows in delivery of negative ions and pumps hydrogen ions within the internal membrane and consequently creates a hydrogen electrochemical gradient. The produced electrochemical gradient generates a proton motive force (PMF) that helps to drive the hydrogen ions back to the inner membrane in presence of ATP synthase. The ATP synthase includes principal subunits: the F0 subunit, which affords a channel for the passage of hydrogen ions back across the internal membrane; and the F1 subunit, which allows the synthesis of ATP from ADP + Pi. As soon as H^+ moves by the F0 subunit, a part of the subunit rotates in the cell membrane. Conformational changes are induced due to rotation within the F1 subunit. It activates the ATP production doing and thus changing the available free energy of the hydrogen electrochemical gradient into the energy of a chemical bond [12, 17].

Electron Transport and Synthesis of ATP

Most of the cell ATP is manufactured using the mechanism of cellular respiration within the mitochondria of each cell of animal and plant cell. The route of ATP synthesis within the cited organelles is almost the same and represents an evolutionarily conserved mechanism. Primarily the mechanism involved two coordinated routes: electron transfer and ATP synthesis (each is initiated *via* cellular membrane binding with proteins). Transportation of electrons through an electron transport chain takes place which consists of numerous protein complexes related to membrane. With the movement of electrons between the protein molecules simultaneously, positive ions are derived across the membrane. The hydrogen ion particles at that point move down the electrochemical angle over the film going through a particular protein complex called ATP synthase. ATP synthase encapsulates the free energy created because of hydrogen angle to work the creation of ATP from ADP and inorganic phosphate. The association of electron transport, proton siphoning, and ATP blend are alluded to as chemiosmotic coupling. In mitochondria, this cycle is a definitive advance of cell breath and called oxidative phosphorylation. In chloroplasts, this cycle is frequently called the light reactions of photosynthesis.

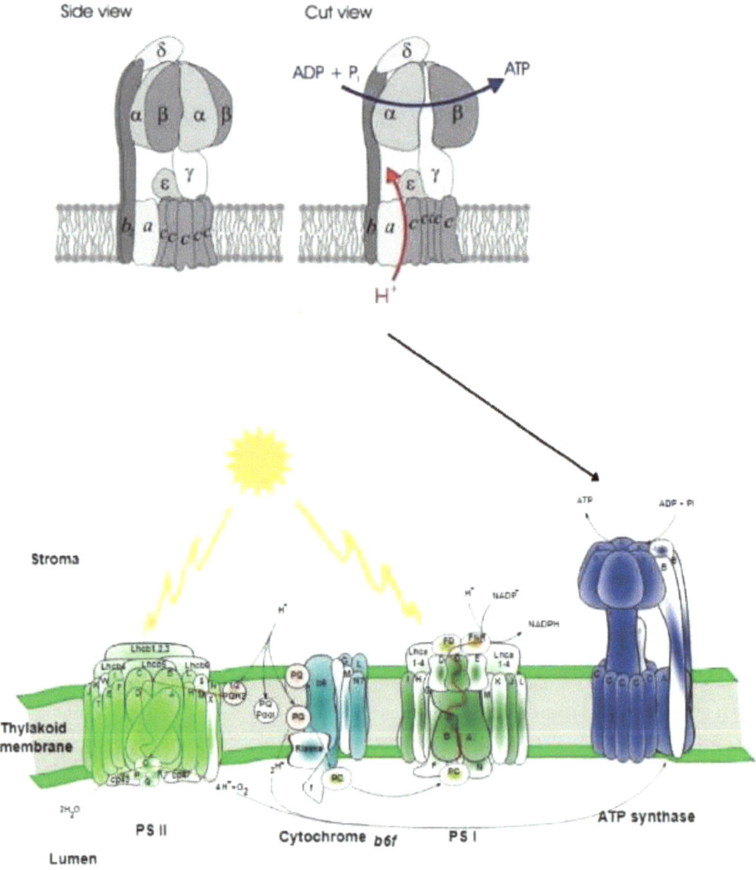

Fig. (23). ATP synthase and electron transport. Wikimedia commons. https://images.app.goo.gl/ai3vSPSgkBR8mY1U6. Wikimedia commons. https://images.app.goo.gl/2VCKaWKL2xbWgPEv7.

The ATP synthase (as shown in Fig. **23**) consists of two subunits: F0 and F1. F0 and F1 subunits consist of different proteins. The F0 subunit consisted of simple membrane layered proteins: 1 copy of protein a; 2 copies of protein b; and between 12-14 copies of protein c. These proteins make the channel for the passage of protons to move across the membrane. The F1 subunit consisted of five different proteins: 3 copies of *alpha*, 3 copies of *beta*, 1 copy each of *gamma*, *delta* and *epsilon*. As protons move over the membrane, the c subunits and the associated *delta* and *epsilon* subunits make the rotation. This rotation prompts conformational changes within *beta* subunits and catalyses the production of ATP.

DNA REPLICATION

In DNA replication, new duplicates of DNA are formed from parent DNA. When two strands of DNA are separated, each DNA standard acts as a precursor for the formation of a new strand. This process is called DNA replication (as shown in Fig. **24**).

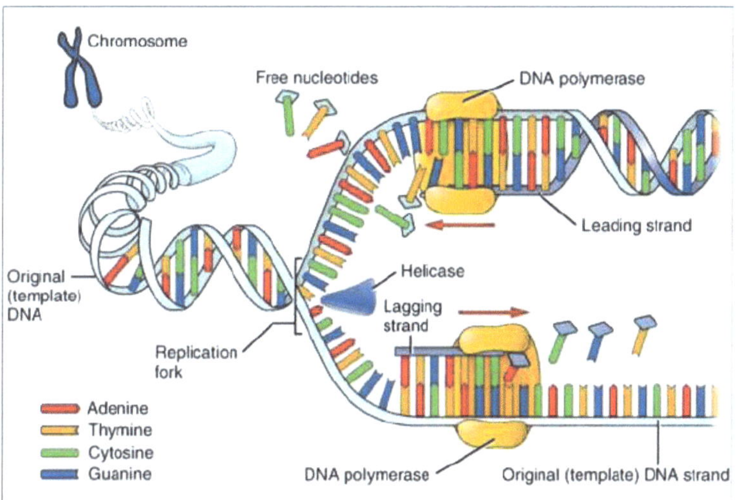

Fig. (24). DNA replication. image credit: Wikimedia Commons https://images.app.goo.gl/A4qe676KVqepRbid6.

The shape of DNA allows it to perform crucial features for the cell:

- Encoding the facts want to assemble and direct the cell
- Transmission of facts from technology to technology

In a way to transmit hereditary data, DNA is duplicated. DNA replication takes location at the S (syntheses) stage cycle of the cell. This simply continues if the

G1 checkpoint is passed, which ensures that the chromosomes have pleasantly isolated as a result of mitosis. An easy way, at some point of DNA replication, the division of the two strands of the DNA unit takes place. Since the two new strands of DNA consolidate each other. The way of replication is a more noteworthy complex way than essentially coordinating nucleotide bases. Replicating DNA involves the collaboration of a grouping of catalysts and administrative cycles; all put away in test by tough blunder checking. DNA replication begins when the compound helicase "loosens up" a little package of the DNA helix, disengaging the two strands. This purpose of the division is known as the replication fork. The two strands are kept detached by single abandoned restricting proteins (SSB) which tie onto every one of the strands. An enzyme called the DNA polymerases are able to make the new DNA strand. Hence, the DNA polymerases can start integrating the unused strand, the catalyst primase joins a short (~60nucleotides) plan of RNA. The DNA polymerases then intensify this preliminary moving along each strand from the 3′ end up to the 5′ end and including nucleotides to the 3′ hydroxyl gathering of the past nucleotide base [5, 18].

TRANSCRIPTION

Translations are the primary stage for gene expression. In this stage, one specific fragment of DNA gets replicated to RNA (particularly mRNA) in the presence of chemical RNA polymerase. DNA and RNA are both nucleic acids. The steps of transcription are as shown below.

1. RNA polymerase, along with one or more common transcription reasons, ties up to DNA promoter.
2. It makes a double transcription bubble that leads to the detachment of the two strands of the DNA helix. During this step, hydrogen bonds are broken down in the joint of complementary DNA nucleotides.
3. RNA polymerase adds up RNA nucleotides. These nucleotides form the complementary nucleotides present on one DNA strand.
4. RNA sugar-phosphate foundation is formed with the help of RNA polymerase which makes RNA strand.
5. Hydrogen bonds present between RNA–DNA helix are broken to release the fresh formation of the RNA strand.
6. If the cell includes the nucleus as the core, the RNA can proceed which includes polyadenylation, capping, and splicing.
7. The RNA may remain within the nucleus or escape to the cytosol through the nuclear pore complex.

The extended DNA translated to an RNA atom is known as a record unit and encodes at least one set of each information. If quality encodes of a protein, the record creates courier RNA (mRNA) which is thusly, set out as a format for the protein blend by the method of interpretation. Rather, the translated quality can be permitted to encode for either non-coding RNA, (for example, microRNA), ribosomal RNA (rRNA), move RNA (tRNA), or other enzymatic RNA molecules called ribozymes. Altogether, RNA makes a difference and synthesizes, regulates, control and handle proteins; it thus plays an essential part inside a cell.

In virology, occasionally, the genome of a negative sense single stranded RNA (ssRNA -) virus can template for a positive sense single stranded RNA (ssRNA+). This is because the positive sense strand contains the data needed to translate the viral protein for viral replication afterward. It is processed by a viral RNA replicase.

Transcription is a technique in which DNA is replicated (*transcribed*) to mRNA which in turn moves the instructions required for protein manufacturing.

The transcription process is carried out in two wide steps. Initially, pre-messenger RNA (as shown in Fig. **15**) is made, in the presence of RNA polymerase enzymes. This process is dependent on Watson-Crick base pairing and the consequent single strand of RNA is the inverse complement of the parent DNA sequence. At this point, the pre-messenger RNA is "edited" to deliver the required mRNA molecule in an operation called *RNA splicing*.

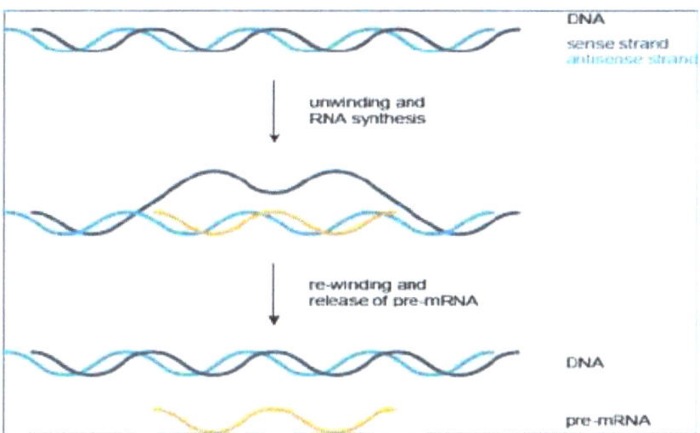

Fig. (25). Trancription; RNA (orange) and DNA (blue). Wikimedia Commons. https://images.app.goo.gl/q1po3U1ccsj8vQGr7.

Formation of pre-mRNA

A pre-mRNA is produced through the transcription of a region of DNA from a linear chromosome. The formation of the pre-mRNA by the transcription of DNA replication takes place in the nucleus (Fig. **25**). In contrast to DNA replication, each strand is duplicated, one strand is interpreted. The strand that has the quality is called the sense strand, simultaneously the integral strand is the antisense strand. The mRNA built during record is a copy of the sense strand; anyway, it's far from the antisense strand which is translated. Ribonucleotide triphosphates (NTPs) line up along the antisense DNA strand, with Watson-Crick base matching (A sets with U). The RNA polymerase gets together with the ribonucleotides to shape a pre-messenger RNA particle, this is integral to a locale of the antisense DNA strand. The record closes while the RNA polymerase compound arrives at a trio of bases this is perused as a "stop" signal. The DNA molecule re-winds to re-shape the twofold helix [5].

TRANSLATION

The messenger RNA constructed during transcription is moved out of the nucleus into the surrounding cytoplasm, then to the ribosome (the cell's protein). Now it coordinates protein synthesis. Messenger RNA requires the help of transfer RNA in protein synthesis. The method of translating the sequence located on a messenger RNA (mRNA) with the help of tRNA during protein synthesis is called *translation*. Each str*etc*h of mRNA (triplet) is called a *codon*, and one codon incorporates a particular amino acid. As the mRNA passes *via* the ribosome, every codon interacts with the *anticodon* of a particular transfer RNA (tRNA) molecule through following Watson-Crick base pairing. This tRNA molecule includes an amino acid at its 3′-terminus that is incorporated into the developing protein chain. The tRNA is then expelled from the ribosome. Below Fig. (**26**) indicates the steps included in protein synthesis.

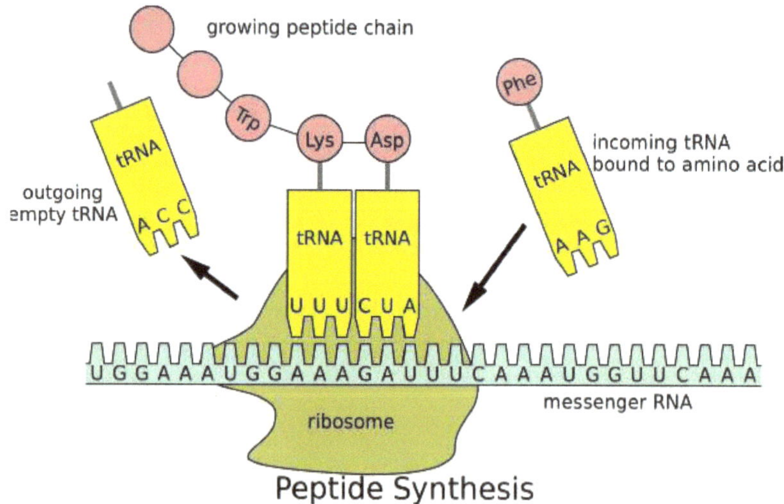

Fig. (26). Protein synthesis. https://www.atdbio.com.

Translation (a) and (b) tRNA molecules occupy the two binding sites of the ribosome, *via* hydrogen bonding to the mRNA; (c) formation of a peptide bond among the two amino acids to shape dipeptide, however, the tRNA is left uncharged; (d) the uncharged tRNA molecule departs the ribosome, the ribosome moves one codon to the right (the dipeptide is translocated from one binding location to the other); (e) subsequent tRNA binds; (f) formation of a peptide bond among the two amino acids to shape a tripeptide; (g) the uncharged tRNA departs the ribosome.

Transfer RNA

Each amino acid possesses its very own top-notch tRNA (or set of tRNAs). Such as the tRNA (Fig. **27**) for phenylalanine (tRNAPhe) is not like from histidine (tRNAHis). Each amino acid is related to its tRNA *via* the 3′-OH organization to shape an ester which reacts with the α-amino group of the terminal amino-acid in the developing protein chain to shape a new amide bond (peptide bond) at some stage in protein synthesis as shown below Fig. (**28**). The response of esters with amines is usually expanded extensively within the ribosome [5, 18].

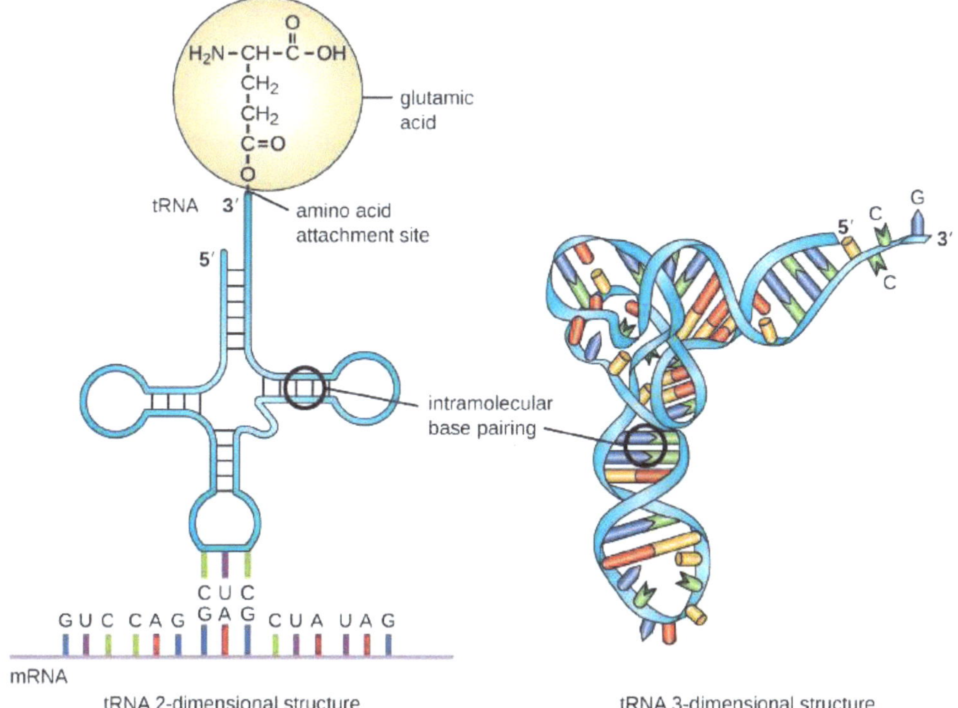

Fig. (27). Transfer RNA. image credit: wikimedia commons. https://images.app.goo.gl/u7dtvcHuPsyoBm2M8.

Fig. (28). Formation of polypeptide.

Protein Synthesis

Growing polypeptide chains react with the 3′-end of the charged tRNA as shown in Fig. (**28**). The amino acid is transmitted from the tRNA to the respective protein. Each transferred RNA molecule possesses a noticeable tertiary shape which is accepted through the enzyme aminoacyl-tRNA synthetase and provides the proper amino acid to the 3′-end of the unchanged tRNA. The presence of changed nucleosides is dominant in stabilizing the tRNA shape. Some of those changes are shown in Fig. (**29**).

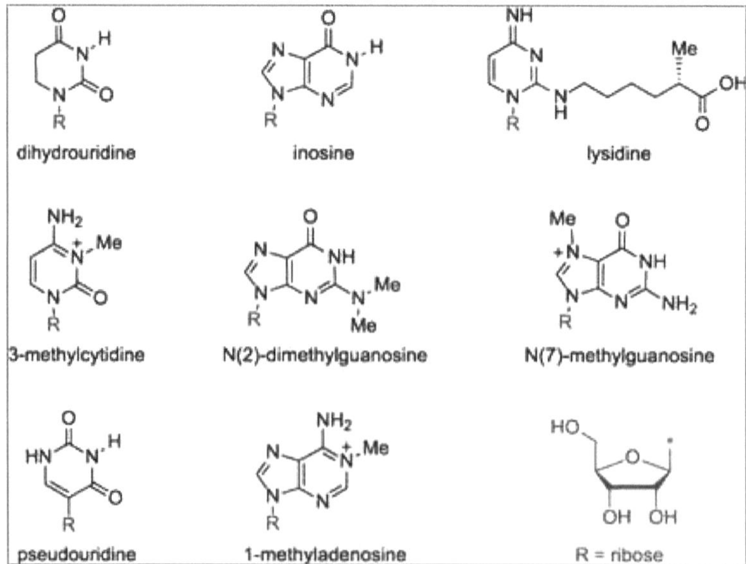

Fig. (29). Chemical structure of modified nucleosides.

The method through which DNA is duplicated to RNA is called **transcription**, and the process by which RNA is utilized to deliver proteins is called **translation** as shown in Fig. (**30**).

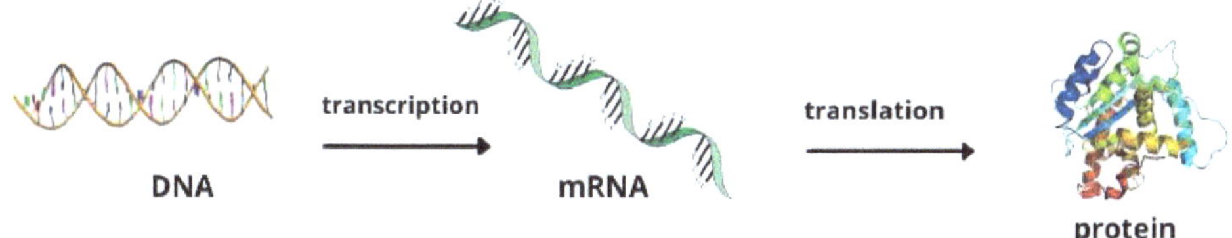

Fig. (30). Transcription and Translation. https://www.atdbio.com.

RECOMBINANT DNA TECHNOLOGY

This technology is a procedure that changes the phenotype of a life form by inserting recombinant DNA molecule into a host organism to produce new genetic combinations.

The method includes the addition of an external part of DNA into the genomic structure which consists of a gene of our interest. The technique by which a gene is incorporated into a recombinant gene is called recombinant DNA technology. It consists of the selection of the preferred gene form management into the host observed *via* the right vector with which the gene needs to be coordinated and recombinant DNA formed.

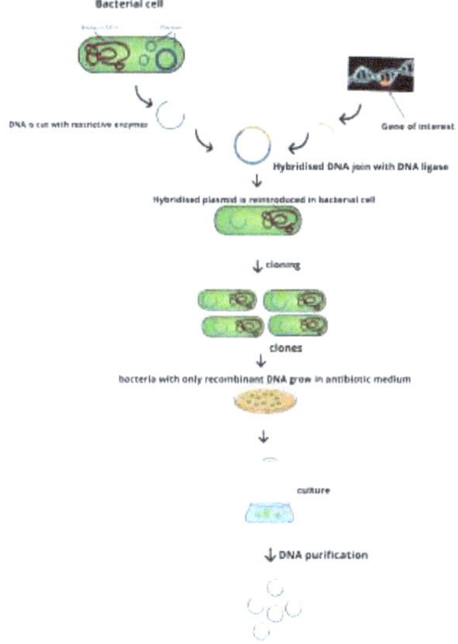

Fig. (31). Recombinant DNA technology. https://ocw.mit.edu.

This recombinant DNA is brought into the host. It is maintained within the host and carried into the offspring's as shown in Fig. (**31**).

Instruments for Recombinant DNA Innovation*:*

The tools mostly embrace the following:

1. Restriction enzymes – Restriction enzymes have two properties useful in recombinant DNA technology. First, they cut DNA into fragments of a size suitable for cloning. Second, many restriction enzymes make staggered cuts that create single-stranded sticky ends conducive to the formation of recombinant DNA. These enzymes are of two kinds, explicitly endonucleases and exonucleases. The endonucleases cut inside the DNA strand however the exonucleases remove the nucleotides from the closures of the strands. The limitation endonucleases are arrangement explicit which are usually palindrome successions and cut the DNA at a specific point. They examine the length of DNA and make the cut at the specific area called the limitation site. The particular qualities and the vectors are cut by similar limitation chemicals to get the integral clingy notes, subsequently making crafted by the ligases easy to tie up the necessary quality to the vector.
2. The vectors offer help with conveying and coordinating the predetermined quality. These are the extraordinary vehicles that convey forward the predetermined quality into the living creatures. Plasmids and bacteriophages are the preeminent regular vectors involved in recombinant DNA innovation as they have the particular characteristic of multiplying and dividing in a very short period of time.
3. The host is a living creature – into which the recombinant DNA is presented. The host is the target instrument of recombinant DNA innovation present inside the vector containing the predetermined DNA with the help of the proteins. There are various ways by which these recombinant DNAs are embedded into the host, these include: specifically–microinjection, biolistic or quality gene gun, substitute cooling and warming, utilization of calcium particles, and so forth [19, 20].

Recombinant DNA Technology Steps

1. Cut the gene at the acknowledgment area – It has just also been dealt with before in the proteins utilized in recombinant DNA innovation.
2. Polymerase chain response – It is a response to open up the quality if the right quality of gene has been cut using the limitation protein.

3. Addition of the necessary recombinant DNA into the host body–As said in apparatuses of recombinant DNA innovation, there are different manners by which this can be refined. The successfully changed cells/life forms convey forward the recombinant quality to the offspring.

Application of Recombinant DNA Technology

1. Application of recombinant DNA innovation in Agriculture – for example, production of Bt Cotton to shield the plants from ball worms.
2. Application in drugs – Production of Insulin by DNA recombinant innovation is a basic model.
3. Gene Therapy – Utilized as an endeavor to address the quality imperfections which offer ascent to heredity illnesses.
4. Clinical finding – ELISA is a model where the utilization of recombinant DNA innovation is seen which is utilized to identify the presence of HIV in humans.

SUMMARY

Bioenergetics is the branch of biochemistry that focuses on how cells transform energy, often by producing, storing, or consuming adenosine triphosphate (ATP). Adenosine triphosphate (ATP) is the key "energy currency" for organisms. The aim of metabolic and catabolic strategies is to synthesize ATP from available beginning materials (from the environment) and to break- down ATP into adenosine diphosphate (ADP) and inorganic phosphate with the aim of utilizing it in organic strategies. Metabolism is the set of life-maintaining chemical alterations inside the cells of living organisms. It also can seek advice from all chemical reactions occurring in living organisms, together with digestion and the transport of materials into and among extraordinary cells, known as intermediary metabolism or intermediate metabolism. Carbohydrate metabolism denotes the diverse biochemical tactics responsible for the synthesis and degradation of carbohydrates in residing organisms.

Glucose is the most energetic carbohydrate in living organisms. Lipids also have plenty of uses. One of the most vital uses is the garage of energy in a compact form. Lipid metabolism is the synthesis and degradation of lipids in cells, it is an important preservation of fat for energy. These fats are acquired from eating meals. Protein metabolism denotes the diverse biochemical approaches for the synthesis of proteins and amino acids and the breakdown of proteins (and other massive molecules) by means of catabolism. Nucleic acid metabolism is a methodology by means of which nucleic acids (DNA and RNA) are synthesized and degraded. Nucleic acids are made by polymers of nucleotides. Nucleotide synthesis is an anabolic mechanism typically related to the chemical response of phosphate, a pentose sugar, and a nitrogenous base. The relation between photosynthesis and cellular respiration is such that the by-product of one process is the reactants of the other. Photosynthesis includes the usage of energy from sunlight, water and carbon dioxide to supply glucose and oxygen. Cellular respiration makes use of $C_6O_6H_{12}$ and oxygen to supply CO_2 and H_2O. Oxidative phosphorylation takes place within the mitochondria. Photosynthesis and cellular respiration are linked *via* a vital relationship. This relationship permits the existence of life. The products of one method are the reactants of the other. Notice that the equation for cell respiration is the direct contrary of photosynthesis:

Cellular Respiration: $C_6H_{12}O_6 + 6O_2 \rightarrow 6CO_2 + 6H_2O$

Photosynthesis: $6CO_2 + 6H_2O \rightarrow C_6H_{12}O_6 + 6O_2$

DNA is made of a double helix possessing complementary strands. During replication, these strands are separated. Each strand of the unique DNA molecule then serves as a template for the manufacturing of its counterpart, a manner referred to as *semiconservative replication*. As a result of semi-conservative replication, the brand new helix might be composed of a unique DNA strand as well as a newly synthesized strand. DNA replication begins whilst the enzyme helicase "unwinds" a small part of the DNA helix, keeping apart the two strands. This factor of separation is referred to as the *replication fork*. The strands are separated *via* means of single-stranded binding proteins (SSB) which bind onto each of the strands. A group of enzymes referred to as the *DNA polymerases* are answerable for growing the brand new DNA strand, but they can't begin the brand new strand off. Therefore, earlier than the DNA polymerases can begin synthesizing the brand new strand, the enzyme primase attaches a short (~60nucleotides) collection of RNA referred to as a *primer*. The DNA polymerases then expand this primer, moving alongside every strand from the 3′ give up to the 5′ give up and including nucleotides to the 3′hydroxyl group of the preceding nucleotide base. The order of nucleotides is retained by means of matching complementary nucleotides at the template strand.

Transcription is the initial and main step of gene expression or protein expression, in which a particular segment of DNA is duplicated to RNA (mainly mRNA) in the presence of enzyme RNA polymerase. In the process of transcription, a DNA sequence is perused by an RNA polymerase, which generates a complementary, antiparallel RNA strand called a primary transcript. Recombinant DNA (rDNA) molecules are designed by scientific lab practices of hereditary recombination like molecular cloning to combine hereditary material from numerous sources and make orders that are not available within the genome. Herbert Boyer first accomplished recombinant DNA in a surviving organism on this earth in 1973. He was from the University of California at San Francisco, and Stanley Cohen from Stanford University made use of *E. coli* restriction enzymes to load foreign DNA into plasmids.

Recombinant DNA molecules also are so often called chimeric DNA, due to the fact they could be created by means of materials from exceptional species consisting of the mythology, chimera.

KEYWORDS

Glycolysis

Glycolysis is a metabolic process at the start of the chain of reactions within the process of cellular respiration – production of cellular energy. It occurs in the presence or absence of oxygen to enable aerobic and anaerobic cellular respiration. The glycolysis pathway converts one glucose (sugar) molecule into two pyruvate molecules; this ten-step conversion occurs in the presence of specific enzymes in the cell cytosol. Glycolysis is sometimes called the Embden-Meyerhof-Parnas or EMP pathway after the scientists that first proposed this mechanism.

The final product of glycolysis is pyruvate in aerobic settings and lactate in anaerobic conditions. Pyruvate enters the Krebs cycle for further energy production.

The reverse of glycolysis; Gluconeogenesis means a new synthesis of glucose. It is the reverse of glycolysis. The body makes glucose in the liver (and also in the kidney).

The Cycle of Citric Acid

is also known as the tricarboxylic acid cycle (TCA **cycle**) or the Krebs **cycle**—is a series of chemical reactions used by all aerobic organisms to generate energy through the oxidation of acetate-derived from carbohydrates, fats, and proteins-into carbon dioxide.

Ketosis

is a metabolic state characterized by elevated levels of ketone bodies in the blood or urine. Physiologic ketosis is a normal response to low glucose availability, such as low-carbohydrate diets or fasting, that provides an additional energy source for the brain in the form of ketones.

Oxidative Phosphorylation

is the process in which ATP is formed as a result of the transfer of electrons from NADH or $FADH_2$ to O_2 by a series of electron carriers. This process, which takes place in mitochondria, is the major source of ATP in aerobic organisms.

The electron transport chain is a series of electron transporters embedded in the inner mitochondrial membrane that shuttles electrons from NADH and FADH2 to molecular oxygen. In the process, protons are pumped from the mitochondrial matrix to the intermembrane space, and oxygen is reduced to form water.

Photosynthesis

cannot be possible if plants don't have the green tint given by chlorophyll. It is a process that takes place in the presence of sunlight, CO_2 and water with favorable conditions like temperature and it is completed through a metabolic pathway. The plant cells undergo *photophosphorylation* to generate ATP after the glucose product formation.

Catabolism

Catabolism is a metabolic process in which large molecules split by removing energy. This process is found in glycolysis and the citric acid cycle.

Photorespiration

Biochemical studies indicate that photorespiration consumes ATP and NADPH, the high-energy molecules made by light reactions. Thus, photorespiration is a wasteful process because it prevents plants from using their ATP and NADPH to synthesize carbohydrates.

Recombinant DNA

(rDNA) molecules are **DNA** molecules formed by laboratory methods of genetic **recombination** (such as molecular cloning) that bring together genetic material from multiple sources, creating sequences that would not otherwise be found in the genome.

LONG QUESTIONS

1. What are the laws of bioenergetics? How is thermodynamics involved in it?
2. Explain glycolysis aerobic and anaerobic in detail. Also mention energetic's and regulation involved in it.
3. Briefly explain recombinant DNA technology.

4. What is the mevalonate pathway, lipid synthesis and degradation?
5. What are nucleic acids?
6. How is the balance between protein synthesis and degradation achieved?
7. What is the main difference between DNA and RNA with respect to the nitrogenous bases present in their nucleotides?
8. What is the main difference between aerobic and anaerobic respiration?
9. How are cellular respiration and photosynthesis related?
10. How to explain DNA replication, translation and transcription?
11. Write the different steps of reaction that convert a polysaccharide into ATP. What are the major reactions, the intermediate compounds formed and the biochemical pathways that take place?
12. For which types of cells is glycolysis the whole some source of energy? Write some advantages and disadvantages if a cell relies solely on glycolysis for energy?
13. The process of glycolysis generates many products and requires numerous enzymes and other molecules to generate the various phases leading up to the final products. Which steps require the breakdown of ATP to ADP? Which steps involve building up ADP into ATP?
14. What cellular organelles are responsible for facilitating energy production? Which specific energy-production pathways are handled by these organelles?
15. Explain why lipoproteins are needed to transport triglycerides in aqueous body fluids. Name the 5 major lipoproteins and explain how they are similar and how they are different.

SHORT-QUESTIONS

1. The first and the second laws are more vital with respect to the biological perspective. Justify.
2. Why is activation energy required even for the reactions which have a negative ΔG value?
3. What constitutes the group lipid?
4. What are proteins and an amino acid?
5. What is the significance of the glycolysis pathway?
6. What is the difference between photosynthesis and cellular respiration?
7. What is pre-messenger-RNA?
8. What is oxidative phosphorylation?
9. How many types of glycolysis and glycolytic pathway phase.
10. What is the preparatory phase?
11. What is free energy?
12. What are high-energy compounds? Write about few examples with a suitable structure.
13. What is Electron Transport Chain? Explain the mechanism of *etc.* at

mitochondrial Matrix?

14. What is photophosphorylation?
15. Explain cyclic and non-cyclic photophosphorylation stages in light phage?
16. Expiation the purine catabolism.
17. Define photosynthesis.
18. What are raw materials for photosynthesis?
19. Give the structure of nucleotides.
20. What is the chemiosmotic coupling of electron transport?

MULTIPLE CHOICE QUESTIONS

1. If ΔH for a given reaction is equal to zero, then ΔG° is equal to

a) -TΔS°

b) TΔS°

c) -ΔH°

d) $\ln k_{eq}$

2. Human beings are can not digest.

a) Starch

b) complex carbohydrates

c) denatured proteins

d) Cellulose

3.The most common monomer of carbohydrates is a molecule of

a) Glucose

b) Maltose

c) Amylose

d) Amylopectin

4. The six most common atoms in organic molecules are

a) C, H, O, He, Ca and S

b) C, H, O, N, P and S

c) C, H, O, Mg, Mn and S

d) C, H, O, N, P and K

5. The structure of a protein can be denatured by

a) Heat

b) The presence of oxygen

c) The polar bonds of water molecules

d) The presence of carbon dioxide gas

6. Which of the enzyme has the lowest turn over number

a) Lysozyme

b) Carbonic anhydrase

c) PEP carboxylase

d) Phasphorilase

7. Which of the following enzyme is not proteinaceous in nature

a) Urease

b) Peptidase

c) Ribozyme

d) Phosphatase

8. Which of the following statement is incorrect regarding RNA

a) RNA is single-stranded

b) RNA is the only genetic material in viruses

c) RNA is the genetic material in some viruses

d) all of these

9. Nitrogenous bases in RNA include

a) AGCT

b) AGCU

c) ACT

d) Depends on the environment

10. RNA is found in

a) Nucleolus

b) Cytoplasm

c) Mitochondria

d) All of these

11. The adjacent nucleotides are joined by

a) Ionic bond

b) Hydrogen bond

c) Phosphodiester bond

d) all of these

12. DNA is

a) Positively Charged

b) Negatively charged

c) Neutral

d) None of these

13. Which of the following statement is true

a) The two DNA strands are parallel and complementary

b) The two DNA strands are anti-parallel and non-complementary

c) The two DNA strands are antiparallel and complementary

d) None of these

14. The diameter of the DNA helix is

a) 10 A

b) 21 A

c) 1.0 A

d) 2.0 A

15. Who all got the Nobel Prize for discovering the structure of DNA double helix

a) Watson and Crick

b) Watson, Crick and Wilkins

c) Watson, Crick and Franklin

d) Watson, Crick and Pauling

16. Watson and Cricks double-helical DNA is a

a) A-DNA

b) B-DNA

c) Z-DNA

d) D-DNA

17. Lipids are important constituents of

a) Nucleus

b) Ribosomes

c) Both a and b

d) Biological membranes

18. All the following are storage polysaccharides except

a) Starch

b) Cellulose

c) Glycogen

d) Dextran

19. Bonds that are cleaved by oxygen in the burning process of fuel are

a) C-H Bonds

b) D-H bonds

c) M-H bonds

d) O-H bonds

CONSENT FOR PUBLICATION

Not Applicable.

CONFLICT OF INTEREST

The author confirms that this chapter contents have no conflict of interest.

ACKNOWLEDGEMENTS

Authors are thankful to Department of Chemistry, North Eastern Regional Institute of Science and Technology,Nirjuli, Itanagar for providing facilities to accomplish the work.

REFERENCES

[1] Nicholls D G, Ferguson SJ. Bioenergetics 3rd ed. Academic Press. ISBN 0-12-518124-8.

[2] Green DE, Zande HD. Universal energy principle of biological systems and the unity of bioenergetics. Proc Natl Acad Sci USA 1981; 78(9): 5344-7. [http://dx.doi.org/10.1073/pnas.78.9.5344] [PMID: 6946475]

[3] Elson D. Metabolism of nucleic acids (Macromolecular DNA and RNA. Annu Rev Biochem 1965; 34: 449-86. [Internet]. [http://dx.doi.org/10.1146/annurev.bi.34.070165.002313] [PMID: 14321176]

[4] Vance JE, Vance D. Biochemistry of Lipids, Lipoproteins and Membranes. 4th Ed. Amsterdam: Elsevier. ISBN 978-0-444-51139-3..

[5] Watson JD. Recombinant DNA: Genes and Genomes: A Short Course. San Francisco: W.H. Freeman 2007.

[6] Lehninger AL. Bioenergetics: the molecular basis of biological energy transformations. 2nd ed., Addison-Wesley 1971.

[7] Photosynthesis McGraw-hill encyclopedia of science & technology 13. New York: McGraw-Hill 2007.

[8] Olson JM. Photosynthesis in the archean era. Photosyn Res 2006; 88(2): 109-17. [http://dx.doi.org/10.1007/s11120-006-9040-5]

[9] Owen OE. Ketone bodies as a fuel for the brain during starvation https:// doi. org/2005. [http://dx.doi.org/10.1002/bmb.2005.49403304246]

[10] Jenkins DJ, Jenkins AL, Wolever TM, Thompson LH, Rao AV. Simple and complex carbohydrates. Nutr Rev 1986; 44(2): 44-9.

[http://dx.doi.org/10.1111/j.1753-4887.1986.tb07585.x] [PMID: 3703387]

[11] Jenkins DJ, Jenkins AL, Wolever TM, Thompson LH, Rao AV. Simple and complex carbohydrates. Nutr Rev 1986; 44(2): 44-9.
[http://dx.doi.org/10.1111/j.1753-4887.1986.tb07585.x] [PMID: 3703387]

[12] Jr. Jones Maitland. Organic Chemistry 1st ed.., 1998.

[13] Food and Nutrition Board. Dietary Reference Intakes for Energy, Carbohydrate, Fiber, Fat, Fatty Acids, Cholesterol, Protein and Amino Acids. Washington, D.C: The National Academies Press 2002/2005; p. 769.

[14] Christie WW, Han X. Lipid Analysis: Isolation, Separation, Identification and Lipidomic Analysis. 4 th Ed. The Oily Press 2010.

[15] Solomon EP, Berg LR, Martin DW. Biology. Cengage Learning. p. 52. ISBN 978- 0534278281– *via* google.books.com

[16] Calvin M. Forty years of photosynthesis and related activities. Photosynth Res 1989; 21(1): 3-16.
[http://dx.doi.org/10.1007/BF00047170] [PMID: 24424488]

[17] Berg JM, Tymoczko JL, Stryer L. Biochemistry. 7th ed., W.H. Freeman & Company 2010.

[18] Dahm R. Discovering DNA: Friedrich Miescher and the early years of nucleic acid research. Hum Genet 2008; 122(6): 565-81. [Internet].
[http://dx.doi.org/10.1007/s00439-007-0433-0] [PMID: 17901982]

[19] Lobban PE, Kaiser AD. Enzymatic end-to end joining of DNA molecules. J Mol Biol 1973; 78(3): 453-71.
[http://dx.doi.org/10.1016/0022-2836(73)90468-3] [PMID: 4754844]

[20] Jenkins DJ, Jenkins AL, Wolever TM, Thompson LH, Rao AV. Simple and complex carbohydrates. Nutr Rev 1986; 44(2): 44-9.
[http://dx.doi.org/10.1111/j.1753-4887.1986.tb07585.x] [PMID: 3703387]

[21] Available from. https://en.wikipedia.org/wiki/Translation_(biology)

[22] Available from. https://www.wikiwand.com/en/Nucleic_acid_metabolism

SUBJECT INDEX

A

Meera Yadav & Hardeo Singh Yadav (Eds.)

B

C

D

E

F

G

H

I

J

K

L

M

N

O

P

R

S

T

V

W

X

www.ingramcontent.com/pod-product-compliance
Lightning Source LLC
LaVergne TN
LVHW070116110826
845147LV00002B/130

* 9 7 8 1 6 8 1 0 8 8 4 9 5 *